PHYSICS THE EASY WAY

Robert L. Lehrman

Science Department
Roslyn High School
Roslyn, New York

BARRON'S

Barron's Educational Series, Inc.
Woodbury, New York / London / Toronto / Sydney

All inquiries should be addressed to:
Barron's Educational Series, Inc.
113 Crossways Park Drive
Woodbury, New York 11797

Library of Congress Catalog Card No. 84-6357

Paper Edition
International Standard Book No. 0-8120-2658-6

Library of Congress Cataloging in Publication Data
Lehrman, Robert L.
    Physics the easy way.

    1. Physics. I. Title.
QC23.L36 1984   530   84-6357
ISBN 0-8120-2658-6 (pbk.)

PRINTED IN THE UNITED STATES OF AMERICA
67   410   98765

Contents

Introduction

So you have decided to study physics. Anyone who would like to understand the mechanism of the universe and the ways in which we investigate it must come to this decision sooner or later. Physics is the most fundamental of the sciences, the place where the most ubiquitous question—WHY?—comes to a stop and peters out into the murky waters of speculative philosophy.

Physics represents the human urge to make sense of the universe, to write the rules of its workings in exact and comprehensible terms. Professional physicists find that they must resort to higher mathematics in order to formulate their theories and explain their findings. For this reason, most of the new developments in physics, the advancing front of scientific knowledge, are well beyond the comprehension of the beginning student. But the basics on which the great superstructure of physical theory is built are accessible to anyone with a reasonable grasp of simple algebra. These fundamental laws have, for the most part, been established for many years, and most of the basic theory in this book could have been written a century ago.

This is by no means the case with the practical applications of these theories. Technology is advancing so rapidly that some of the statements made here may soon be out of date. This is no cause for alarm; properly armed with the fundamentals, you should be able to keep up fairly well with many of the technological advances based on a particular body of theory.

Although this book limits itself to simple mathematics, a knowledge of algebra and geometry is indispensable. To understand any part of physics, you must solve problems. You will have to be able to manipulate and solve simple linear equations, to set up and solve proportionalities, and to take square roots. You will have to know, for Chapters 1, 2, 11, and 14, how to deal with the sines, cosines, and tangents of angles. Scientific notation will be useful throughout and absolutely essential in Chapter 15. In this day, it would be folly to do all this arithmetic without a calculator; a simple, four-function type is adequate, and a scientific calculator, with trigonometric functions and scientific notation, will be even more useful.

In your pursuit of this work, I can wish you nothing better than the uplift in spirit that flows from perceiving the beauty inherent in the patterns according to which the universe behaves.

Robert L. Lehrman

CHAPTER 1

Notions of Motion

It is the job of physics to explain why things happen. But the first task is to describe *what* happens. Here we are going to use algebra to write mathematical descriptions of the way things move. If you need a special name for the mathematical description of motion, you can call it *kinematics*.

We must begin by defining our variables carefully. You no doubt already have some idea of what is meant by *speed*. Suppose that, as you start on a trip, you note that your watch reads 4:10 and the odometer of your car says 45,350 miles. When you arrive at your destination, your watch says 5:40 and the odometer reads 45,422 miles. Now you do the following calculation:

distance traveled: 45,422 − 45,350 = 72 miles

time elapsed: 5:40 − 4:10 = 1:30 = 1.50 hours

speed = 72 miles/1.50 hours = 48 mi/hr

Now, you do not know how fast you were going at any one time. On the highway, you were doing 55, or perhaps more if no one was watching; when you stopped for a red light, your speed was zero. What you have calculated is the *average* speed for the whole trip.

1

Your calculation was correct, because it agrees with the standard kinematic definition of average speed, which is

$$v_{av} = \frac{\Delta s}{\Delta t}$$

(Equation 1-1)

Let's decode this equation:

v_{av} stands for average speed

s stands for distance

t stands for time

Δ is the Greek capital letter delta

The symbol Δ calls for some explanation. It can be translated as "change in" or "difference of." Thus Δs stands for the difference between the original odometer reading and the final reading. Therefore Δs is the distance traveled during this particular trip. Similarly, Δt is the time elapsed during the trip. The equation says that average speed is the distance traveled divided by the elapsed time. The unit of speed in the International System (*Système Internationale* in French, or SI) is the meter per second (m/s), which is a little over 2 miles per hour. (See Appendix 1.)

Sample Problem

1-1 If it takes you 45 seconds to run 340 meters, what is your average speed?

Solution   According to Equation 1-1,

$$v_{av} = \frac{\Delta s}{\Delta t} = \frac{340 \text{ m}}{45 \text{ s}} = 7.6 \text{ m/s}$$

Sample Problem

1-2 How long does it take a skier to travel 650 meters, going 18.5 m/s?

Solution   Equation 1-1 can be transformed into

$$\Delta t = \frac{\Delta s}{v_{av}} = \frac{650 \text{ m}}{18.5 \text{ m/s}} = 35 \text{ s}$$

Note how the units behave:

$$\frac{\text{m}}{\text{m/s}} = \cancel{\text{m}}\left(\frac{\text{s}}{\cancel{\text{m}}}\right) = \text{s}$$

Sample Problem

1-3 At 9:00 A.M. you start on a walk. At 10:30 A.M. you are 4.6 miles from your starting point. In SI units, what was your average speed? (There are 1,600 meters to a mile.)

Solution   Distance traveled, Δs, is

$$4.6 \cancel{\text{ mi}} \times \frac{1,600 \text{ m}}{1 \cancel{\text{ mi}}} = 7,360 \text{ m}$$

Elapsed time, Δt, is

$$1.5 \cancel{\text{ hrs}} \times \frac{3,600 \text{ s}}{1 \cancel{\text{ hr}}} = 5,400 \text{ s}$$

From Equation 1-1,

$$v_{av} = \frac{\Delta s}{\Delta t} = \frac{7,360 \text{ m}}{5,400 \text{ s}} = 1.36 \text{ m/s}$$

Your speedometer does not indicate averages; it tells you how fast you are going *now*. This is your *instantaneous speed*. If the speedometer reading does not change during the trip, you are said to be going at *constant speed*. In this case the constant value of the speed will be equal to your average speed for the whole trip.

Core Concept

Average speed is distance traveled divided by the elapsed time:

$$v_{av} = \frac{\Delta s}{\Delta t}$$

Try This What is the average speed of a runner who finishes the 1,500-meter race in 3 minutes 30 seconds?

1-2. Vectors: Which Way?

Your speedometer gives you very incomplete information. You cannot rely on it to tell you where you will end your trip, because it says nothing about *which way* you are going. The quantity that tells you that is called *velocity*.

Velocity is one of the many physical quantities that have direction. Quantities of this kind are called *vectors*. In contrast, quantities that have no directional property, such as time, temperature, energy, and volume, are called *scalars*. A scalar quantity is completely specified when you give its magnitude and units. A vector quantity is incomplete unless its direction is also stated. Speed, as distinguished from velocity, is a scalar. Vectors are usually denoted by boldface letters. Thus, **v** means velocity. The corresponding scalar—the magnitude without the direction—is denoted by an ordinary letter, for example, v for speed.

The most basic of all vectors, the quantity that defines the properties of a vector, is *displacement*. Displacement is simply distance and direction from some starting point or origin of coordinates. Suppose you start from your home and walk 2 miles. Where are you? Well, which way did you go? If you walked straight north, your displacement is 2 miles north.

Vectors can be added, but they do not obey the same rules of mathematics as scalars. If you walk 2 miles north and then 3 miles east, you have walked 5 miles. But your displacement from your starting point, as shown in Figure 1-1, is 3.6 miles at 56° east of north. This quantity is called the *vector sum* of the other two vectors.

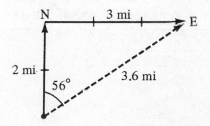

FIGURE 1-1

Look at the vectors labeled **A** and **B** in Figure 1-2. They might represent displacements, velocities, electric fields, or any other physical quantity that has direction. The vector **A** points up and a little to the right. The vector **B** is longer, indicating that its magnitude is greater than that of **A**. Also, its direction is different—down and to the right.

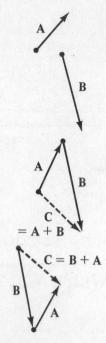

FIGURE 1-2

To perform a *vector addition,* just imagine that these two vectors represent displacements. You start at the tail end of **A** and walk the distance and direction that this vector represents. Then, from your new position, you walk the distance and direction specified by **B**. Your total displacement, the vector sum of **A** and **B**, can be found by drawing the vector **B** with its tail at the head of **A**. Then, an arrow starting at the tail of **A** (your original starting point) and drawn to the head of **B** (your endpoint) is the vector sum of **A** and **B**, written as **A** + **B**. When the vectors represent quantities other than displacements, the length of each vector must be proportional to the magnitude of the quantity. Thus a vector 2 cm long would represent an electrical attraction twice as strong as a vector 1 cm long, for example.

You might have elected to add the two vectors in the opposite order. The last diagram in the figure represents the vector sum **B** + **A**, and it gives the same sum as **A** + **B**. Maybe it looks different because it is in a different configuration, but if you examine it carefully, you will see that it is the same length as **A** + **B** and points in the same direction. Therefore it is the same vector.

When vectors are at right angles to each other, they can be added very easily by means of the simple trigonometry rules. Examples are given in the sample problems.

Sample Problem

1-4 You walk 7 miles south and then 3 miles west. What is your displacement from your starting point?

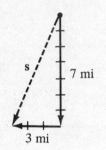

FIGURE 1-3

Solution   The sum of the two displacements is represented in the diagram of Figure 1-3. The vector pointing south is 7 units long and the one pointing west is 3 units long. The vector sum is found by the theorem of Pythagoras:

$$(7 \text{ mi})^2 + (3 \text{ mi})^2 = s^2$$

$$s = 7.6 \text{ mi}$$

The direction is found by using the tangent function:

$$\frac{3 \text{ mi}}{7 \text{ mi}} = \tan \theta$$

$$\theta = 23°$$

The displacement is 7.6 mi 23° W of S.

Sample Problem

1-5 A child is playing with a toy car on the floor of a train that is moving eastward. While the train travels 12.0 m, the child pushes the car 2.6 m northward on the floor of the train. What is the resulting displacement of the car?

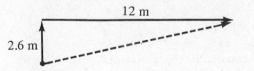

FIGURE 1-4

Solution   The two displacements can be added in either sequence. Figure 1-4 shows the vector addition. The magnitude of the total displacement is

$$s = \sqrt{(2.6 \text{ m})^2 + (12.0 \text{ m})^2} = 12.3 \text{ m}$$

The angle has a tangent of

$$\frac{12.0\ m}{2.6\ m} = 4.615$$

so the direction of the vector sum is 78° E of N.

Core Concept

Vectors are quantities with both magnitude and direction, and they can be added geometrically.

Try This Find the vector sum of 3.5 m east and 5.8 m south. Give both magnitude and direction of the sum.

1.3. Velocity Vectors

Sometimes you have two different velocities at the same time. Your total velocity is then the vector sum of the separate velocities. It is found by vector addition, using the same rules as in adding displacement vectors.

Suppose, for example, a boat can travel 4.0 m/s in still water. It is in a river that flows southward at 5.5 m/s, as shown in Figure 1-5. If the boat heads eastward, directly across the river, what are the direction and magnitude of its total velocity?

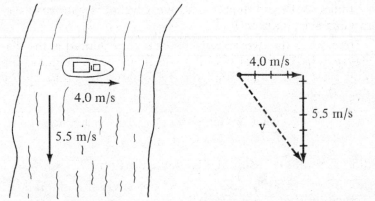

4.0 m/s

5.5 m/s

4.0 m/s

5.5 m/s

v

FIGURE 1-5

The vector diagram in Figure 1-5 shows how the two velocity vectors can be added. You can add them in either sequence; in the diagram, the order shown is

boat's velocity + river's velocity = total velocity

Since these velocities are vectors, they must be added by the rules of vector mathematics, that is, as shown in the vector diagram. We get the answer by solving the triangle. The total velocity of the boat is the hypotenuse of the triangle. Using the theorem of Pythagoras, we obtain

$$\mathbf{v}_{total}^2 = (4.0\ m/s)^2 + (5.5\ m/s)^2$$

from which $\mathbf{v}_{total} = 6.8$ m/s

The boat will obviously not travel directly eastward, but at some angle south of east. To find the angle, use the tangent function:

$$\tan \theta = \frac{5.5\ m/s}{4.0\ m/s}$$

which gives $\theta = 54°$. The total velocity of the boat is therefore 6.8 m/s 54° south of east.

More examples of the addition of velocity vectors are given in Sample Problems 1-6, 1-7, and 1-8.

Sample Problem

1-6 A plane is headed directly east at 340 mi/hr when the wind is from the south at 45 mi/hr. What is its velocity with respect to the ground?

FIGURE 1-6

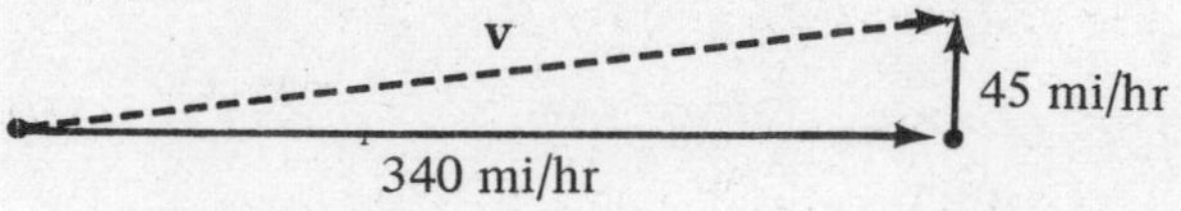

$$\mathbf{v}$$

45 mi/hr

340 mi/hr

Solution The vector diagram for adding these two velocities is shown in Figure 1-6. The hypotenuse of a right triangle is found this way:

$$\mathbf{v}^2 = \left(340\frac{\text{mi}}{\text{hr}}\right)^2 + \left(45\frac{\text{mi}}{\text{hr}}\right)^2$$

$$\mathbf{v} = 343 \text{ mi/hr}$$

and the angle is

$$\tan \theta = \frac{45 \text{ mi/hr}}{340 \text{ mi/hr}}$$

$$\theta = 8°$$

Total velocity is 343 mi/hr at 8° N of E.

Sample Problem

1-7 You are exercising by running around the deck of an ocean liner traveling west at 18.5 m/s. What is your total velocity if you are running athwartships, going south at 7.3 m/s?

Solution Since the two velocities are perpendicular to each other, they add by the theorem of Pythagoras:

$$\mathbf{v} = \sqrt{(18.5 \text{ m/s})^2 + (7.3 \text{ m/s})^2} = 19.9 \text{ m/s}$$

The direction west of south is the angle whose tangent is the ratio of the west velocity to the south velocity:

$$\tan \theta = \frac{18.5 \text{ m/s}}{7.3 \text{ m/s}}$$

$$\theta = 68° \text{ W of S}$$

Sample Problem

1-8 You want to drive your motorboat directly across a stream that flows at

FIGURE 1-7

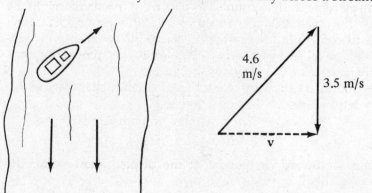

4.6 m/s

3.5 m/s

$\mathbf{v}$

3.5 m/s, and you know that your boat can do 4.6 m/s in still water. Find (a) the angle upstream at which you must point the boat; (b) the resulting speed of the boat in the cross-stream direction.

Solution As shown in Figure 1-7, the velocity of the boat is now the hypotenuse of a right triangle and the resulting velocity is its short leg.

(a) To find the angle, note that we know the hypotenuse of the triangle and the side opposite the angle we are looking for. Then

$$\sin \theta = \frac{\text{opp}}{\text{hyp}} = \frac{3.5 \text{ m/s}}{4.6 \text{ m/s}}$$

so
$$\theta = 50°$$

(b) To find the resulting speed:

$$\mathbf{v}^2 + (3.5 \text{ m/s})^2 = (4.6 \text{ m/s})^2$$

$$\mathbf{v}^2 = (4.6 \text{ m/s})^2 - (3.5 \text{ m/s})^2$$

$$\mathbf{v} = 3.0 \text{ m/s}$$

Core Concept

Velocities are vectors and can be added by the same mathematical rules that govern displacement vectors.

Try This In a plane that goes 620 mi/hr, the pilot wants to fly directly north when the wind is blowing at 110 mi/hr from the east. In what direction must he head the plane?

1.4. A Vector in Parts

In some cases, a vector tells us more than we really want to know. If you get stopped for going 75 mi/hr in a 55 mi/hr zone, the cop is not concerned with whether the direction of your motion is east or north. The only part of the vector that is of any interest is its magnitude.

In other cases, we want to know only what effect the vector has in a particular direction. For example, if you want to know how your time zone changes as you travel, your displacement northward is of no significance; you need to know only how far east or west you have come. And if you are concerned with the height of the sun in the sky at noon, it is only your north-south displacement that counts. Typically, a problem might be stated this way: If I travel 350 miles in a direction 30° west of north, how much is my displacement in the northward direction? We want to ignore any movement along an east-west axis. The quantity we are looking for is the *northward component* of the displacement vector.

It is easily found, as shown in Figure 1-8. From the definition of the cosine, we can write

$$\cos 30° = \frac{s_N}{350 \text{ mi}}$$

where s_N is the northward component of the displacement vector. Its value is

$$s_N = (350 \text{ mi})(\cos 30°) = 303 \text{ mi}$$

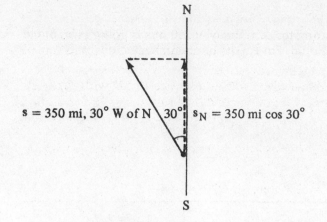

$s = 350$ mi, $30°$ W of N $\quad 30°\quad$ $s_N = 350$ mi cos $30°$

FIGURE 1-8

If you are worried about the time of sunrise, you want the component of your displacement vector on an east-west axis. The principle is the same—see Figure 1-9. This time the displacement vector is making an angle of $60°$ with the axis, so the equation is

$$s_W = (350 \text{ mi})(\cos 60°) = 175 \text{ mi}$$

This process works on any kind of vector, including velocity vectors. We can write a perfectly general rule for finding the component of any vector on any axis:

$$P_x = P \cos \theta_x \qquad \text{(Equation 1-4)}$$

where **P** is a vector (as indicated by boldface type), P_x is its component on the x axis, and θ_x is the angle between the vector and the axis.

FIGURE 1-9

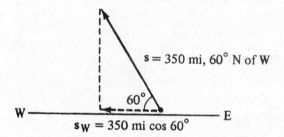

$s = 350$ mi, $60°$ N of W

$60°$

W ———————————— E

$s_W = 350$ mi cos $60°$

Sample Problem

1-9 You are driving along at 20 m/s on a road that heads at an angle of $25°$ W of N. At what rate are you going (a) north; (b) west?

Solution   (a) Since your velocity makes an angle of $25°$ with the north axis, your velocity northward is

$$v_N = (20 \text{ m/s})(\cos 25°) = 18 \text{ m/s}$$

(b) The angle that your velocity makes with the west axis is $(90° - 25°) = 65°$. Therefore your westward velocity is

$$v_W = (20 \text{ m/s})(\cos 65°) = 8.5 \text{ m/s}$$

It should be noted that $\cos (90° - 25°) = \sin 25°$. In general, the component of any vector on an axis perpendicular to the axis given is the vector times the sine of the angle. Thus the west component is

$$v_W = (20 \text{ m/s})(\sin 25°) = 8.5 \text{ m/s}$$

Sample Problem

1-10 An airplane descending to the runway at 250 m/s is going at an angle of 22° with the horizontal. Find (a) its horizontal velocity; (b) its rate of descent.

Solution   (a) Since the airplane makes an angle of 22° with the horizontal, the horizontal component of its velocity is

$$v_{horiz} = (250 \text{ m/s})(\cos 22°) = 232 \text{ m/s}$$

(b) The rate of descent is its vertical velocity, at right angles to the horizontal, so

$$v_{vert} = (250 \text{ m/s})(\sin 22°) = 94 \text{ m/s}$$

Core Concept

The component of a vector on any axis is found by multiplying the vector by the cosine of the angle that the vector makes with the axis:

$$P_x = \mathbf{P} \cos \theta_x$$

Try this   You are going 40 mi/hr up a hill that makes an angle of 22° with the horizontal. How fast are you going (a) horizontally; (b) vertically?

1-5. Acceleration

Things get more complicated when you change your velocity. You do it in your car by stepping on the gas or the brakes, or by turning the steering wheel. Whenever velocity is changing—in either magnitude or direction—the motion is said to be *accelerated*.

Let's take the simplest case first. Suppose you are trying out your new car. You are satisfied to find that you can get from 0 to 60 mi/hr in 9.2 seconds. What you have just measured is the *acceleration* of the car.

Acceleration is the rate at which velocity changes. You can find the average acceleration of the car this way:

$$\text{acceleration} = \frac{60 \text{ mi/hr} - 0}{9.2 \text{ sec}} = 6.5 \text{ mi/hr/sec}$$

The average acceleration is 6.5 miles per hour per second. In other words, the speed increases by 6.5 mi/hr in each second of the trip.

This leads to an algebraic definition of average acceleration:

$$\mathbf{a}_{av} = \frac{\Delta v}{\Delta t} \qquad \text{(Equation 1-5)}$$

In words: average acceleration is the change in velocity divided by the length of time required to make that change.

Let's take a specific case. Suppose a car is traveling at 12 m/s and increases its speed uniformly to 30 m/s, taking 15 seconds to do so. "Uniformly" means that the acceleration of the car is constant, so it must be equal to the average acceleration during that time. The question is, What is the value of that acceleration?

First of all, $\Delta v = 18$ m/s, representing an increase from 12 m/s to 30 m/s. Therefore

$$a = \frac{\Delta v}{\Delta t} = \frac{18 \text{ m/s}}{15 \text{ s}} = 1.2 \text{ m/s}^2$$

Note that the SI unit of acceleration is the meter per second per second, or meter per second squared. In each second, the speed of the car increased by 1.2 m/s.

See the sample problems for other examples of the use of this relationship.

Sample Problem

1-11 What is the acceleration of a rocket ship in outer space that takes 5.0 s to increase its speed from 1,240 m/s to 1,300 m/s?

Solution   The change in its speed is

$$\Delta v = 1{,}300 \text{ m/s} - 1{,}240 \text{ m/s} = 60 \text{ m/s}$$

Then

$$a = \frac{\Delta v}{\Delta t} = \frac{60 \text{ m/s}}{5.0 \text{ s}} = 12 \text{ m/s}^2$$

Sample Problem

1-12 How much does the speed of a car increase if it accelerates uniformly at 2.5 m/s^2 for 5 s?

Solution   Equation 1-5 gives the definition of average acceleration:

$$a_{av} = \frac{\Delta v}{\Delta t}$$

Since the acceleration is constant, it is equal to a_{av}. Multiply the equation through by Δt to get

$$\Delta v = a \, \Delta t$$

$$\Delta v = \left(2.5 \, \frac{\text{m}}{\text{s}^2}\right)(5 \text{ s}) = 7.5 \text{ m/s}$$

Sample Problem

1-13 A car is going 8.0 m/s on an access road into a highway, and then accelerates at 1.8 m/s^2 for 7.2 s. How fast is it then going?

Solution   Equation 1-5 can be transformed to

$$\Delta v = a \, \Delta t = \left(1.8 \frac{\text{m}}{\text{s}^2}\right)(7.2 \text{ s}) = 13.0 \text{ m/s}$$

This is the *increase* in the speed of the car. Since it started at 8.0 m/s, its new speed is

$$8.0 + 13.0 = 21 \text{ m/s}$$

Core Concept

Acceleration is the rate of change of velocity:

$$\mathbf{a}_{av} = \frac{\Delta \mathbf{v}}{\Delta t}$$

Try This   In outer space a rocket ship is traveling at the enormous speed of 2,800 m/s. What is its acceleration if it increases its speed uniformly and is going 2,840 m/s after 25 s?

When motion is accelerated, velocity varies as a function of time. Like any function, this one can be made visual by representing it as a graph. Such a graph is shown in Figure 1-10.

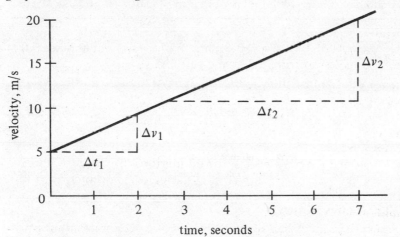

FIGURE 1-10

What does this graph tell you? First of all, note that at time 0 the moving object is going 5 m/s. Time 0 is purely arbitrary; it is the moment when we start our stopwatch. Two seconds later, the velocity has increased to 8 m/s, and it keeps on increasing for the entire 7 seconds plotted in the graph. We do not know what it was doing before we started our clock, or after the first 7 seconds.

The graph can tell us the acceleration of the object. From Equation 1-5,

$$a = \frac{\Delta v}{\Delta t}$$

In the first 2 seconds, $\Delta t = 2$ s and $\Delta v = 8$ m/s $- 5$ m/s $= 3$ m/s. The acceleration is therefore 1.5 m/s^2.

It is clear from the figure that some other time interval would give the same value of acceleration. From simple geometry,

$$\frac{\Delta v_1}{\Delta t_1} = \frac{\Delta v_2}{\Delta t_2}$$

The acceleration is the same regardless of which time interval we elect in order to calculate it. This is true because the graph is a straight line.

In any straight-line graph, the ratio of the increment on the vertical axis to the corresponding increment on the horizontal axis is called the *slope* of the graph. Whenever the graph is a straight line, the ratio of the two variables is a constant. In this case, the graph shows that the acceleration of the object, for the 7 seconds plotted, did not change.

Core Concept

The slope of a velocity-time graph is the acceleration of the object.

Try This Determine the slope of the graph of Figure 1-10 in the interval between 3 and 6 s.

1-7. Starting Up

There is a lot of physics we can do with a simple form of accelerated motion if we can handle it algebraically. The motion is that which is *uniformly accelerated* and *starting at rest,* like the new car you tried out in Section 1-5.

How far did your car travel in going from 0 to 60 miles per hour in 9.2 seconds? It will be easier to solve this problem in SI units; 60 mi/hr is about 26 m/s. Now, we know that

$$v_{av} = \frac{\Delta s}{\Delta t} \qquad \text{(Equation 1-1)}$$

Therefore, if we want to find the value of Δs, we can multiply both sides of this equation by Δt to get

$$\Delta s = v_{av}\, \Delta t$$

So we need to know the average speed of the car during the trip. It is surely *not* 26 m/s! That was its speed at the end of the trip, but for the rest of the time it was going slower. In fact, if it started at 0 and increased its speed uniformly to 26 m/s, it must have averaged just half its final speed, or 13 m/s. The distance is therefore

$$\Delta s = \left(13\frac{m}{s}\right)(9.2\,s) = 120 \text{ m}$$

It will be handy to have a simple set of equations to use whenever we have to deal with motion that is uniformly accelerated in a straight line starting at rest. Let's find some equations.

First, let's define our symbols. Let a be the acceleration. Since it is constant, it is equal to a_{av}. Let s be the distance from the starting point. If we designate the starting point as position zero, then s is equal to Δs, the distance traveled. Similarly, start the clock when the motion begins; then the time $t = \Delta t$. And, since the initial velocity is 0, the change in velocity Δv is the same as the final velocity, which we will call v.

The first equation we need is simply an adaptation of the definition of average acceleration, Equation 1-5:

$$a = \frac{v}{t} \qquad \text{(Equation 1-7a)}$$

Next, we need a way to find the distance traveled. From Equation 1-1, this is

$$s = v_{av}t$$

In going from rest to speed v, the average speed is $v/2$. And, from Equation 1-7a, $v = at$. Therefore

$$s = \left(\frac{at}{2}\right)t$$

This gives us a useful standard relationship:

$$s = \frac{1}{2}at^2 \qquad \text{(Equation 1-7b)}$$

Look at the sample problems to see how this equation is applied.

Sample Problem

1-14 How far does a car travel if it starts at rest and accelerates at 3.0 m/s² for 6.5 s?

Solution Since it starts at rest and accelerates uniformly, Equation 1-7b applies:

$$s = \frac{1}{2}at^2$$

$$s = \frac{1}{2}\left(3.0 \ \frac{m}{s^2}\right)(6.5 \ s)^2 = 63 \ m$$

Sample Problem

1-15 What is the acceleration of a rocket that accelerates uniformly from rest and travels 650 m in the first 12 s?

Solution Since the acceleration is uniform and starts from rest, Equation 1-7b applies. It can be written as

$$a = \frac{2s}{t^2} = \frac{2(650 \ m)}{(12 \ s)^2} = 9.0 \ m/s^2$$

Sample Problem

1-16. How long does it take a car, starting from rest, to travel 240 m if its acceleration is 1.90 m/s²?

Solution From Equation 1-7b,

$$t = \sqrt{\frac{2s}{a}} = \sqrt{\frac{2(240 \ m)}{1.90 \ m/s^2}} = 16 \ s$$

Note how the units work out:

$$\sqrt{\frac{m}{m/s^2}} = \sqrt{\frac{\not{m} \cdot s^2}{\not{m}}} = s$$

We need one more. From Equation 1-7a,

$$t = \frac{v}{a}$$

Substituting this in Equation 1-7b gives

$$s = \frac{1}{2}a\left(\frac{v}{a}\right)^2$$

which is easily converted to

$$v = \sqrt{2as} \qquad\qquad \textbf{(Equation 1-7c)}$$

The sample problems show how this expression can be applied.

Sample Problem

1-17 How far does a car travel if it starts at rest and accelerates at 3.0 m/s² until it reaches a speed of 22 m/s?

Solution Since the car starts at rest and accelerates uniformly, Equation 1-7c can be applied:

$$v = \sqrt{2as}$$

Square both sides and divide by 2*a* to get

$$s = \frac{v^2}{2a}$$

$$s = \frac{(22 \ m/s)^2}{2(3.0 \ m/s^2)} = 81 \ m$$

Check the units:

$$\frac{(m/s)^2}{m/s^2} = \left(\frac{m^2}{\cancel{s^2}}\right)\left(\frac{\cancel{s^2}}{m}\right) = m$$

This set of useful equations can be applied in another circumstance as well. They work for an object that slows down from speed v to rest, changing its speed uniformly as it goes.

Sample Problem

1-18 A car going 22 m/s has its brakes jammed on and leaves skid marks 45 m long. What was its acceleration?

Solution   Slowing down is also acceleration, but in the opposite direction from speeding up. (See Section 1-9.) Since the car comes to rest, Equation 1-7c applies. Then

$$a = \frac{v^2}{2s} = \frac{(22 \text{ m/s})^2}{2(45 \text{ m})} = 5.4 \text{ m/s}^2$$

Core Concept

Problems involving objects that accelerate uniformly, starting or ending at rest, can be solved with a set of three simple equations:

$$a = \frac{v}{t}; \qquad s = \frac{1}{2}at^2; \qquad v = \sqrt{2as}$$

Try This   What is the acceleration of a rocket-driven sled that travels 360 m in 5.0 s, starting from rest and accelerating uniformly?

Try This   How fast is a car going if it starts at rest and accelerates uniformly at 2.8 m/s^2 while traveling 220 m?

1-8. Falling Freely

When something is dropped, its velocity starts at zero and increases. It is accelerated. It is a remarkable fact that in a vacuum all objects—rocks, books, feathers, raindrops, dust particles—have the same acceleration. If dropped together, all will accelerate uniformly and strike the ground at the same time. This rule does not hold in everyday experience only because the resistance of the air tends to hold things back—and it holds feathers more effectively than rocks.

The acceleration of an object in free fall (in a vacuum) does not depend on the nature of the object. It depends only on where the object happens to be. The acceleration is uniform, and its value is the *acceleration due to gravity*, represented by the letter g. On earth, g is about 9.8 m/s^2, or 32 ft/s^2.

The value is far different on other planets—3.3 m/s^2 on Mars, 25.6 m/s^2 on Jupiter; and when our astronauts walked on the moon, anything they dropped accelerated downward at a mere 1.67 m/s^2. Even on earth, the acceleration due to gravity varies a little. It is about 9.78 m/s^2 at the equator, increasing to 9.83 m/s^2 at the poles. It also decreases with altitude, dropping from 9.80 m/s^2 at the surface (at latitude 40°) to 9.79 m/s^2 at altitude 10 miles.

Whatever the local value of g, it can be used in Equations 1-7 to find out how long it takes something to hit the ground or how fast it will be going when it hits. Just remember that for an object in free fall, $a = g$.

Sample Problems 1-19 and 1-20 show how to proceed. A word of caution, however: If the object is very light, or if it has a lot of flat surface, or if it falls a great distance, air resistance becomes significant, and the equations will tell you only lies.

Sample Problem

1-19 A high-wire artist missteps and falls 9.2 m to the ground. What is her speed on landing?

Solution: The motion is uniformly accelerated and starts at rest, so from Equation 1-7c

$$v = \sqrt{2as} = \sqrt{2\left(9.8\frac{m}{s^2}\right)(9.2\ m)} = 13\ m/s$$

Sample Problem

1-20 A cat falls out of a tree, dropping 16 m to the ground. How long is the cat in the air?

Solution: Since the cat started at rest and accelerated uniformly, Equation 1-7b applies; $a = g$, so

$$s = \frac{1}{2}gt^2$$

Multiply through by $2/g$ and take the square root of both sides to get

$$t = \sqrt{\frac{2s}{g}}$$

$$t = \sqrt{\frac{2\ (16\ m)}{9.8\ m/s^2}} = 1.8\ s$$

Core Concept

An object in free fall accelerates uniformly, and the value of the acceleration due to gravity depends only on the object's location in space.

Try This You drop a rock from a bridge, and it hits the water 2.3 s later. Find (a) the height of the bridge; (b) the velocity of the rock when it hits.

1-9. The Direction of Acceleration

Any change of velocity—speeding up, slowing down, or turning a corner—is an acceleration. We distinguish these different kinds of accelerated motion by assigning a *direction* to the acceleration. In other words, acceleration is a vector.

If you are traveling in a straight line and speed up, your change in velocity ($\Delta \mathbf{v}$) is in the direction you are going. Then your acceleration is also in that direction. This is the first rule for the direction of an acceleration: when the acceleration is in the same direction as the velocity, the result is an increase in speed.

If, on the other hand, you are slowing down, then the change in your velocity is *opposite* to the velocity itself; if you are going east at 30 mi/hr and slow down to 20 mi/hr, $\Delta \mathbf{v} = 10$ mi/hr west. Since acceleration is always in the direction of the change in velocity, we get a second rule: when acceleration is in the opposite direction from velocity, the result is a decrease in speed.

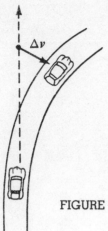

Δv

Now suppose you do not change speed, but are making a turn. This is a change in only the direction of velocity; since velocity is a vector, this is surely a change in velocity—an acceleration. Figure 1-11 shows how to find the direction of this acceleration. The dotted line show what the path of the car would be if its motion were not accelerated—that is, if it traveled at constant speed in a straight line. The arrow is a vector representing the change in the velocity of the car; that is, the vector that must be added to the old velocity to get the new velocity. The acceleration is in the same direction. When the car turns to the right, its acceleration is to the right, perpendicular to its velocity.

FIGURE 1-11

Core Concept

When an object speeds up, its acceleration is in the same direction as its velocity; when it slows down, its acceleration is in the opposite direction; when it changes direction with no change in speed, its acceleration is perpendicular to the velocity, toward the side to which the object is turning.

Try This In each of the following cases, state the direction of the acceleration:

1. A rising elevator is coming to rest.
2. A car going west speeds up.
3. A car going east slows down.
4. A bicycle going southward starts to turn left.

1-10. More Free Falls

To put a rock into free fall, you do not necessarily just drop it. You can also throw it—up, down, or sideways. Whichever you do, the rock is in free fall once it leaves your hand, and it must therefore accelerate downward at the rate of the acceleration due to gravity—9.8 m/s^2.

FIGURE 1-12

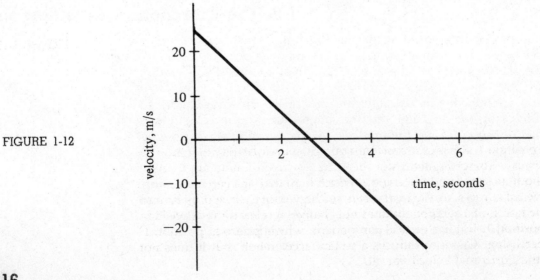

Suppose, for example, you throw the rock straight upward and it leaves your hand going 25 m/s. Since its acceleration is downward, opposite to its velocity, it must slow down. After 1 second, it is going (25 m/s − 9.8 m/s) = 15.2 m/s. It loses another 9.8 m/s of velocity during the next second. Eventually, it comes to rest and starts downward. Then it speeds up, since its acceleration is now in the same direction as its velocity.

Look at the graph of Figure 1-12 to see how this accelerated motion takes place. At time 0, the velocity is 25 m/s. When 2.6 seconds have passed, the velocity has dropped to zero—the rock has come to rest at the top of its flight. From that point on, the velocity is negative, meaning that the rock is now on the way down. It stops at the point from which it was thrown, going 25 m/s. But through the entire flight, the slope of the graph—the acceleration—has not changed. Even when the rock is at the top of its flight, at rest, its velocity has not stopped changing, at the rate of 9.8 m/s².

Sample Problem

1-21 You throw a baseball straight up, and it leaves your hand at 15 m/s. What is its velocity 2.0 s later?

Solution First, let's find the amount of speed the baseball loses as it rises. From Equation 1-5,

$$\Delta v = g\ \Delta t$$

$$\Delta v = \left(9.8\ \frac{m}{s^2}\right)(2.0\ s) = 19.6\ m/s$$

Since this is a loss of speed, it must be subtracted from the initial speed:

$$v_{final} = 15\ m/s - 19.6\ m/s = -4.6\ m/s$$

The negative sign indicates that the ball has reversed its direction of travel.

Now suppose you stand on the roof and throw the rock horizontally. Figure 1-13 shows what happens. The path of the rock is a parabola; its velocity is shown by vectors at six positions of its trajectory.

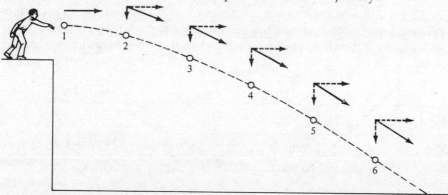

FIGURE 1-13

At position 1, it has just been thrown and its velocity is horizontal. A little later, it is going slightly faster, heading downward at an angle. As it travels, both its speed and the angle at which it is heading keep increasing.

Now take a look at the *components* of the velocity, shown as broken lines. The horizontal component does not change; it is the same at position 6 as at position 1. But the vertical component, which is zero at position 1, keeps increasing. Gravity produces a *vertical* acceleration, but it does not change the horizontal velocity at all.

Looking at the vertical motion only, we see that the rock starts at rest and accelerates uniformly at rate g. The time required for the rock to hit the ground is exactly the same as if it were dropped from rest. The horizontal motion has nothing to do with its vertical fall.

How far does the rock travel horizontally? You can figure this out by calculating how long it takes the rock to hit the ground, just as if it were simply dropped. During that time, it is traveling horizontally at constant speed, so you can easily figure out how far it goes. See Sample Problem 1-22.

Sample Problem

1-22 A rock is thrown horizontally at 25 m/s from a roof that is 15 m high. How far does the rock go horizontally before hitting the ground?

Solution   First, let's find out how long it takes the rock to fall to the ground. Since the rock has no initial vertical velocity and accelerates uniformly, Equation 1-7b applies; $a = g$, so

$$s = \frac{1}{2}gt^2$$

which can be written as

$$t = \sqrt{\frac{2s}{g}} = \sqrt{\frac{2\,(15\text{ m})}{9.8\text{ m/s}^2}} = 1.75\text{ s}$$

During that time the rock is traveling horizontally at a constant speed, so, from Equation 1-1,

$$\Delta s = v\,\Delta t = \left(25\ \frac{\text{m}}{\text{s}}\right)(1.75\text{ s})$$

$$\Delta s = 44\text{ m}$$

Core Concept

Acceleration of an object in free fall depends only on gravity; it is always vertical regardless of which way the object is moving.

Try This   A baseball is thrown from the roof of a building at 20 m/s in a horizontal direction. It strikes the ground 2.8 s later. Find (a) how far from the building the baseball lands; (b) the height of the building.

1-11. Going in Circles

When an object is traveling in a circular path at constant speed, its direction is constantly changing. Therefore this is an example of accelerated motion.

The motorcyclist in Figure 1-14 is going around a circular track of radius r, and his speed is v. The direction of his velocity is tangential; at the moment, he is at the northernmost point of the track and traveling eastward. Let T stand for the *period* of his motion, that is, the length of time it takes him to go around once. Since the distance around the track is $2\pi r$, his speed is

$$v = \frac{2\pi r}{T} \qquad \text{(Equation 1-11a)}$$

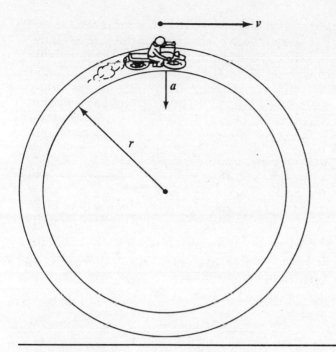

FIGURE 1-14

Sample Problem

1-23 What is the period of the motion of a runner going 9.2 m/s on a circular track whose radius is 22 m?

Solution From Equation 1-11a,

$$T = \frac{2\pi r}{v} = \frac{2\pi\,(22\text{ m})}{9.2\text{ m/s}} = 15\text{ s}$$

As the motorcyclist travels, he is constantly turning to the right, and his speed is constant. This means that his acceleration is to the right and perpendicular to his motion, always toward the center of the circle. The acceleration is said to be *centripetal*, which simply means toward the center of a circle.

The magnitude of centripetal acceleration can be calculated from this formula:

$$a_c = \frac{v^2}{r} \qquad \text{(Equation 1-11b)}$$

See the sample problems for examples.

Sample Problem

1-24 What is the acceleration of a motorcycle going 28 m/s on a circular track whose radius is 140 m?

Solution From Equation 1-11b,

$$a_c = \frac{v^2}{r} = \frac{(28\text{ m/s})^2}{140\text{ m}} = 5.6\text{ m/s}^2$$

Sample Problem

1-25 The clothes in a rotary drier whose radius is 0.25 m are accelerated toward the center of the drier at 22 m/s². How fast are the clothes moving?

Solution From Equation 1-11b,

$$v = \sqrt{ar} = \sqrt{\left(22\tfrac{\text{m}}{\text{s}^2}\right)(0.25\text{ m})} = 2.3\text{ m/s}$$

When acceleration has constant magnitude and is kept always perpendicular to velocity, motion is in a circle at constant speed and the centripetal acceleration can be calculated:

$$a_c = \frac{v^2}{r}$$

Try This The motorcyclist of Figure 1-14 is on a track whose radius is 150 m, and he makes one complete circuit every 30 s. Find the direction and magnitude of his (a) velocity and (b) acceleration when he is at the southernmost point of the track.

1-12. Going Into Orbit

Suppose you could build a tower several miles high and mount a horizontal cannon at its top, as shown in Figure 1-15. The cannonballs will fall to earth, always accelerating downward in free fall. The greater the speed, the farther a ball will travel before it strikes. The fastest ball in the picture will never hit the ground at all. It keeps falling toward the earth, but it never lands; the earth curves away from it as it falls. It is in orbit.

FIGURE 1-15

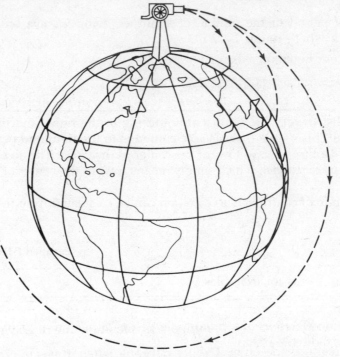

Objects in orbit around the earth—artificial satellites, Spacelabs, the moon—are in free fall. They must obey the usual rule of free fall: their acceleration depends only on where they are. A satellite orbiting the earth at low altitude has a downward acceleration of 9.8 m/s². Farther away, the acceleration is less. For the moon, it is a great deal less.

We do not build towers to put satellites in orbit, but we get the satellites up there. They are lifted by rockets. Once the desired altitude has been reached, the last-stage rocket is pointed horizontally and fired, placing the satellite into a horizontal path. Its acceleration is down, perpendicular to its velocity. If the velocity is exactly right, the direction of the satellite will change just enough as it travels to keep the velocity always perpendicular to the acceleration. The path will then be a circle.

For the satellite to go into a circular orbit, it must conform to the condition for uniform circular motion, Equation 1-11b. If it is in orbit not far above the surface of the earth, the radius of the orbit, r, is the radius of the earth and the acceleration, a, is $g = 9.8$ m/s^2. Solving the equation tells us just what speed must be given to the satellite to put it into circular orbit. The calculation in Sample Problem 1-26 shows that it has to go about 7,900 m/s. At that speed, the satellite circles the earth every 85 minutes.

Sample Problem

1-26 An artificial satellite is to be put into orbit a short distance above the surface of the earth. What speed must it be given to make it go into a circular orbit? (The radius of the earth is 6.4×10^6 m.)

Solution The satellite's acceleration must be g, since it is in free fall. If this is to produce a circular orbit, it must satisfy Equation 1-11b, so

$$g = \frac{v^2}{r}$$

which can be written as

$$v^2 = gr$$

$$v^2 = \left(9.8 \ \frac{m}{s^2}\right)(6.4 \times 10^6 \ m)$$

$$v = 7,900 \ m/s$$

Sample Problem

1-27 What is the acceleration due to gravity at the surface of a planet with a radius of 1.6×10^6 m if a satellite near the surface makes one circular orbit every 1,700 s?

Solution The acceleration due to gravity must be the centripetal acceleration of the satellite. We can find its speed from Equation 1-11a:

$$v = \frac{2\pi r}{T} = \frac{2\pi(1.6 \times 10^6 \ m)}{1,700 \ s} = 5,910 \ m/s$$

Now we can find the acceleration:

$$a = \frac{v^2}{r} = \frac{(5,910 \ m/s)^2}{1.6 \times 10^6 \ m} = 22 \ m/s$$

Core Concept

An object in orbit is in free fall with centripetal acceleration due to gravity.

Try This How fast was the Lunar Orbiter traveling when it was in orbit around the moon, near its surface? The radius of the moon is 3.5×10^6 m, and the acceleration due to gravity at its surface is 1.67 m/s^2.

Summary Quiz

For each of the following, fill in the missing word or phrase:

1. Distance traveled divided by elapsed time gives _____ .
2. The SI unit of velocity is the _____ .
3. A quantity that has magnitude but no direction is called a _____ .
4. A _____ has magnitude and direction.

5. Depending on their relative directions, the vector sum of two vectors 6 m and 2 m long must be somewhere between _____ m and _____ m long.
6. If a plane is flying in a direction 30° west of north, the northward component of its velocity is the velocity times the _____ of 30°.
7. The rate of change of velocity is called _____ .
8. The SI unit of acceleration is the _____ .
9. If a car starts at rest and accelerates uniformly, the distance it travels is proportional to the _____ of the time it travels.
10. All objects in free fall at a given place have the same _____ .
11. If a car is going northward and the driver jams on its brakes, the direction of its acceleration is _____ .
12. If a car is going northward and starts to turn left, the direction of its acceleration is _____ .
13. When a baseball is hit to center field, only the _____ component of its velocity changes as it travels.
14. The rate at which the velocity of the baseball changes is _____ .
15. When an object is going in a circular path at constant speed, the direction of its acceleration is _____ .
16. The acceleration of the moon is toward the _____ .
17. When an object is in orbit, its acceleration is produced by _____ .

Problems

1. A bicycle averages 4.5 m/s while traveling for 10 min. How far does it travel?
2. What is the average speed of a car that travels 4.6×10^4 m in 1 hr? Give your answer in SI units.
3. You are on an ocean liner that is going eastward at 12.0 m/s, and you run southward at 3.6 m/s. Find the magnitude and direction of your resulting velocity.
4. A pilot wants to fly her plane directly eastward when the wind is from the north at 55 mi/hr. If the air speed is 230 mi/hr, in what direction must she head her plane?
5. You walk 2.0 mi north, then 4.5 mi east, then 6.2 mi south. What is your displacement from your starting point?
6. You drive a car 45 mi in a direction north 30° W. How much farther west are you?
7. A sailor's compass tells him that he is traveling N 55° W, and his sextant tells him that at the end of 6.0 hr he is 35 mi farther north. How fast is he going?
8. What is the acceleration of a car that speeds up from 12 m/s to 30 m/s in 15 s?
9. If a car can accelerate at 3.2 m/s², how long will it take to speed up from 15 m/s to 22 m/s?
10. How far does a motorcycle travel if it starts at rest and is going 22 m/s after 15 s?
11. What is the acceleration of a car that gets to a speed of 18 m/s from rest while traveling 240 m?
12. A ball is dropped from a window 24 m high. How long will it take to reach the ground?
13. An arrow is fired straight up, leaving the bow at 15 m/s. If air resistance is negligible, how high will the arrow rise?
14. A firefighter drops from a window into a net. If the window is 34 m above the net, at what speed does the firefighter hit the net?

15. A trained acrobat can safely land on the ground at speeds up to 15 m/s. What is the greatest height from which the acrobat can fall?

16. An elevator descending at 4.4 m/s is accelerated upward at 1.5 m/s^2 for 2.0 s. What is its velocity at the end of that time?

17. A toy train is traveling around a circular track 2.0 m in radius, and it makes a complete circuit every 4.5 s. Find (a) its velocity; (b) its acceleration.

18. A carousel is considered safe if no rider is accelerated at more than 3.0 m/s^2. What is the greatest permissible speed of a rider at the outer edge of a carousel that is 6.5 m in radius?

19. A car will skid if its acceleration as it makes a turn is more than 3.5 m/s^2. If the car is traveling at 20 m/s, what is the radius of the smallest circle it can travel in without skidding?

20. In the spin cycle of a washing machine, the clothes must be accelerated at 75 m/s^2 to squeeze the water out of them. If the radius of the basket is 30 cm, how many revolutions must it make per minute?

21. A baseball is thrown horizontally from a window that is 22 m high. If the initial speed of the ball is 18 m/s, find (a) how long the ball takes to reach the ground; (b) how far from the building it lands.

22. The radius of Mars is 3.4×10^6 m, and the acceleration due to gravity near its surface is 3.3 m/s^2. How fast would a satellite have to go in order to be in orbit near the surface of Mars?

23. An astronaut discovers a strange planet that has a small satellite orbiting near its surface, making a complete circuit every 40 min. If the radius of the planet is 2.2×10^6 m, what is the acceleration due to gravity at its surface?

CHAPTER 2

Forces: Push and Pull

2-1. What Is a Force?

Every now and then, the whole structure of physical theory has to be reorganized. This happens when the theory comes into conflict with observations of the real world. Then someone with an analytical mind looks at the theory and says, "That can't be right!"

It happened in the sixteenth century, when the Italian professor Galileo examined the accepted answer to the question of what makes things move. The Greek philosopher Aristotle had given the answer centuries before: things keep moving, he said, as long as there are forces acting on them. When the horse stops pulling, the wagon stops moving. The harder the horse pulls, the faster the wagon goes. The spear was a problem. What keeps it going after it leaves the soldier's hand? Aristotle said that a current of air displaced by the spear and coming back behind it keeps pushing the spear forward.

Galileo looked at this theory and said, "That can't be right." The theory could not account for the endless motion of the planets through airless space. By adopting a whole new set of assumptions, Galileo was able to calculate, for the first time, the flight of a cannonball.

The trouble, said Galileo, is that the problem has not been correctly stated. We do not have to explain why something keeps on moving, once started. Anything in motion will keep on moving in a straight line forever, unless something is done to it. The thing we have to explain is the *change* in velocity, and the "something" we have to apply to cause the change is called a *force*.

25

This is surely a much more reasonable proposition today than it was in Aristotle's time. When you take your foot off the gas pedal, your car does not suddenly come to a stop. It coasts on, only gradually losing its velocity. If you want the car to stop, you have to do something to it. That is what your brakes are for: to exert a force that decreases the car's velocity. A spacecraft illustrates this point vividly, for Voyager has been coasting through the solar system for years, studying, in turn, the surfaces of Mars, Jupiter, Saturn, and Neptune. Nothing is pushing it. When we want it to speed up, slow down, or change direction, we send it signals that fire its control rockets.

Core Concept

A force is something that can change the velocity of an object by making it start, stop, speed up, slow down, or change direction.

Try This Tell what effect the air has on a flying baseball according to (a) Aristotle; (b) Galileo.

2-2. Kinds of Force: Friction

FIGURE 2-1

Your car has three controls whose function is to change its velocity: the gas pedal, the brake pedal, and the steering wheel. The brakes make use of the same force that stops your car if you just let it coast: friction.

Sliding friction is a force that is generated whenever two objects are in contact and there is relative motion between them. When something is moving, friction *always* acts in such a direction as to retard the relative motion. Thus the direction of the force of friction on a moving object will always be directly opposite to the direction of the velocity. The brake shoes slow down the rotation of the wheels, and the tires slow the car until it comes to rest.

In the simplest case, the force of friction can be measured quite easily by means of a spring scale. (The spring scale can be used to measure all kinds of forces. Basically, it measures the force pulling on its shackle.) A spring scale is calibrated in force units—pounds, or (in the SI) *newtons*. A newton (abbreviated N) is a rather small unit of force; it takes 4.45 newtons to equal 1 pound.

Figure 2-2 shows a spring scale being used to measure the force of sliding friction between a brick and the horizontal surface on which it is resting. The scale is pulling the brick along at constant speed. In this condition, the brick is said to be in *equilibrium*. Since its velocity is not changing, it follows that the net force acting on it is zero.

However, the spring scale is surely exerting a force, which is indicated on its face. This force is pulling the brick to the right. The brick can remain in equilibrium only if there is an equal force pulling it to the left, so that the net force acting is zero. In the situation shown, the only force pulling the brick to the left is the frictional force between the brick and the surface. Therefore the reading on the spring scale is equal to the force of friction.

FIGURE 2-2

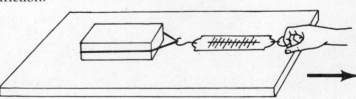

Anyone who has ever moved furniture around, or slid a box across the floor, knows that the frictional force is greater when the furniture or the box is heavier. The force of friction does not change much as the object speeds up, but it depends very strongly on how hard the two surfaces are pressed together.

When a solid moves through a liquid, such as a boat moving through the water, there is a frictionlike force retarding the motion of the solid. It is called *viscous* (that's pronounced VISKus, not VISHus) *drag*. Like friction, it always acts opposite to velocity. Unlike friction, however, it increases greatly with speed, and it depends more on the shape of the object and on the nature of the liquid than on the object's weight. Gases also produce viscous drag, and you are probably most familiar with this as the air resistance that acts on a car at high speed.

Friction and viscous drag are forces that act to retard motion.

Try This Explain why an outboard motor has to work harder to keep a boat going at 10 mi/hr than at 5 mi/hr.

2-3. Gravity

Probably the first law of physics that everyone learns is this: If you drop something, it falls. Since its velocity keeps on changing as it falls, there must be a force acting on it all the while. That is the force we call *gravity*.

You can measure gravity by balancing it off with a spring scale. When you hang something on a spring scale and read the scale in pounds or newtons, you usually call the force of gravity acting on the object by a special name. You call it the *weight* of the object.

Weight is not a fixed property of an object; it varies with location. A person who weighs 160 pounds at the North Pole will check in at 159.2 pounds at the equator. If he should step on a scale on the moon, it would read only 27 pounds. Everything weighs less where the acceleration due to gravity is smaller. Weight, in fact, is directly proportional to the acceleration due to gravity, or

$$w \propto g$$

Obviously, weight also depends on something else, since things have different weights even if all are at the same place. Weight depends on how much stuff there is in the object. If you buy 10 pounds of sugar, you expect to get twice as much as if you buy 5 pounds. And if you take both sacks to the moon, one will still weigh twice as much as the other. The amount of sugar, the *mass* of the sugar, did not change when it was brought somewhere else. And the more sugar you have, the more it weighs.

Weight, then, is proportional both to mass and to the acceleration due to gravity:

$$w = mg \qquad \text{(Equation 2-3)}$$

This equation works nicely, without introducing any constants, if the units are carefully defined. Mass is measured in kilograms; the kilogram is one of the basic units of the SI. By definition, when you multiply the mass in kilograms by the acceleration in meters per second squared, the weight comes out in newtons. Sample Problem 2-1 shows how this relationship is used. Note that the definition tells us that $1 \text{ N} = 1 \text{ kg} \cdot \text{m/s}^2$.

Sample Problem

2-1 If a hammer has a mass of 2.5 kg, how much does it weigh (a) on earth; (b) on Mars?

Solution (a) Equation 2-3 tells us that

$$w = mg$$

and we know that g on earth is 9.8 m/s^2. Therefore

$$w = (2.5 \text{ kg})\left(9.8 \ \frac{\text{m}}{\text{s}^2}\right)$$

$$w = 24.5 \ \frac{\text{kg} \cdot \text{m}}{\text{s}^2} = 24.5 \text{ N}$$

(b) On Mars:

$$w = (2.5 \text{ kg})\left(3.3 \ \frac{\text{m}}{\text{s}^2}\right) = 8.3 \text{ N}$$

Sample Problem

2-2 A 62-kg astronaut lands on a strange planet. He drops his phasor, and finds that it falls the 3.5 m from his doorway to the ground in 1.6 s. Find (a) the acceleration due to gravity; (b) the astronaut's weight.

Solution (a) From Equation 1-7b,

$$a = \frac{2s}{t^2} = \frac{2(3.5 \text{ m})}{(1.6 \text{ s})^2} = 2.7 \text{ m/s}^2$$

$$\text{(b) } w = mg = (62 \text{ kg})\left(2.7\frac{\text{m}}{\text{s}^2}\right) = 167\frac{\text{kg} \cdot \text{m}}{\text{s}^2} = 167 \text{ N}$$

Core Concept

Weight, the force of gravity, is the product of mass and the acceleration due to gravity:

$$w = mg$$

Try This What is the mass of a girl who weighs 340 N on earth?

2-4. Elastic Recoil

The basic feature of a solid, as opposed to a liquid or a gas, is that it has a definite shape. It resists changes in its shape and, in so doing, exerts a force against whatever force is applied to it.

Look, for example, at the meter bar supported at its ends, shown in Figure 2-3. If you push down on it, you bend it, and you can feel it pushing back on you. The harder you push, the more the bar bends and the harder it pushes back. It bends just enough to push on you with the same force that you exert on it. The force it exerts is called *elastic recoil*.

FIGURE 2-3

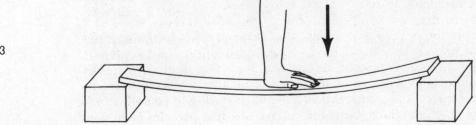

The same thing happens when you stand on the floor, or on the ground. You can't see the floor bend, but it bends just the same, although not very much. Even a feather resting on the floor bends it a little. The elastic recoil force needed to support the feather is very small, so the amount of bending is too small to detect.

A rope does not resist bending, but it certainly does resist stretching. When a rope is stretched, it is said to be in a state of *tension*. The tension can be measured by cutting the rope and inserting a spring scale into it, as shown in Figure 2-4. The spring scale will read the tension in the rope, in pounds or newtons. If you pull with a force of 30 N, for example, the tension in the rope is 30 N, and that is what the scale will indicate. Something, such as the elastic recoil of the wall to which the scale is attached, must be pulling on the other end with a force of 30 N. The tension in the rope is the same throughout, and is equal to the elastic recoil force that the rope exerts at its ends.

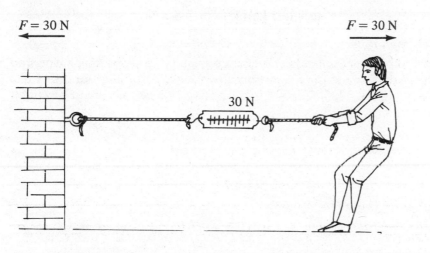

$F = 30 \, \text{N}$ $F = 30 \, \text{N}$

30 N

FIGURE 2-4

Core Concept

A force applied to a solid distorts the shape of the solid, causing it to exert a force back on the force that distorted it.

Try This   In a tug-of-war, each team exerts a force of 350 pounds on the rope. What is the tension in the rope?

2-5. Some Other Forces

If you take a deep breath and dive into a pool, you will have a lot of trouble keeping yourself submerged. Something keeps pushing you up. That force is called *buoyancy*.

The force of buoyancy acts in an upward direction on anything submerged in a liquid or a gas. Buoyancy is the force that makes ships float and helium-filled balloons rise. A rock sinks because its weight is larger than the buoyancy of the water. A submerged cork rises because the buoyancy is more than its weight. When it reaches the surface, some of it emerges, but the rest is still under water. The amount under water is just enough to produce a buoyant force equal to its weight, so it stays put.

There are some other familiar forces, and others not so familiar. You know about magnetism, which attracts iron nails to a red horseshoe. You have met electric force, which makes a nylon shirt cling to you when you try to take it off, or which refuses to release the dust particles from your favorite record. Airplanes stay up because of the lift force generated by the flow of air across their wings. A rocket takes off because of the force generated by the gases expanding in it. We will meet other forces from time to time, but we have enough to get along with for now.

Core Concept

There are many other kinds of force, including buoyancy, which acts upward on anything submerged in a liquid or a gas.

Try This An object submerged in water has a mass of 6 kg, and the buoyancy force acting on it is 75 N. Will the object rise or fall in the water?

Try This Figure 2-5 shows a boat that has sunk to the bottom of a lake and is being dragged out by a man pulling on its painter. There is also a diagram indicating eight directions of space. From these eight, select the one that indicates the direction of each of the following forces acting on the boat: (a) friction; (b) viscous drag; (c) buoyancy; (d) weight; (e) elastic recoil of the rope; (f) elastic recoil of the lake bottom.

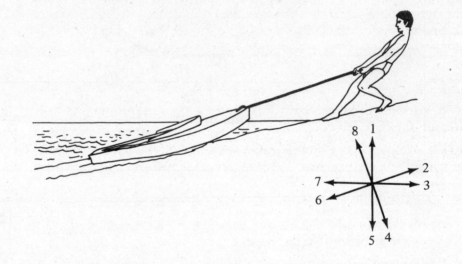

FIGURE 2-5

2-6. Action and Reaction

The batter steps up to the plate and takes a healthy swing, sending the ball into left field. The bat has exerted a large force on the ball, changing both the magnitude and the direction of its velocity. But the ball has also exerted a force on the bat, slowing it down. The batter feels this when the bat hits the ball.

Next time up, he strikes out. He has taken exactly the same swing, but exerts no force on anything. (Air doesn't count.) The batter has discovered that it is impossible to exert a force unless there is something there to push back. Forces exist *only* in pairs. When object *A* exerts a force on object *B*, then *B* must exert a force on *A*. The two forces are sometimes called *action* and *reaction*, although which is which is often rather arbitrary. The two parts of the interaction are equal in magnitude, are opposite in direction, and act on different objects.

Each of the six forces acting on the boat of Figure 2-5 is half of an interaction pair. Friction of the lake bottom pulls the boat in direction 6; the boat pulls the sand of the lake bottom in direction 2. Elastic recoil of the lake bottom supports the boat; elastic recoil of the boat depresses the lake bottom. The viscous drag of the water pulls the boat in direction 6, and the boat pulls some water along with it, in direction 2. Tension in the rope pulls the boat in direction 2; the boat stretches the rope in direction 6.

The weight of the boat is also one half of an interaction. It is the gravitational influence of the whole earth that pulls the boat down. If the law of interaction is valid, the gravity of the boat must be pulling the earth up. This effect cannot be detected in the context under discussion, but that does not invalidate the law. If we calculated the effect the force has on the earth, we would find that it is far too small to detect.

The law of action and reaction leads to an apparent paradox, if you are not careful how it is applied. The horse pulls on the wagon. If the force of the wagon pulling the horse the other way is the same, as the law insists, how can the horse and wagon get started?

The error in the reasoning is this: If you want to know whether the horse gets moving, you have to consider the forces acting *on the horse*. The force acting on the wagon has nothing to do with the question. The horse starts up because the force he exerts with his hooves is larger than the force of the wagon pulling him back. And the wagon starts up because the force of the horse pulling it forward is larger than the frictional forces holding it back. To know how something moves, consider the forces acting *on it*. The action and reaction forces *never* act on the same object.

2-7. Balanced Forces

If a single force acts on an object, the velocity of the thing must change. If two or more forces act, however, their effects may eliminate each other. This is the condition of equilibrium, in which there is no net force and the velocity does not change. We saw such a condition in the case in which gravity is pulling an object down and the spring scale, used for weighing it, is pulling it upward.

An object in equilibrium may or may not be at rest. A parachutist, descending at constant speed, is in equilibrium. His weight is just balanced by the viscous drag on the parachute, which is why he put it on in the first place. A heavier parachutist falls a little faster; his speed increases until the viscous drag just balances his weight.

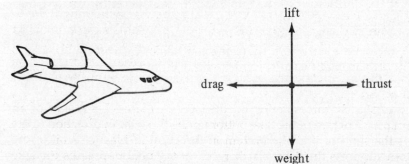

FIGURE 2-6

Balancing the vertical forces is not enough to produce equilibrium. An airplane traveling at constant speed, as in Figure 2-6, is in equilibrium under the influence of four forces, two vertical and two horizontal. Vertical: gravity (down) is just balanced by the lift produced by the flow of air across the wing. Horizontal: viscous drag is just balanced by the thrust of the engines. Both the vertical and the horizontal velocities are constant.

The brick of Figure 2-2, resting on a tabletop and being pulled along at constant speed, is another example. Vertical: the downward force of gravity is balanced by the upward force of the elastic recoil of the tabletop. Horizontal: the tension in the spring scale, pulling to the right, is balanced by the friction pulling it to the left, opposite to the direction of motion.

Core Concept

If an object is in equilibrium—at rest or moving at constant speed in a straight line—the total force acting on it in any direction is exactly equal in magnitude to the force in the opposite direction.

Try This The mass of the brick in Figure 2-2 is 1.6 kg, and the spring scale reads 4 N. Find (a) the force of friction on the brick; (b) the elastic recoil of the tabletop.

2-8. Components of a Force

The crate of Figure 2-7 is being dragged along the floor by means of a rope, which is not horizontal. The rope makes an angle θ to the floor.

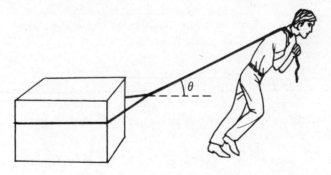

FIGURE 2-7

The tension in the rope, acting on the crate, does two things to it. First, it drags the crate across the floor. Second, it tends to lift the crate off the floor. The smaller the angle θ, the larger the effective force that is dragging the crate, and the smaller the effective force that is lifting it. When $\theta = 0$, the entire force is dragging and there is no lifting at all. Conversely, when $\theta = 90°$, the entire force is lifting the crate.

How can you find out how much force is being used to drag the crate? Force is a vector, and it obeys the same mathematical rules as velocity vectors and displacement vectors, which you used in Chapter 1. The dragging force is the component of the tension in the rope acting parallel to the floor, that is, the horizontal component. According to Equation 1-4, it is found as follows:

$$F_{\text{horiz}} = \mathbf{F} \cos \theta$$

Suppose, for example, that the tension in the rope is 250 N and the angle the rope makes with the ground is 25°. Then the dragging force, as shown in Figure 2-8, is

$$F_{horiz} = (250 \text{ N})(\cos 25°) = 227 \text{ N}$$

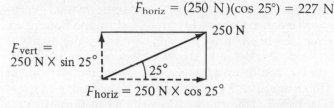

$F_{vert} =$
$250 \text{ N} \times \sin 25°$

250 N

25°

$F_{horiz} = 250 \text{ N} \times \cos 25°$

FIGURE 2-8

And, as before, the other component, lifting the crate, is

$$F_{vert} = (250 \text{ N})(\sin 25°) = 106 \text{ N}$$

This is not enough to get the crate off the floor, but it relieves the floor of some of the weight.

How much is the friction in this situation? If the crate is moving at constant speed, the force to the right must equal the force to the left. Since the force pulling to the right is the horizontal component of the tension in the rope, the friction must be the same—227 N.

And how much is the elastic recoil? The upward forces must equal the downward forces. The only downward force is the weight of the crate, say, 500 N. The *total* upward force must then be 500 N. But 106 N of this is provided by the upward component of the tension in the rope. The rest— 394 N—is the elastic recoil of the floor.

Sample Problem

2-3 Using a window pole that makes an angle of 23° with the window, you push up on the pole with a force of 85 N to close the window. Find (a) the effective force that is pushing the window up; (b) the force pushing the window against its sash.

Solution (a) Since the force makes an angle of 23° with the vertical, its vertical component is (85 N)(cos 23°) = 78 N.
(b) The horizontal component is (85 N)(sin 23°) = 33 N.

Core Concept

The effect that a force has in any direction can be found by calculating its component in that direction.

Try This You push a lawn mower with a force of 160 N, exerted directly along its shaft. The shaft makes an angle of 30° with the ground. Find (a) how much force is moving the lawn mower; (b) how much force is pushing the lawn mower into the ground.

2-9. Equilibrium with Several Forces

For the purpose of making a complete analysis of the forces acting on an object, a *vector diagram* is a useful device.

Figure 2-9 is a vector diagram showing the forces on a brick being dragged along a tabletop. Four forces act: gravity (weight), friction, elastic recoil of the tabletop, and tension in the cord. Each force is represented by a vector, drawn at the correct angle and with its length proportional to the force. For the purpose of analysis, all vectors are represented by components along a pair of axes perpendicular to each other. In this case, we elect to use horizontal and vertical axes, since three of the forces are already on these axes. To do the analysis, we have to resolve the tension vector into its components on the vertical and horizontal axes. Sample Problem 2-4 shows how this vector diagram can be used to solve for unknown forces.

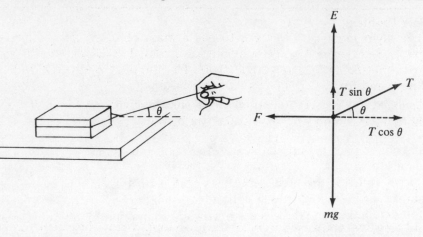

FIGURE 2-9

Sample Problem

2-4 The 2.5-kg brick of Figure 2-9 is being pulled by a cord that makes an angle of 20° with the horizontal and has 7 N of tension in it. Find (a) the force of friction; (b) the elastic recoil of the tabletop.

Solution (a) On the horizontal axis, the friction must be equal to the horizontal component of the tension, so

$$F = (7 \text{ N})(\cos 20°)$$

$$F = 6.6 \text{ N}$$

(b) On the vertical axis, the downward force (weight) must equal the sum of upward forces, so

$$m\mathbf{g} = \mathbf{E} + \mathbf{T} \sin \theta$$

$$\mathbf{E} = m\mathbf{g} - \mathbf{T} \sin \theta$$

$$\mathbf{E} = (2.5 \text{ kg})\left(9.8 \ \frac{\text{m}}{\text{s}^2}\right) - (7 \text{ N})(\sin 20°)$$

$$\mathbf{E} = 24.5 \text{ N} - 2.4 \text{ N} = 22.1 \text{ N}$$

Another example is the lawn mower. In this case, there are four forces acting: gravity, friction, the force along the shaft, and the elastic

FIGURE 2-10

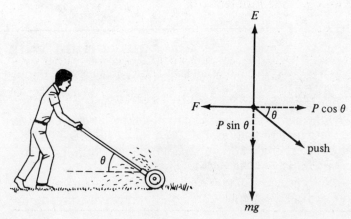

recoil of the ground. The force exerted by the poor fellow doing the work is resolved into vertical and horizontal components. See Sample Problem 2-5.

Sample Problem

2-5 The lawn mower of Figure 2-10 has a mass of 22 kg, and is being pushed at constant velocity against a frictional resistance of 150 N, with the shaft making an angle of 30°. Find (a) the compression in the shaft; and (b) the elastic recoil of the ground.

Solution (a) The compression in the shaft is the push force, and its horizontal component must equal the frictional resistance:

$$F = P \cos \theta$$

$$P = \frac{F}{\cos \theta} = \frac{150 \text{ N}}{\cos 30°}$$

$$P = 173 \text{ N}$$

(b) Now we can use the value of the push force to find the elastic recoil:

$$E = m\mathbf{g} + P \sin \theta$$

$$E = (22 \text{ kg})\left(9.8 \ \frac{m}{s^2}\right) + (173 \text{ N})(\sin 30°)$$

$$E = 302 \text{ N}$$

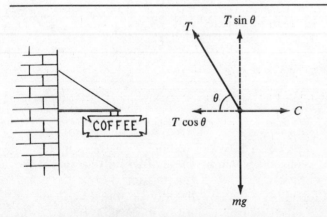

FIGURE 2-11

Here is one more example: What is the tension in the rope holding up the sign in Figure 2-11? This can be found by analyzing the forces acting *on the bar*. The rope pulls the bar up and to the left; the weight of the sign pulls it straight down; the elastic recoil of the wall pushes it to the right. If the weight of the bar itself is too small to worry about, the solution to the problem is as shown in Sample Problem 2-6.

Sample Problem

2-6 If the sign of Figure 2-11 weighs 240 N and the angle the rope makes with the bar is 55°, how much is the tension in the rope?

Solution One equation is enough to get this answer. On the vertical axis:

$$m\mathbf{g} = T \sin \theta$$

$$T = \frac{m\mathbf{g}}{\sin \theta} = \frac{240 \text{ N}}{\sin 55°}$$

$$T = 293 \text{ N}$$

Core Concept

To calculate the forces acting on an object, first analyze them into components along two axes at right angles to each other.

Try This   A helium-filled balloon weighs 25 N, and is acted on by four forces, as shown in Figure 2-12. These forces are its weight, the buoyant force of the air, the wind pushing the balloon to the right, and the tension in the rope that is holding it down. The rope makes an angle of 20° with the vertical, and the tension in it is 16 N. Find the buoyancy and the force exerted by the wind.

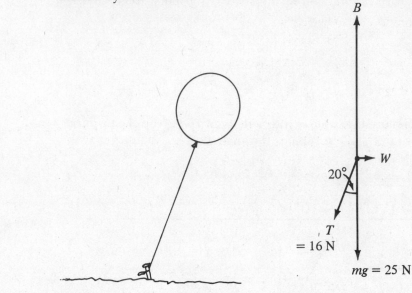

FIGURE 2-12

2-10. The Inclined Plane

A wagon rolls downhill, propelled only by its own weight. But gravity pulls straight down, not at an angle downhill. What makes the wagon go is a *component* of its weight, a part of its weight acting downhill, parallel to the surface the wagon rests on.

A component of a force can act in any direction, not just vertically or horizontally. On the inclined plane, the weight of the wagon has two different effects: it acts *parallel* to the surface of the hill, pushing the wagon downhill; and it acts perpendicular (or *normal*) to the surface, pushing the wagon into the surface. As the hill gets steeper, the parallel component becomes larger and the normal component decreases.

FIGURE 2-13

Figure 2-13 shows the wagon and a vector diagram of the components of its weight. Geometrically, the components are found by dropping perpendiculars from the end of the vector to the two axes. The angle marked θ in the vector diagram is equal to the slope of the hill. This can easily be shown by a little geometry; both angles θ are complements of the two angles marked ϕ. Sample Problem 2-7 shows how the two components of the weight are calculated.

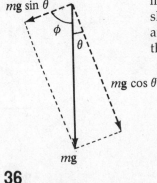

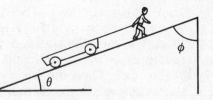

36

2/FORCES: PUSH AND PULL

Sample Problem

2-7 The wagon of Figure 2-13 weighs 40 lb, and the angle that the hill makes with the horizontal is 35°. Find (a) the force pushing the wagon downhill; (b) the force pushing the wagon into the surface.

Solution (a) Parallel (downhill) component of weight:

$$F_P = \mathbf{w} \sin \theta$$

$$F_P = (40 \text{ lb})(\sin 35°)$$

$$F_P = 23 \text{ lb}$$

(b) Normal component of weight:

$$F_N = \mathbf{w} \cos \theta$$

$$F_N = (40 \text{ lb})(\cos 35°)$$

$$F_N = 33 \text{ lb}$$

Sample Problem

2-8 A well-oiled, frictionless wagon with a mass of 75 kg is pulled uphill, using a force of only 110 N. What is the angle that the hill makes with the horizontal?

Solution As before, the component of weight parallel to the plane is the weight times the sine of the slope of the plane. The weight of the wagon (Equation 2-3) is

$$m\mathbf{g} = (75 \text{ kg})(9.8 \text{ m/s}^2) = 735 \text{ N}$$

Then

$$\sin \theta = \frac{110 \text{ N}}{735 \text{ N}} \quad \text{and} \quad \theta = 8.6°$$

Sample Problem

2-9 A plank will break if a force of 350 N is applied to its center. What is the largest weight it can support if it is tilted to an angle of 35°?

Solution At the breaking point, the normal component of the weight is 350 N, which is equal to the weight times the cosine of the angle. Then

$$\mathbf{F}_N = \mathbf{w} \cos \theta$$

so

$$\mathbf{w} = \frac{\mathbf{F}_N}{\cos \theta} = \frac{350 \text{ N}}{\cos 35°} = 430 \text{ N}$$

When the wagon is resting on the surface, the elastic recoil of the surface is just enough to cancel the normal component of the wagon's weight. If the wagon is to stay in equilibrium, you have to pull on it, uphill, to prevent it from running away. If there is no friction, the uphill force needed is the same whether the wagon is standing still, or going either uphill or downhill at constant speed.

The situation is different if the wagon is moving and there is friction. If the wagon is going uphill, you have to pull harder, because the friction is working against you, holding it back. The total force you need to keep the wagon going is then equal to the parallel component of the weight plus the friction. On the other hand, if you are lowering the wagon down the hill, holding the rope to keep it from running away from you, friction is acting uphill, helping you to hold the wagon back. Then the force you must exert is the parallel component of the weight *minus* the friction.

When an object is on an inclined plane, it is propelled down the plane by a force equal to $w \sin \theta$, and a force $w \cos \theta$ pushes it into the surface.

Try This The 120-kg wagon of Figure 2-13 is being pulled up a 20° slope. Find (a) the tension in the rope; (b) the elastic recoil of the surface.

2-11. Making Things Turn

Imagine a wheel of fortune like the one shown in Figure 2-14. The operator gives it a spin, and everyone stands around to see where it comes to rest.

FIGURE 2-14

As the wheel spins around a fixed axle, it has not been moved out of its position. It is in a state of *translational* equilibrium because the operator's force was canceled by the elastic recoil of the axle. The push put the wheel into *rotation*, but did not make it translate from one place to another.

If there were no friction on the wheel, it would keep spinning forever. The only thing that slows it up and eventually brings it to rest is the friction in the axle and the rubbing of all the nails against the marker. Anything that spins with a constant angular velocity is in a state of *rotational equilibrium*, and will keep on spinning unless something is done to it. The earth has been doing it for several billion years, and will still be turning on its axis long after you and I are forgotten.

The operator of the wheel of fortune set it spinning by applying a force to it. But whether a force makes a change in the angular velocity of something depends on where and how that force is applied. It takes a *torque* to change angular velocity, and a force may or may not produce a torque.

Consider, for example, the floating life ring of Figure 2-15, which has two strings attached to it. In Figure 2-15a the strings are being pulled in opposite directions with equal forces, so the ring is in translational equilib-

FIGURE 2-15

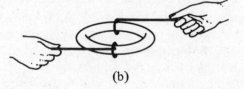

(a)   (b)

rium and will not move. However, in Figure 2-15b the two strings are not in line, and the ring will rotate. While there is no net force on it, there certainly *is* a net torque.

What determines the amount of torque? Suppose that the operator of the wheel of fortune wants to set it spinning as rapidly as possible. How would he apply his force? He would not have to think about it, but there are three precautions he would automatically take:

1. Push as hard as possible; torque is proportional to the force he applies.
2. Apply the force at the rim of the wheel; torque is proportional to the distance from the center of rotation at which the force is applied.
3. Apply the force tangentially.

A push directly toward or away from the center of rotation produces no torque, no matter how hard the push. The force must be—or at least have a component that is—perpendicular to the radial distance to the center of rotation.

These three considerations can be summarized algebraically:

$$\tau = F \perp r \qquad \text{(Equation 2-11)}$$

where τ (the Greek letter tau) is the torque, F is the force, and r is the radial distance from the center of rotation to the point of application of the force. The symbol $\perp$ reminds us that only the tangential force counts.

Torque is the product of a force and a distance, so it is measured in pound-feet or, in SI, in newton-meters (abbreviated as $N \cdot m$). A torque wrench is used by mechanics when they have to tighten a nut by some specified amount. It is calibrated in pound-feet. Sample Problem 2-10 is a calculation of torque in the system favored by scientists.

Sample Problem

2-10 The radius of the wheel of fortune in Figure 2-14 is 1.2 m, and the operator applies a force of 45 N tangentially to get it spinning. What torque has he supplied?

Solution

$$\tau = F \perp r$$

$$\tau = (45 \text{ N})(1.2 \text{ m})$$

$$\tau = 54 \text{ N} \cdot \text{m}$$

Sample Problem

2-11 A 32-kg child sits on a seesaw. If she is 2.2 m from the pivot, what is the torque that her weight exerts, making the seesaw rotate around the pivot?

Solution   Her weight is a vertical force, perpendicular to her distance from the pivot, so

$$\tau = Fr = mgr = (32 \text{ kg})(9.8 \text{ m/s}^2)(2.2 \text{ m}) = 690 \text{ N} \cdot \text{m}$$

Sample Problem

2-12 A torque of 30 N · m is required to turn the steering wheel of a car. If the radius of the wheel is 26 cm, how much force is needed?

Solution   The force will be applied tangentially, perpendicular to the radius, so $\tau = Fr$ and

$$F = \frac{\tau}{r} = \frac{30 \text{ N} \cdot \text{m}}{0.26 \text{ m}} = 115 \text{ N}$$

Torque, the product of radius vector and tangential force, changes the rotational state of objects:

$$\tau = \mathbf{F} \perp \mathbf{r}$$

Try This A 45-kg woman sits 2.5 m from the center of rotation of a seesaw. How much torque does her weight produce around the pivot point?

2-12. Balanced Torques

If an object is acted on by more than one torque, the effects of the torques may cancel each other and leave the object in rotational equilibrium.

Consider, for example, the winch shown in Figure 2-16. When you turn the crank, the rope winds up on the shaft, and a bucket raises water from the well. If the crank is turning at constant angular velocity, it is in equilibrium, and the net torque on it must be zero.

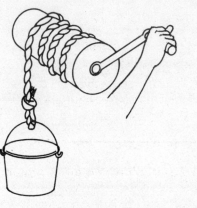

FIGURE 2-16

Imagine what would happen if you let go of the handle. The bucket would fall, unwinding the rope while it turned the crank in the counterclockwise direction. The weight of the bucket would exert a torque on the winch, counterclockwise, equal to the product of the weight and the radius of the shaft.

While you raise the bucket, you cancel this torque by applying a clockwise torque to the handle. This is equal to the force you apply times the length of the crank. The two torques are equal in magnitude, but opposite in direction so they add up to zero, and the winch remains in equilibrium. Sample Problem 2-13 shows how this principle can be applied.

Sample Problem

2-13 The bucket of Figure 2-16, with the water in it, has a mass of 6.0 kg; the radius of the drum on which the rope is wound is 12 cm; the length of the crank arm is 35 cm. How much force is needed to lift the bucket?

Solution The torque exerted by the weight of the bucket on the drum is the weight times the radius of the drum:

$$\tau_{bucket} = m\mathbf{gr}$$

$$\tau_{bucket} = (6.0 \text{ kg})\left(9.8 \frac{\text{m}}{\text{s}^2}\right)(0.12 \text{ m})$$

$$\tau_{bucket} = 7.06 \text{ N} \cdot \text{m}$$

Since the torque on the crank must be the same,

$$\tau_{crank} = \mathbf{Fr}$$

$$F = \frac{\tau}{\mathbf{r}}$$

$$F = \frac{7.06 \text{ N} \cdot \text{m}}{0.35 \text{ m}} = 20 \text{ N}$$

Sample Problem

2-14 You are using a crowbar 2.4 m long, pivoted at 0.20 m from the end, to lift, at that end, a rock that weighs 750 N. How much force do you have to exert?

Solution Since the rock is 0.20 m from the pivot, the torque it exerts around the pivot is

$$\mathbf{Fr} = (750 \text{ N})(0.20 \text{ m}) = 150 \text{ N} \cdot \text{m}$$

To lift the rock in equilibrium, you must exert an equal torque in the opposite direction. Your force is 2.2 m from the pivot, so

$$\mathbf{F}(2.2 \text{ m}) = 150 \text{ N} \cdot \text{m} \quad \text{and} \quad \mathbf{F} = 68 \text{ N}$$

In another example, two children are sitting on one side of a seesaw, and Daddy is on the other, as in Figure 2-17. Now there are *two* counterclockwise torques, produced by the weights of the children. Daddy's torque, applied clockwise, has to equal the sum of the counterclockwise torques. Sample Problem 2-15 shows how this works out.

Sample Problem

2-15 The seesaw of Figure 2-17 is 12 ft long and is pivoted at its center. George weighs 40 lb and is sitting at one end; his sister Alice, weighing 32 lb, is 1 ft from him. Where does their 155-lb daddy have to sit to balance the seesaw?

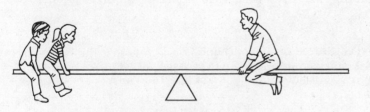

FIGURE 2-17

Solution Write an equation stating that the sum of counterclockwise torques (George's and Alice's) must equal Daddy's clockwise torque. All radius vectors, including Daddy's unknown *r*, *must* be measured from the same point—in this case, the pivot:

$$(40 \text{ lb})(6 \text{ ft}) + (32 \text{ lb})(5 \text{ ft}) = (155 \text{ lb})r$$

$$240 \text{ lb} \cdot \text{ft} + 160 \text{ lb} \cdot \text{ft} = (155 \text{ lb})r$$

$$r = 2.6 \text{ ft}$$

So Daddy must sit 2.6 ft *from the pivot*.

Here's one more example. A pile of books rests on a plank that is supported on either end by a chair, as in Figure 2-18. The books are not in the center. How much of the weight of the books is supported by each of the chairs on which the plank rests? You solve this by imagining that the

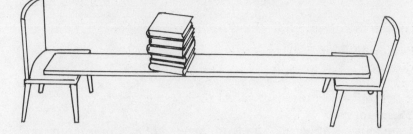

FIGURE 2-18

right-hand chair is pushing up on the plank, making it try to turn counter-clockwise around the left-hand chair. The counterclockwise torque produced by the elastic recoil of the right-hand chair must be canceled by the clockwise torque, which comes from the weight of the books. See Sample Problem 2-16.

Sample Problem

2-16 The mass of the books of Figure 2-18 is 35 kg; the plank is 3.7 m long, and the books are centered 1.2 m from the left-hand chair. What is the elastic recoil of each chair?

Solution Consider the left-hand chair to be a pivot around which the whole plank is trying to turn. Then the weight of the books is producing a clockwise torque around that pivot, and the elastic recoil of the other chair is producing a counterclockwise torque. These two torques produce rotational equilibrium, so they must be equal to each other. Then:

$$(m\mathbf{g}r)_{books} = (Er)_{right\text{-}hand\ chair}$$

$$(35\ kg)\left(9.8\ \tfrac{m}{s^2}\right)(1.2\ m) = (3.7\ m)E_{right}$$

$$412\ N \cdot m = (3.7\ m)E_{right}$$

$$E_{right} = 111\ N$$

To get the elastic recoil of the left-hand chair, note that the books are also in translational equilibrium, so the total upward force must equal the total downward force:

$$m\mathbf{g} = E_{right} + E_{left}$$

$$(35\ kg)\left(9.8\ \tfrac{m}{s^2}\right) = 111\ N + E_{left}$$

$$E_{left} = 232\ N$$

Core Concept

An object is in rotational equilibrium when the counterclockwise torques acting on it are equal to the clockwise torques.

Try This A light, horizontal plank 3.5 m long rests with one end on a rock and the other on a scale. A woman is standing on the plank, 1.0 m from the rock, and the scale reads 120 N. How much does the woman weigh?

42
2/FORCES: PUSH AND PULL

Summary Quiz

For each of the following, fill in the missing word or phrase:

1. Application of a force changes the _____ of an object.
2. The direction of the force of sliding friction is always _____ to that of velocity.
3. An object moving at constant speed is in a state of _____ .
4. The SI unit of force is the _____ .
5. It takes about 4.45 N to equal a _____ .
6. The force that retards a solid moving through a liquid is called _____ .
7. The force of gravity acting on an object is called the object's _____ .
8. The amount of material in an object is called the object's _____ .
9. Mass times acceleration due to gravity equals _____ .
10. One $kg \cdot m/s^2$ = 1 _____ .
11. When the shape of a solid is distorted, it produces a force called _____ .
12. When a rope is stretched, the _____ in it results in an elastic recoil force at each end.
13. A solid submerged in a liquid experiences an upward force called _____ .
14. If you stand next to a wall and push it northward with a force of 30 lb, it will push you with a force of _____ lb.
15. When you stand on a floor, your weight is balanced by the _____ force of the floor.
16. If a 50-lb girl on roller skates stands on a hill where the slope is 30°, the gravitational force pushing her down the hill is _____ .
17. To find the component of a force on any axis, multiply the force by the _____ of the angle the force makes with the axis.
18. If it takes a force of 200 N to move a wagon up a frictionless hill at constant speed, the force needed to let the wagon roll downhill at constant speed is _____ .
19. Application of a _____ changes the angular velocity of an object.
20. Radius times tangential forces equals _____ .
21. The SI unit of torque is the _____ .
22. To turn a crank 0.30 m long, you exert a force of 100 N. The torque you apply to the crank is _____ .
23. An object is in rotational equilibrium when the sum of all _____ acting on it is zero.

Problems

1. What is the mass of a dog that weighs 75 N?
2. An astronaut with all her equipment has a mass of 95 kg. How much will she weigh on the moon, where the acceleration due to gravity is 1.67 m/s^2?
3. A rope is attached to a 35-kg rock to lift the rock from the bottom of the lake. If the buoyancy on the rock is 50 N and the viscous drag is 25 N, how much is the tension in the rope?
4. How much is the viscous drag acting on a rocket-driven sled that is going at constant speed against a frictional force of 22,000 N when the thrust of the engine is 31,000 N?
5. A sled is being pulled along a horizontal road at constant speed by means of a rope that makes an angle of 25° with the horizontal. If the friction between the sled and the snow is 85 N, how much is the tension in the rope?

6. On a camping trip, you stretch a rope between two trees and hang your knapsack from the middle of it, to keep it safe from bears. The mass of the knapsack is 32 kg, and each half of the rope makes an angle of 40° with the horizontal. Find (a) the amount of weight supported by each half of the rope; (b) the tension in the rope.

7. A 35-kg child is on a swing supported by two ropes. A baby sitter is holding the swing back so that the ropes make an angle of 25° with the vertical. How much is the tension in each rope?

8. In a sign supported as shown in Figure 2-11, the tension in the rope is 350 N. How much does the sign weigh if the angle between the rope and the wall is 40°?

9. A 20-kg pile of books is resting on a plank tilted so that it makes an angle of 20° with the ground. How much force do the books exert normally against the plank?

10. A force of 40 lb is needed to push a wagon up a 35° slope. How much does the wagon weigh? (Neglect friction.)

11. A crowbar 2.0 m long is used to pry a rock out of the ground, by pivoting the crowbar 0.5 m from the rock and pushing down on the other end. The end is pushed with a force of 250 N. Find (a) the torque being applied to the crowbar by the person doing the job; (b) the force being applied to the rock.

12. A light scaffold is supported by a rope at each end, and a painter is standing 1.0 m from the left end. If the painter weighs 600 N and the scaffold is 2.5 m long, how much is the tension in each of the ropes?

13. A winch, consisting of a crank that turns a shaft with a rope wrapped around it, is being used to lift cargo into a boat. The shaft has a diameter of 3.0 cm, and the crank is 45 cm long. If a load of cargo weighs 1200 N, how much force is needed to lift it?

14. A 7.0-kg bowling ball is placed on a ramp sloped at 15°. Find (a) the force that propels the ball down the ramp; (b) the force that compresses the surface of the ramp.

15. You have a rope attached to a cart that weighs 450 N, and you are lowering the cart down a 25° slope. If the friction is 75 N, how hard do you have to pull on the rope to prevent the cart from running away from you?

16. A plank 4.0 m long has its left end resting on a rock and the other supported by a rope, in a horizontal position. If the greatest tension the rope can stand without breaking is 350 N, how far from the rock can a 55-kg girl walk out on the plank before the rope breaks?

CHAPTER 3

Things That Flow

3-1. Phases of Matter

To the ancient Greeks, the universe was composed of four elements: fire, earth, water, and air. To us, these "elements" can be used to define everything that composes the universe as we know it: fire stands for energy, and the others represent matter in its three phases—solid, liquid, and gas.

The objects of our world are in the solid phase. Each has its own special size and shape, its characteristic mass, weight, volume, and so on.

Liquids and gases are fluids, things that flow. They have no shape of their own, but will assume the shape of whatever container they are in. While a given sample of water, air, carbon dioxide, or milk has a definite, measurable mass, it does not have the solid's capacity to resist changes in its shape by exerting an elastic recoil force.

A given sample of a liquid has a definite volume, whereas a sample of a gas does not. If you pour a pint of coffee from a pot into three cups, it is still a pint. The coffee settles into the cup, and its volume is defined by the cup and by the upper, horizontal surface. It is this surface that distinguishes a liquid from a gas.

It is quite possible to pour a heavy, visible gas like chlorine, for example, from one container into another. It will not stay in the container, however, unless you put a cap on it. A gas has no upper surface and will spread out indefinitely. The volume of a gas is the volume of the container it is in.

45

A solid has a definite shape and a definite volume; a liquid has only a definite volume; a gas has neither.

Try this Tell whether each of the following properties is a characteristic of a given sample of a solid, of a liquid, and of a gas: (a) mass; (b) volume; (c) weight; (d) elasticity; (e) surface; (f) shape; (g) density; (h) color; (i) length.

3-2. Density

Which is heavier, milk or alcohol?

The word "heavy," unfortunately, has at least three different meanings. If you say that a rock is too heavy to lift, you are talking about the *weight* of the rock. When "heavy" is used in that sense, there is no answer to the question about the milk and the alcohol. A gallon of either weighs more than a drop of the other.

The question as asked is clearly intended to refer to a general property of the kind of substance—milk or alcohol. It is not a question about any particular sample, so it cannot be asking about the weight. It is a question not about weight, but about *density*, which is a property of the kind of material, regardless of the size of the sample.

Density tells you to what extent the mass of a substance is concentrated. Lead is a high-density material because a great deal of mass is concentrated into a small volume. Air, on the other hand, has a very low density.

It is easy to compare the densities of two liquids, or of a solid and liquid. If something is less dense than the liquid, it will float; if more dense, it will sink. Wood is less dense than water. A toothpick will float, and so will a thousand-pound pine log. Both have the same density.

The exact definition of density is the ratio of mass to volume:

$$D = \frac{m}{V}$$

(Equation 3-2)

In the SI, the unit of density is the kilogram per cubic meter (kg/m^3), but this is rarely used. The commonly used unit is the gram per cubic centimeter (g/cm^3). This is a very convenient unit because most ordinary liquids and solids have densities that are given by numbers of reasonable size in this unit. The density of water is just 1 g/cm^3, varying a little as the temperature changes. Gases have much lower densities. See the sample problems.

Sample Problem

3-1 What is the density of a liquid if 75 cm^3 of it has a mass of 93 g?

Solution From Equation 3-2,

$$D = \frac{m}{V}$$

$$D = \frac{93 \text{ g}}{75 \text{ cm}^3}$$

$$D = 1.24 \text{ g/cm}^3$$

Sample Problem

3-2 A "heavy" oil has a density of 0.91 g/cm^3. What is the mass of 2.5 liters of this oil? (A liter is 1,000 cm^3.)

Solution From Equation 3-2,

$$m = DV = \left(0.91 \tfrac{g}{cm^3}\right)(2,500 \ cm^3) = 2,300 \ g \ or \ 2.3 \ kg$$

Sample Problem

3-3 How big a bottle is needed to hold 5.0 kg of mercury, which has a density of 13.6 g/cm^3?

Solution From Equation 3-2,

$$V = \frac{m}{D} = \frac{5,000 \ g}{13.6 \ g/cm^3} = 370 \ cm^3$$

Beware of the third meaning of "heavy." A heavy oil does not have a high density. In fact, it will float on water. "Heavy" in this sense refers to gooeyness, or viscosity.

Density is mass per unit volume:

$$D = \frac{m}{V}$$

Try This What is the volume of 650 g of ether, which has a density of 0.62 g/cm^3?

3-3. Pressure

The effect that a force has on a surface depends on how the force is applied. A man in spiked shoes dents the turf or the floor that he stands on. Put the same man in ordinary shoes and he does no damage. Yet the force he applies to the floor, his weight, is the same in both cases.

The difference is that the spikes concentrate all the force into a small area. The spikes do not change the total force, but they greatly increase the *force per unit area.* This quantity, the force per unit area, is called *pressure.* This is how it is defined mathematically:

$$p = \frac{F}{A} \qquad \text{(Equation 3-3)}$$

The bottom of a size 10 shoe is about 11 inches long and 4 inches wide, so the total area of two such shoes is $2(11 \ in)(4 \ in) = 88 \ in^2$. In a pair of size 10 shoes, the pressure a 180-pound man exerts on the ground is

$$p = \frac{F}{A} = \frac{180 \ lb}{88 \ in^2} = 2.0 \ lb/in^2$$

which is read as "two point zero pounds per square inch." If the man walks in snow, this pressure is enough to compress the snow, and he sinks into it. He can put on skis, which have a much larger area, to reduce the pressure he exerts on the snow.

There are many units of pressure in use, and they can be confusing. A convenient unit, one that can serve as a reference for the others, is the *atmosphere* (atm), defined as the average pressure exerted by the air at sea level. It is about 14.7 lb/in^2. The SI unit is the newton per square meter

(N/m^2), which is called a *pascal* (Pa). This is a very small unit, and the *kilopascal* (kPa), equal to 1,000 Pa, is coming into use. It takes a little over 100 kPa to equal 1 atmosphere. Meteorologists have been using the *millibar*, which is one tenth of a kPa.

Sample Problem

3-4 What pressure is exerted on the snow by a 180-lb skier if his skis are 6 ft long and 5 in wide?

Solution   The area of the skis is

$$2 \times (72 \text{ in}) \times (5 \text{ in}) = 720 \text{ in}^2$$

To get the pressure, we have Equation 3-3

$$p = \frac{F}{A}$$

$$p = \frac{180 \text{ lb}}{720 \text{ in}^2}$$

$$p = 0.25 \text{ lb/in}^2$$

Sample Problem

3-5 A boy and a sled have a combined mass of 38 kg. The runners of the sled are 1.60 m long and 1.2 cm wide. Find the pressure exerted on the snow in (a) pascals; (b) atmospheres.

Solution   (a) The force exerted on the snow is the weight of the boy on the sled:

$$mg = (38 \text{ kg})(9.8 \text{ m/s}^2) = 372 \text{ N}$$

The area of contact is $2(1.60 \text{ m})(0.012 \text{ m}) = 0.0384 \text{ m}^2$. Then

$$p = \frac{F}{A} = \frac{372 \text{ N}}{0.0384 \text{ m}^2} = 9,700 \text{ Pa}$$

(b) Converting the units, with 100,000 Pa = 1 atm, we have

$$(9,700 \text{ Pa})\left(\frac{\text{atm}}{10^5 \text{ Pa}}\right) = 0.097 \text{ atm}$$

Sample Problem

3-6 Soft snow can be compressed by about 3,000 Pa of pressure. What is the smallest area that a pair of snowshoes must have if they will enable a 70-kg person to walk over the snow without sinking in?

Solution   The force on the snow is the person's weight $= mg = (70 \text{ kg})(9.8 \text{ m/s}^2) = 686 \text{ N}$. Then, from Equation 3-3,

$$A = \frac{F}{p} = \frac{686 \text{ N}}{3000 \text{ N/m}^2} = 0.23 \text{ m}^2$$

Core Concept

Pressure is force per unit area:

$$p = \frac{F}{A}$$

Try This   What is the pressure exerted on the ground by a camera and tripod weighing 7.5 lb if each leg of the tripod has a tip whose area of contact with the ground is 0.05 in^2? (Watch it! The pressure is much more than you think!)

A liquid or a gas exerts pressure against any object immersed in it. This is called *hydrostatic* pressure. Your eardrums are sensitive to changes in the hydrostatic pressure on them, and you feel these changes when you dive deep into the water or when you rise suddenly in an elevator or an airplane.

We all live at the bottom of a sea of air, and this air is always exerting pressure on everything. The classic method of measuring this pressure—still widely used—is with a *mercury barometer* (Figure 3-1). It is an easy thing to make. The method is as follows (see warning below): Take a glass tube about a meter long and fill it with mercury. Hold your finger over the open end and push it into an open bowl of mercury. When you remove your finger, the level of mercury in the tube will drop, leaving a vacuum in the space above the mercury. The mercury in the tube is supported by the pressure of the air on the open surface of the mercury in the bowl, transmitted through the liquid. The height of the mercury level in the tube, above that in the bowl, is a measure of the pressure of the air. (Don't try this; mercury vapor is poisonous, and special precautions must be taken.)

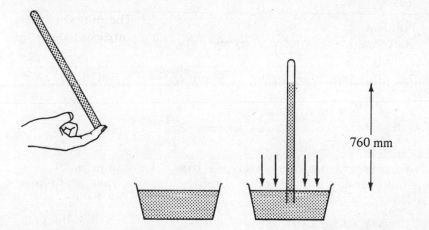

760 mm

FIGURE 3-1

The first barometer of this sort used water in the tube, but this called for a column over 34 feet high. The change to mercury made it possible to use a shorter tube, since mercury has more than 13 times the density of water.

Let's figure out how high the mercury column will stand in the barometer when the air pressure outside is 1 standard atmosphere, or 101.3 kPa. Equilibrium is reached when the pressure at the bottom of the tube, caused by the weight of all the mercury above it, is equal to the atmospheric pressure holding up the column. Therefore we want to know what length of mercury column produces a pressure of 101.3 kPa at the bottom.

The volume of the mercury is the length of the mercury column (h) times its cross-section area (A):

$$V = hA$$

From Equation 3-2, the mass of the mercury is

$$m = VD = hAD$$

and from Equation 2-3, its weight is

$$w = mg = hADg$$

The force that the weight of the mercury exerts at the bottom is just its weight: $F = w$. Then, from Equation 3-3,

$$p = \frac{F}{A} = \frac{w}{A} = \frac{hADg}{A}$$

so, finally,

$$p = hDg \qquad \text{(Equation 3-4)}$$

To find the height of the mercury column, we had better get the density into SI units first:

$$D = \left(13.6\frac{g}{cm^3}\right)\left(\frac{100 \text{ cm}}{m}\right)^3\left(\frac{kg}{1,000 \text{ g}}\right) = 13,600 \text{ kg/m}^3$$

Now we can solve Equation 3-4 for the height of the column:

$$h = \frac{p}{Dg} = \frac{101,300 \text{ N/m}^2}{(13,600 \text{ kg/m}^3)(9.8 \text{ m/s}^2)}$$

$$h = 0.760 \text{ m}$$

Pressures are often expressed in terms of the height of the mercury column they will support, and 1 standard atmosphere is 760 mm of mercury.

Sample Problem

3-7 How tall would a barometer tube have to be if it were filled with alcohol, density 0.87 g/cm³?

Solution The density of alcohol, in SI units, is 870 kg/m³. Then, from Equation 3-4,

$$h = \frac{p}{Dg} = \frac{101,300 \text{ N/m}^2}{(870 \text{ kg/m}^3)(9.8 \text{ m/s}^2)} = 11.9 \text{ m}$$

Sample Problem

3-8 In a partially evacuated chamber, the barometer stands at 210 mm. How much is the pressure in the chamber in (a) atmospheres; (b) pascals?

Solution (a) Since it takes 760 mm of mercury to make 1 atm, the pressure is

$$\left(\frac{210 \text{ mm}}{760 \text{ mm}}\right) \text{atm} = 0.28 \text{ atm}$$

(b) One atmosphere is 101.3 kPa, so the pressure is

$$(0.28)(101.3 \text{ kPa}) = 28 \text{ kPa, or } 28,000 \text{ Pa.}$$

Sample Problem

3-9 If a wall is 3 m high and 7 m long, how much force does the atmosphere exert on each side of it?

Solution The area is $(7 \text{ m}) \times (3 \text{ m}) = 21 \text{ m}^2$. Then, from Equation 3-3,

$$F = pA = \left(101,300\frac{N}{m^2}\right)(21 \text{ m}^2) = 2,100,000 \text{ N}$$

Core Concept

The atmosphere exerts pressure which can be measured in terms of the height of the mercury column it will support in a barometer:

$$p = hDg$$

Try this You have a table whose top measures 3 ft by 6 ft. What is the total force exerted by the atmosphere on the top surface of the table?

All divers know that the pressure on them increases as they go deeper into the water, and it can become dangerous if they go too deep. This pressure is created by the weight of all that water above the diver, just as the pressure at the bottom of a barometer tube is produced by the weight of the mercury.

Notice that in deriving the expression for the pressure at the bottom of a mercury column, Equation 3-4, we found that the width of the column did not matter. The pressure at a depth of 1 foot in Lake Erie is the same as that 1 foot down from the surface of the water in your bathtub. In both cases, you can find the pressure by applying Equation 3-4.

The shape of the bathtub does not matter, either. This can be shown by the well-known Pascal's vases of Figure 3-2. In the narrow tube, the wide tube, the conical tube, and the squiggly tube, the water level at the top is the same. This can only be so because the pressure is the same at the bottom of all the tubes.

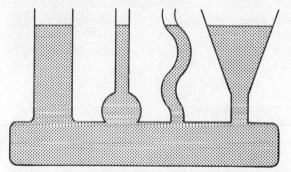

FIGURE 3-2

Hydrostatic pressure at any point is exerted equally in all directions. Imagine a thin slice of water, set horizontal at some arbitrary depth. It is not going anywhere, so the force pushing it up from the bottom must be equal to the force pushing it down from above. You could do the same mental trick with any slice of water at all, no matter how it is oriented. If an object is immersed in water, the force on any surface will always be perpendicular to the surface. Hydrostatic pressure has no direction (it is a scalar), and the direction of the force it exerts on a surface depends only on the orientation of the surface.

In calculating the pressure at any given depth, using Equation 3-4, you must be sure to add the pressure at the surface, if there is any. In the barometer, the top surface of the mercury faces a vacuum, so no such correction is needed. But the pressure on a diver's eardrums is 1 atmosphere before he jumps into the water, and it increases from there as he dives deeper. Therefore, after calculating hDg, you have to add 1 atmosphere, or 100 kPa, or 14.7 lb/in², or 760 mm of mercury, or any other measure of the air pressure at sea level. See Sample Problems 3-10 and 3-11.

Sample Problem

3-10 What is the pressure at a depth of 23 m in water?

Solution   First, we need the density of water in SI units. As before, this is 1,000 times as much as the numerical value in grams per cubic centimeter. Therefore $D = 1,000$ kg/m³. Now, from Equation 3-4,

$$p = hDg = (23 \text{ m})\left(1000\tfrac{\text{kg}}{\text{m}^3}\right)\left(9.8\tfrac{\text{m}}{\text{s}^2}\right) = 225,000 \text{ Pa}$$

This is 225 kPa. To get the actual pressure, we have to add the atmospheric pressure at the surface, which is 101 kPa. Therefore $p = 326$ kPa.

Sample Problem

3-11 Through what vertical distance does a diver move to increase the pressure on herself by 1 atm?

Solution   From Equation 3-4,

$$h = \frac{p}{Dg} = \frac{101,000 \text{ N/m}^2}{\left(1000 \frac{\text{kg}}{\text{m}^3}\right)\left(9.8 \frac{\text{m}}{\text{s}^2}\right)} = 10.3 \text{ m}$$

A diver should know that each 10 meters of descent increases the pressure on the body by about 1 atmosphere. A healthy diver can tolerate a pressure of 10 atmospheres, but must be extremely careful to come up slowly. A sudden reduction in pressure can be fatal, as it causes nitrogen dissolved in the blood to form bubbles that block the circulation.

Core Concept

Hydrostatic pressure in a liquid is directly proportional to the depth of the liquid:

$$p = hDg$$

Try This   A large tank holds salad oil, density 0.9×10^3 kg/m^3, to a depth of 6 m. What is the pressure at the bottom of the tank?

3-6. Flying High

Our atmosphere has no definite top; gases have no surface. As you go higher up, the pressure drops, just as it does in a liquid. But gases compress easily; as the pressure drops, the air simply expands and occupies more space. Its density drops at higher altitudes, gradually fading away to zero in the immensity of outer space.

The pressure–depth relationship, Equation 3-4, works for the atmosphere just as well as the ocean. However, it is not as easy to use. The reason is that the density is decreasing as you rise. There is no single number you can plug into the equation to calculate the pressure. Also, what would you use for h? There is no upper surface to measure from.

With a little advanced mathematics, we can come up with a practical way to determine what happens to the air pressure as a person rises into the atmosphere. The rule is this: For every kilometer of rise into the atmosphere, the pressure drops by about 11%. Thus, at an altitude of 1 km, the pressure is

$$(1 \text{ atm})(0.89) = 0.89 \text{ atm}$$

By the time a person reaches 2 km, it has dropped by an additional 11%:

$$(0.89 \text{ atm})(0.89) = 0.79 \text{ atm}$$

and so on. To find the pressure at any altitude, just multiply the surface pressure by 0.89 once for each kilometer of altitude. In algebraic terms,

$$p = p_0(0.89)^n \qquad \text{(Equation 3-6)}$$

where p_0 is 1 atmosphere and n is the number of kilometers of altitude. See Sample Problem 3-12.

Sample Problem

3-12 What is the atmospheric pressure at an altitude of 3 km?

Solution   From Equation 3-6,

$$p = (1 \text{ atm})(0.89)^3 = 0.70 \text{ atm}$$

Figure 3-3 is a graph showing how the atmospheric pressure varies with altitude. You might check the answer to the Sample Problem against this graph.

This rule for pressure is fairly rough, because the actual pressure in the atmosphere depends on temperature, humidity, winds, and a great many other factors. But the equation is useful to get approximate results.

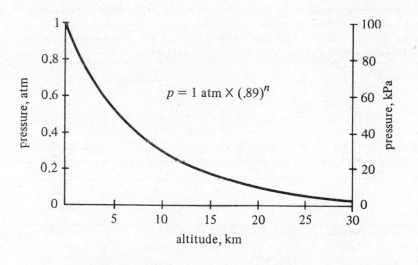

$$p = 1 \text{ atm} \times (.89)^n$$

FIGURE 3-3

Core Concept

Atmospheric pressure drops percentagewise as altitude increases:

$$p = p_0(0.89)^n$$

Try This   How much is the atmospheric pressure, in atmospheres, at an altitude of 5 km?

3-7. Buoyancy

When you take a deep dive, why does the water push you upward?

Buoyancy is caused by a difference in hydrostatic pressure. The water pushes against your body on all sides. But it pushes harder on the lower surface because that is deeper in the water. Since the water is pushing upward from below and downward from above, the upward force is greater than the downward force. The difference between these two forces is the buoyancy.

Let's make it quantitative. Suppose you have a can with a side spout, as shown in Figure 3-4, and you fill the can with water up to the spout. Then you take a rectangular block of some kind and immerse it in the water. The block will displace an amount of water equal to its own volume, and that much water will flow out of the spout.

FIGURE 3-4

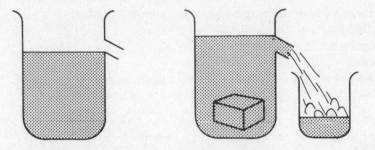

The pressure against the top surface of the block is p_{top}, and the pressure against the bottom surface (upward) is p_{bottom}. Letting h stand for the depth of the water at top and bottom, and D for the density of the water, we can write, from Equation 3-4,

$$p_{\text{top}} = h_{\text{top}}Dg$$

$$p_{\text{bottom}} = h_{\text{bottom}}Dg$$

If the block has length l and breadth b, the area of both top and bottom is bl. Since the force is pA (Equation 3-3),

$$F_{\text{top}} = (blh)_{\text{top}}Dg \quad \text{and} \quad F_{\text{bottom}} = (blh)_{\text{bottom}}Dg$$

Buoyancy is the difference between these two forces:

$$B = F_{\text{bottom}} - F_{\text{top}} = bl(h_{\text{bottom}} - h_{\text{top}})Dg$$

But $(h_{\text{bottom}} - h_{\text{top}})$ is the thickness of the block, so $bl(h_{\text{bottom}} - h_{\text{top}})$ is its volume—and also the volume of the water it displaces. Therefore

$$B = VDg$$

Further more, VD is the mass of the displaced water (Equation 3-2) and mg is its weight (Equation 2-3). Therefore *the force of buoyancy acting on a submerged object is equal to the weight of the fluid displaced.* This is Archimedes' principle, known since ancient Greek times. Sample Problem 3-13 shows this principle in operation.

Sample Problem

3-13 A steel box measuring 40 cm long, 22 cm high, and 35 cm wide is submerged in water. How large is the force of buoyancy exerted on it by the water?

Solution The volume of the box is

$$(40 \text{ cm}) \times (22 \text{ cm}) \times (35 \text{ cm}) = 30{,}800 \text{ cm}^3$$

and this is the volume of water that it displaces, since it is completely submerged. To get the mass of the water, we use Equation 3-2:

$$m = DV$$

$$m = \left(1\frac{\text{g}}{\text{cm}^3}\right)(30{,}800 \text{ cm}^3)$$

$$m = (30{,}800 \text{ g})\left(\frac{\text{kg}}{1{,}000 \text{ g}}\right) = 30.8 \text{ kg}$$

The buoyancy on the box is the weight of this much water. Using Equation 2-3, we have

$$w = mg$$

$$w = (30.8 \text{ kg}) \left(9.8 \tfrac{\text{N}}{\text{kg}}\right)$$

$$w = 302 \text{ N}$$

Sample Problem

3-14 A rock with a mass of 22 kg and a volume of 0.018 m³ is attached to a spring scale, calibrated in newtons, and immersed in water. What does the spring scale read?

Solution The scale must read the difference between the weight of the rock and its buoyancy. Its weight is

$$mg = (22 \text{ kg}) (9.8 \text{ m/s}^2) = 216 \text{ N}$$

Its buoyancy is the weight of the 0.018 m³ of water that it displaces, $w = mg$; but $m = DV$, so

$$w = DVg = \left(\tfrac{1000 \text{ kg}}{\text{m}^3}\right) (0.018 \text{ m}^3) (9.8 \text{ m/s}^2) = 176 \text{ N}$$

Therefore, the scale reads

$$216 \text{ N} - 176 \text{ N} = 40 \text{ N}$$

A geologist in the field makes use of this principle to determine the density of a rock. All he needs is a spring scale calibrated in grams—mass units. This kind of scale will tell the mass of an object, provided that it is used where the acceleration due to gravity is 9.8 m/s². On Mars, it will tell you lies.

Here is how the method works: Hang the rock on the spring scale to find its mass. Then submerge the rock in water and read the scale again. The difference between the two readings gives the mass of the water displaced. Since the density of water is 1 g/cm³, the mass in grams is numerically equal to the volume in cubic centimeters—which is also the volume of the rock. Dividing the mass of the rock by its volume gives its density. (See Sample Problem 3-15.)

Sample Problem

3-15 A rock is suspended from a gram-calibrated spring scale, which reads 155 g when the rock is in air and 83 g when it is in water. What is the density of the rock?

Solution Since the buoyancy equals the weight of water displaced, the mass of displaced water must be (155 g − 83 g) = 72 g. The density of water is 1 g/cm³, so the volume of displaced water is 72 cm³; this must also be the volume of the rock. Then, from Equation 3-2,

$$D = \frac{m}{V}$$

$$D = \frac{155 \text{ g}}{72 \text{ cm}^3} = 2.2 \text{ g/cm}^3$$

Core Concept

When an object is immersed in a fluid, the force of buoyancy on it is equal to the weight of the displaced fluid.

Try This An unknown fluid is poured into an overflow can like that of Figure 3-4, up to the spout. A block of aluminum is suspended from a scale and lowered into the fluid. If the scale reads 25 g when the block is in air and 15 g when the block is in water, and the amount of fluid that flows out of the spout is 8 cm³, what is the density of the fluid?

3-8. It Floats!

If you push a log under water and release it, it will rise to the top and float there. Clearly, this implies that, when the log is submerged, the force of buoyancy acting on it is greater than its weight.

When the log is submerged, it displaces its own volume of water. But that volume of water weighs more than the log does, so the buoyancy is larger than the weight of the log. This is true only if the density of the log is less than the density of the water. Generally, any object will rise in a fluid if the fluid has a greater density than the object.

How high will the log rise? It gets to the surface and sticks out a little. Now, the volume of water it displaces is equal to the volume of only that part of the log that is below the water surface, as in Figure 3-5. The log sits happily at the surface because the buoyancy force on it is equal to its weight. And since the buoyancy is equal to the weight of displaced water, it follows that the log sinks into the water just deeply enough to displace its own weight of water, as in Sample Problem 3-16.

FIGURE 3-5

Sample Problem

3-16 An overflow can is filled up to the spout. Then a chunk of wood is floated on the water, and 330 cm^3 of water flows out of the spout. If the wood is then pushed down until it is completely submerged, an additional 75 cm^3 of water flows out. What is the density of the wood?

Solution   The volume of the wood is the amount of water it displaces when completely submerged: 330 cm^3 + 75 cm^3 = 405 cm^3. Its mass is the same as the mass of water it displaces when floating, which is 330 g. Therefore its density is

$$\frac{m}{V} = \frac{330 \text{ g}}{405 \text{ cm}^3} = 0.81 \text{ g/cm}^3$$

Sample Problem

3-17 If a log floats with 40% of its volume above the surface of the water, what is its density?

Solution   Since the force of buoyancy on the log is equal to its weight, the weight of the log is equal to the weight of the displaced water. But this much water takes up only 60% as much volume as the whole log does, since only 60% of the log is under water. Therefore the density of the log must be 60% of that of the water, or 0.60 g/cm^3.

A steel ship floats, too, even though the density of steel is more than 7 times that of water. The reason is that the ship is not mainly steel; it is mainly air. When the enclosed air is taken into account, the *average* density of the ship is much less than 1 g/cm^3.

A balloon filled with helium or with hot air will rise for the same reason that the log rose to the top of the water. But it cannot rise to the top of the atmosphere because there is no top. Remember, however, that the density of the air decreases with altitude. The balloon rises until its average density is equal to the density of the air outside. If the balloonist wants to go higher, she increases the volume of the balloon by adding more helium or hot air to it. This increases its weight, to be sure; helium has mass. But it increases the buoyancy more, because the displaced air weighs more than the added helium.

Core Concept

A floating object displaces its own weight of fluid.

Try This An ice cube 2 cm on a side floats in water with 0.3 cm of ice projecting above the surface of the water. What is the density of the ice?

3-9. The Skin of a Liquid

Try this: Take a clean razor blade and lay it gently on the surface of a quiet pot of clean water. With a little care, you can make it rest on the surface of the water.

Is the razor blade floating, the way a cork floats? Certainly not; the density of steel is many times that of water. If you look carefully at the blade, you will see that no part of it is under water, as shown in Figure 3-6. The blade is actually resting on the surface. It even depresses the surface a little, as though it were resting on a slightly elastic membrane. If you push it through the surface, it will sink.

FIGURE 3-6

This elastic property of the surface of a liquid is called *surface tension*. Surface tension is responsible for many of the familiar properties of liquids. For example, a thin stream of water flowing out of a tap will break up into droplets because the surface tension makes the water contract into the smallest possible space. Also, it is the surface tension of water that prevents water from penetrating into tiny spaces, such as the pores in dirty clothing. Soap and other detergents promote penetration because they reduce surface tension. You can't float a razor blade on soapy water.

Many insects make use of the surface tension of water. A water strider actually rests on the surface film (Figure 3-7). A mosquito larva lives under water, with its breathing apparatus projecting through the surface. Its body is supported by the tension of the surface film acting on its snorkel.

FIGURE 3-7

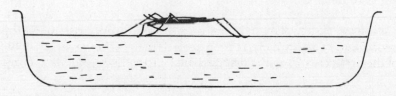

The surface of a liquid acts like an elastic membrane.

Try This   How would you go about drowning mosquito larvae?

3-10. Drag Again

As we saw in Chapter 2, when an object moves through a fluid, a retarding force, called viscous drag, is acting on it. If you are planning to move a boat through the water, or a train through the air, it is important to design your vehicle so as to keep the drag as low as possible. On the other hand, if you fall out of an airplane, you want as much drag as you can get to slow down your fall—hence the parachute.

The amount of viscous drag depends on three factors: (1) the nature of the fluid; (2) the speed of the motion; and (3) the size and shape of the object.

Every fluid has a characteristic *viscosity*. You could compare the viscosities of different fluids by dropping a steel ball into them and measuring the rate at which it falls in each. In air, it will be retarded little, and it will accelerate at very nearly the acceleration due to gravity in a vacuum. Water is more viscous; the ball will not speed up as much. If you drop the ball into a bucket of molasses, it will soon reach a slow, constant speed, and will take a long time to get to the bottom of the bucket. Molasses is a highly viscous fluid.

Viscous drag increases with speed. As the steel ball falls through the fluid, accelerating all the while, it will eventually be going fast enough so that the viscous drag on it is equal to its weight. Then it will be in equilibrium and will continue to fall at constant speed. This speed is called its *terminal velocity*.

If a man falls out of an airplane at a fairly low altitude, he is not in free fall. Figure 3-8 shows what happens. When the man leaves the plane, he is moving slowly, so air resistance is small, and he starts to accelerate at 9.8 m/s^2. Note that the slope of the graph, at the beginning, is 9.8 m/s^2. However, the curve soon begins to level off. The rate at which the man's speed increases is getting smaller, since the increasing air resistance prevents him from accelerating so rapidly. Eventually, at a speed of about 50 m/s, the speed stops increasing altogether. The air resistance is now

FIGURE 3-8

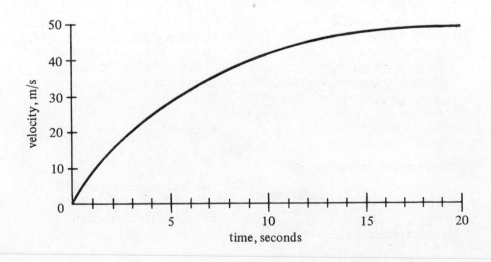

equal to the man's weight, and he continues to fall at constant speed, in equilibrium.

Sky divers, jumping out at very high altitudes, have more complex motions. When they deplane at altitudes of a couple of miles, they often assume the "delta position"—falling head first with a stiff body and arms at the sides. This position produces minimum drag. After about 15 seconds, the diver has fallen about 800 meters and is going at a terminal velocity of 100 m/s. He or she does not dare open the parachute at that speed because the sudden increase in drag could produce a severely injuring jerk on the shroud lines. As the diver falls into denser air, the drag increases, and is increased still further by flattening the body and spreading the arms and legs. The terminal velocity decreases to about 50 m/s. Near the ground, the parachute is opened, increasing the drag still more and reducing the terminal velocity to 15 m/s—still enough to produce quite a jolt on landing.

The viscous drag on the sky diver depends on the shape of his or her body. In the delta position, the air flows around the body smoothly, like the air around the airfoil surface in the picture. This flow is said to be *laminar*, that is, in layers. Laminar flow produces the minimum of viscous drag. If the surface of the object is more irregular, so that the air forms swirls and eddies, the drag is greatly increased. The flow is then said to be *turbulent*.

The engineer who designs a boat to travel through water, or a car or plane to go through the air, makes the shape smooth and rounded, so that the air flow will be laminar. This is especially important if the vehicle is to travel at high speed. A vehicle designed to produce laminar flow as it races through the air or the water is said to be *streamlined*.

Core Concept

Viscous drag is least when the flow of fluid is laminar, and greatest when it is turbulent.

Try This In terms of fluid flow, explain the utility of the shape of (a) a parachute; (b) a fish.

3-11. Pipelines

When water flows through a smooth pipe, the flow is laminar and obeys certain rules. Some of them are obvious, but others are rather surprising.

Let's look at what happens when the pipe gradually narrows, as in Figure 3-9. In the picture, one section of the water has been shown in three different positions. It flows at *A* through a wide pipe; then it passes into *B*, where the pipe is getting narrower; finally, at *C*, it is in a narrow pipe of uniform width.

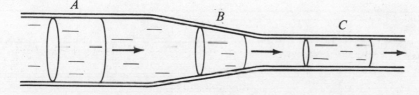

FIGURE 3-9

First of all, notice that this particular piece of water occupies a much longer section of pipe at C than it does at A. Also, it has to take the same length of time to travel its own length in either pipe. Therefore it must be going faster at C than at A. You have probably made use of the fact that the water in a pipe has to speed up as it goes through a constriction; did you ever partly cover the end of a hose (Figure 3-10) to make the water flow out faster?

FIGURE 3-10

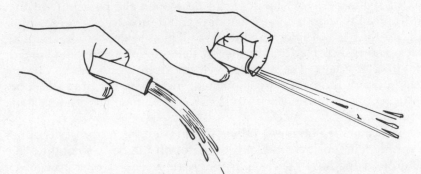

A second rule about flowing water is more unexpected. The section of water at A (Figure 3-9) is flowing along at constant speed, so it must be in a state of equilibrium. Therefore the force pushing it to the left is equal to the force pushing it to the right, just as a cart moves along at constant speed when all the forces on it balance out. This means that, in the wide pipe, the pressure is constant throughout.

Now look at what happens when the water enters section B, where the pipe is gradually constricted. In that region, it is speeding up—after all, it is going faster at C than at A. This means that the force pushing it to the right is greater than the force pushing it to the left. It follows that *the pressure at C must be less than the pressure at A!*

The rule we have just found is called *Bernoulli's theorem*; it states that in a flowing liquid or gas, the pressure is least where the speed is greatest. That this is true is easily demonstrated with a device like that of Figure 3-11. The water in the vertical tubes stands highest in the widest part of the pipe, showing that the pressure is greatest where the water flows slowest.

There are a number of practical applications of Bernoulli's principle. The venturi meter is a device for measuring the rate of flow of a fluid through a pipe; the principle is shown in Figure 3-11. The higher the fluid stands in the vertical tube, the slower the liquid is flowing.

FIGURE 3-11

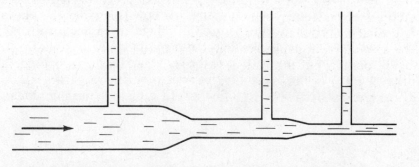

Core Concept

In a liquid or gas in which the flow is laminar, the pressure is least where the speed of flow is greatest.

Try This Sometimes a part of an artery in the body becomes swollen and widens. Explain why this is dangerous.

The familiar atomizer or spray gun also uses the Bernoulli principle (Figure 3-12). The flow of air across the mouth of the vertical pipe is constricted. This reduces the air pressure at that point, making it less than the atmospheric pressure on the open surface of the liquid. Thus the fluid rises up into the vertical tube, just as it does when you use your mouth to reduce the pressure at the top of a drinking straw.

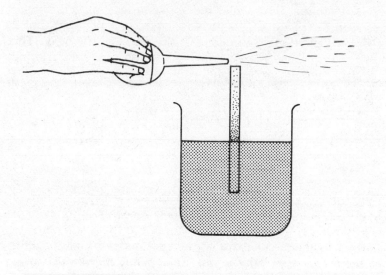

FIGURE 3-12

3-12. What Makes It Fly?

Long ago, nature designed a structure that makes use of Bernoulli's principle to produce lift, enabling birds to fly. It is no coincidence that the cross-sectional shape of an airplane wing is much like that of a bird's wing. Both operate on the basis of the same law of physics.

The principle on which the wing functions is illustrated in Figure 3-13. As the wing—of bird or airplane—passes through the air, there is a laminar flow of the air across both its upper and lower surfaces. But the upper surface has a curve to it. Air, hugging the surface as it flows, must go farther across the upper surface than the lower one. Since the upper and lower streams come together behind the wing, without turbulence, they must be making the trip across the wing surface in equal times. The upper stream must be going faster, and, according to Bernoulli, is therefore under less pressure than the air flowing across the flat lower surface. The reduced pressure on the upper surface means that the upward force on the wing is greater than the downward force.

The amount of lift depends on how fast the air flows across the surface, that is, on how fast the plane is flying. Lift increases with the square of velocity. A plane designed for high speed needs a lot less wing than a slow-flying sport plane. There is a trade-off; the military fighter plane with

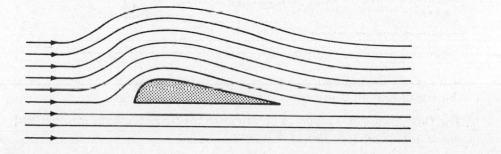

FIGURE 3-13

small wings has to get up a lot of speed to take off, so it needs a powerful engine and a long runway. One compensation is that the small wing produces less drag.

Lift also depends on the *angle of attack* of the wing. This is the angle that the lower wing surface presents to the air flow. In Figure 3-14, the angle of attack has been increased from the position of Figure 3-13, as it would be when the plane is taking off or landing. With the larger angle, there is some loss of laminar flow across the upper surface, as you can see by the turbulence in the picture. This increases drag considerably. But it also increases lift, which is why the pilot tilts the plane in this way. The extra lift comes about because the lower wing surface is now interfering with the flow of air. When the air flow on the lower surface slows down, its pressure increases.

FIGURE 3-14

There is a limit to how much extra lift the pilot can get this way. If he tilts the plane too much, the air flow across the upper surface breaks away from the surface and becomes turbulent, as in Figure 3-15. Lift disappears. In this condition, the plane is said to *stall*. The pilot can recover from a stall if he has enough space and control to speed up sufficiently to get back his lift at a smaller angle of attack.

FIGURE 3-15

When a plane is coming in for a landing, it has to slow down to well below its stall speed at the usual angle of attack. That is why the wing flaps are lowered as the plane approaches the runway, giving the wing cross-section the configuration shown in Figure 3-16.

FIGURE 3-16

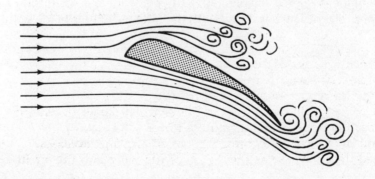

Lift is produced by the reduced air pressure in the fast-flowing air across the upper surface of the wing.

Try This Explain the two important ways in which lowering the wing flaps helps prepare a plane for landing.

Summary Quiz

For each of the following, fill in the missing word or phrase:

1. A fluid with no upper surface is called a _____ .
2. Every solid is characterized by a definite _____ .
3. The ratio of mass to volume is called _____ .
4. A rock will sink in water because its _____ is more than that of water.
5. Pressure is the ratio of force applied to the _____ over which it is applied.
6. The pressure a woman exerts on the ground increases as the heels of her shoes are made _____ .
7. In making a barometer, mercury is the liquid of choice because of its high _____ .
8. In a standard atmosphere, the height of the mercury in a barometer is _____ .
9. At a depth of 10 m in a lake, the pressure is 2 atm. At 20 m, the pressure is _____ .
10. If the downward pressure at some point is 2,200 mb, the upward pressure is _____ .
11. At an altitude of 3.7 mi, the air pressure is about 0.5 atm. At 7.4 mi, it is _____ atm.
12. Difference of pressure on top and bottom of a submerged object produces the force known as _____ .
13. The buoyancy on an object is equal to the _____ of the fluid it displaces.
14. When an object is floating, the _____ of the displaced fluid is equal to that of the floating object.
15. An object will float on a fluid whenever its _____ is less than that of the fluid.
16. The elastic property that tends to hold a liquid together is called _____
17. When an object is moving through a fluid, the amount of viscous drag depends, in part, on the _____ of the fluid.
18. Viscous drag increases as the _____ of the object increases.
19. At the terminal velocity of an object that is falling, the viscous drag is equal to the object's _____ .
20. Viscous drag is least when the flow of fluid around the object is _____ .
21. If an object is _____ , there will be little or no turbulence in the fluid through which it is moving.
22. As a fluid passes into a narrower section of pipe, its speed _____ .
23. As the fluid in a pipe speeds up, its pressure _____ .
24. Lift on an airplane wing results from the fact that the air is flowing _____ across the upper surface than the lower.
25. Both lift and drag increase as the _____ of an airplane increases.
26. Both lift and drag increase as the _____ of the wing into the air increases.

Problems

1. To determine the density of a liquid, an empty graduated cylinder is placed on a balance, and its mass is found to be 120 g. The liquid is then added, up to the 100 cm^3 mark, and the cylinder placed back on the balance. If the balance now reads 265 g, what is the density of the liquid?
2. What is the mass of 1 liter (=1,000 cm^3) of alcohol, whose density is 0.83 g/cm^3?
3. A new plastic is to be made, to be fabricated into little balls that will sink in any liquid whose density is less than 1.20 g/cm^3. What is the volume needed in a container to package 500 g of this plastic?
4. What pressure is exerted on the ground by a 2,500-lb buffalo if each hoof has an area of 4 in^2?
5. What is the total force, in pounds, exerted by the atmosphere on one side of a sheet of paper that measures $8\frac{1}{2} \times 11$ inches?
6. How many standard atmospheres are represented by a pressure of 62 lb/in^2?
7. What is the atmospheric pressure at the altitude at which commercial airliners fly—about 6 mi, or 9.6 km?
8. What is the pressure, in atmospheres, on the ears of a diver who is 65 m deep in the water?
9. A dam is to be made to hold back an artificial lake 110 m deep. Find the pressure that the bottom of the dam must withstand, in pascals (1 Pa = 1 N/m^2). Note that the density of water, in SI units, is 10^3 kg/m^3.
10. What is the pressure at the bottom of a tank of salad oil, density 880 kg/m^3, if the tank is 7.5 m deep? Express your answer in atmospheres; 1 atm = 10^5 Pa.
11. What is the force of buoyancy, in newtons, acting on a rock that has a volume of 650 cm^3 and is immersed in water?
12. An anchor is suspended from a spring scale that reads 350 N, and the reading drops to 245 N when the anchor is lowered into the water. What is the volume of the anchor?
13. To find the density of a rock, a geologist suspends it from a spring scale calibrated in mass units—grams. The scale reads 45 g when the rock is in air, and 32 g when it is in water. What is the density of the rock?
14. A cubical block of wood, 3.0 cm on a side, is floated in an overflow can, and it displaces 18 cm^3 of water. Find the block's (a) mass; (b) density.
15. A rectangular plank is 6.5 cm thick; it floats in water with 1.5 cm projecting above the surface. What is its density?
16. An iceberg floats with 86% of its volume under water. What is the density of ice?

CHAPTER
Making Things Move

4-1. How to Change a Velocity

When one of those space exploration machines is traveling in outer space, it keeps on going in a straight line at constant speed, unless otherwise instructed. The instructions sent by radio take the form, "Fire rocket c for t seconds." The purpose of the firing is to change the velocity of the space-craft—to speed it up, slow it down, or change its direction.

How much does the velocity change? That depends on which rocket fires, and for how long. Rockets vary according to the amount of force they exert on the spacecraft. A powerful rocket fired for a short burst might produce the same amount of change in velocity as a weak rocket fired gently for a long time. The amount by which the velocity changes depends on the product of the force exerted by the rocket and the length of time for which it is exerted. This product is called the *impulse*, represented by P in the definition

$$\mathbf{P} = \mathbf{F} \, \Delta t \qquad\qquad \text{(Equation 4-1a)}$$

In the SI, impulse is expressed in newton-seconds (N · s). The amount that the velocity of an object changes depends directly on the amount of impulse applied to it. An impulse of 500 N · s, for example, could be produced by a force of 100 N operating for 5 s, or by 500 N operating for 1 s. In either case, it produces twice as much change in the velocity of a given object as would an impulse of 250 N · s. This relationship is expressed in the proportionality

$$\Delta v \propto F \, \Delta t$$ (Equation 4-1b)

Sample Problems 4-1, 4-2, and 4-3 show how this relationship can be used.

Sample Problem

4-1 To slow down a car, a braking force of 1,200 N is applied for 10 s. How much force would be needed to produce the same change in velocity in 6 s?

Solution   To produce the same change in velocity, the same impulse is needed. Thus

$$F_1 \, \Delta t_1 = F_2 \, \Delta t_2$$

$$(1{,}200 \text{ N})(10 \text{ s}) = F_2(6 \text{ s})$$

$$F_2 = 2{,}000 \text{ N}$$

Sample Problem

4-2 A frictionless wagon is pushed, from rest, with a force of 60 N for 14 s. If it then strikes a wall and comes to rest in 0.15 s, how much force does the wall exert on it?

Solution   The increase from rest to the final velocity is the same amount of change as the decrease from the final velocity to rest, so the two impulses are equal:

$$(60 \text{ N})(14 \text{ s}) = F_2(0.15 \text{ s})$$

$$F_2 = 5{,}600 \text{ N}$$

Sample Problem

4-3 A spacecraft has two rocket engines, one producing a thrust of 300 N and the other 750 N. Firing the smaller engine for 10 s speeds the ship up from 80 m/s to 95 m/s, and the large engine is then fired for 12 s. How fast is the ship then going?

Solution   The two impulses are

$$\text{impulse}_1 = F \, \Delta t = (300 \text{ N})(10 \text{ s}) = 3{,}000 \text{ N} \cdot \text{s}$$

$$\text{impulse}_2 = F \, \Delta t = (750 \text{ N})(12 \text{ s}) = 9{,}000 \text{ N} \cdot \text{s}$$

So the second impulse must produce three times as much change in velocity as the first. The first change was 95 m/s − 80 m/s = 15 m/s, so the second impulse increased the velocity by 45 m/s, from 95 m/s to 140 m/s.

Core Concept

Change in velocity of an object is proportional to the product of force and the time during which the force acts:

$$\Delta v \propto F \, \Delta t$$

Try This   A spacecraft is going 350 m/s; a retro rocket that provides a force of 520 N for 10 s slows it down to 300 m/s. Then another retro rocket fires, with a force of 130 N for 4 s. How fast is the craft then going?

The effectiveness of an impulse in producing a change in velocity depends on some property of the object on which the impulse acts. An impulse of 50 N · s provided by a bow might send an arrow flashing through the air, but it would have little effect on a ferryboat.

Expression 4-1b is written simply as a proportionality, which is fine as long as you do not change the object you are dealing with. But the actual value of the ratio of impulse to change in velocity varies according to the object. We might want to rewrite the equation this way:

$$k \, \Delta v = F \, \Delta t$$

in which k is some property of the object. The constant k is much larger for a ferryboat than for an arrow; if the same impulse is applied to both, Δv is much smaller for the ferryboat.

Just what is that constant k, anyway? We can find out by applying a little algebra to some facts that we already know. First, divide the previous equation by Δt:

$$F = k \, \frac{\Delta v}{\Delta t}$$

But Equation 1-5 tells us that $\Delta v / \Delta t$ is the acceleration of the object. Therefore

$$F = ka$$

Now suppose we take our arrow, or our ferryboat, or anything else, and just drop it so that it falls freely. We learned in Chapter 1 that any object in such a circumstance falls with acceleration g, the acceleration due to gravity. Therefore, for a falling object,

$$F = kg$$

But the force that makes the thing fall is none other than the force of gravity, its weight, which is mg according to Equation 2-3. Therefore,

$$mg = kg$$

and the constant of proportionality k, which converts Expression 4-1b into an equation, is the mass, m, of the object. Ferryboats have more mass than arrows. Now we can write the equation that, more than any other, lies at the heart of the science of mechanics:

$$\textbf{F} = m\textbf{a} \qquad \text{(Equation 4-2)}$$

This is *Newton's second law of motion.* Sample Problems 4-4, 4-5, and 4-6 are applications of this law. Note that there is no difficulty with units as long as we stick to the SI, since $1 \text{ N} = 1 \text{ kg} \cdot \text{m/s}^2$.

Sample Problem

4-4 What force is needed to accelerate a 60-kg wagon from rest to 5.0 m/s in 2.0 s?

Solution   First, find the acceleration from Equation 1-5:

$$a = \frac{\Delta v}{\Delta t} = \frac{5.0 \text{ m/s}}{2.0 \text{ s}} = 2.5 \text{ m/s}^2$$

Then apply Equation 4-2:

$$F = ma = (60 \text{ kg})(2.5 \text{ m/s}^2) = 150 \text{ kg} \cdot \text{m/s}^2 = 150 \text{ N}$$

Sample Problem

4-5 A frictionless wagon going at 2.5 m/s is pushed with a force of 380 N, and its speed increases to 6.2 m/s in 4.0 s. What is its mass?

Solution The acceleration of the wagon is

$$a = \frac{\Delta v}{\Delta t} = \frac{6.2 \text{ m/s} - 2.5 \text{ m/s}}{4.0 \text{ s}} = 0.93 \text{ m/s}^2$$

From Equation 4-2,

$$m = \frac{F}{a} = \frac{380 \text{ N}}{0.93 \text{ m/s}^2} = 410 \text{ kg}$$

Sample Problem

4-6 What acceleration would be given to a 7.5-kg bowling ball being swung with a propelling force of 120 N?

Solution From Equation 4-2,

$$a = \frac{F}{m} = \frac{120 \text{ N}}{7.5 \text{ kg}} = 16 \text{ m/s}^2$$

Core Concept

Force equals mass times acceleration ($F = ma$).

Try This What is the mass of a frictionless sled that will be accelerated at 3.0 m/s² by a force of 130 N?

4-3. Cooperating Forces

Spaceships in outer space and things falling freely have one thing in common: Each has only a single force acting on it. We rarely deal with this sort of situation. When we push something, there are other forces acting as well. The F in Equation 4-2 is not any one of these forces; it is the vector sum of all of them. If they all add up to zero, then the object will not accelerate; it will be in equilibrium. F is the *net force* on the object.

Consider, for example, a 350-kg rocket about to take off vertically from its launching pad. You decide that you would like it to zoom upward with a substantial acceleration—say, $a = 8$ m/s². So you apply Equation 4-2 to this acceleration and conclude that you need an engine with a thrust of $ma = 2,800$ N. If you design your rocket engine that way, you will be severely disappointed. Your rocket weighs $mg = 3,430$ N; 2,800 N will never get it off the ground. The thrust you need is 2,800 N *more than* the weight of the rocket. Sample Problem 4-7 shows how the needed thrust must be calculated.

Sample Problem

4-7 What thrust is needed to fire a 350-kg rocket straight up with an acceleration of 8.0 m/s²?

Solution The net force needed to produce this acceleration is

$$F = ma = (350 \text{ kg})(8.0 \text{ m/s}^2) = 2,800 \text{ N}$$

However, there is a downward force acting on it as well, equal to

$$w = mg = (350 \text{ kg})(9.8 \text{ m/s}^2) = 3,430 \text{ N}$$

The net force is the thrust minus the weight, or

$$\text{thrust} - 3{,}430\text{ N} - 2{,}800\text{ N}$$

so the rocket engine must product a thrust of 6,230 N.

Here's another example: The child and sled of Figure 4-1 have a mass of 40 kg. The acceleration of the sled is produced by the net horizontal force. The horizontal component of the tension in the tow rope pulls the sled to the right, while friction is pulling it to the left. The net force is the difference—and that is what accelerates the sled. See Sample Problems 4-8 and 4-9.

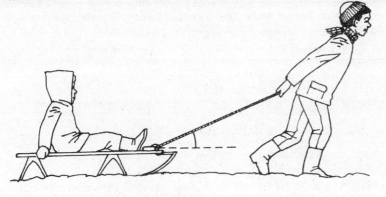

FIGURE 4-1

Sample Problem

4-8 What is the acceleration of the child on the sled, with combined mass 40 kg, if the friction is 60 N and the rope is being pulled with a force of 170 N at an angle of 35° with the ground?

Solution The horizontal component of the force pulling the sled forward is (from Equation 1-4)

$$T_{\text{hor}} = \mathbf{T}\cos\theta = (170\text{ N})(\cos 35°) = 139\text{ N}$$

To get the net force, subtract the force to the right from the frictional force to the left:

$$F = 139\text{ N} - 60\text{ N} = 79\text{ N} = 79\text{ kg} \cdot \text{m/s}^2$$

From Equation 4-2,

$$a = \frac{F}{m} = \frac{79\text{ kg} \cdot \text{m/s}^2}{40\text{ kg}} = 2.0\text{ m/s}^2$$

Sample Problem

4-9 A 2.0-kg weather balloon is released and begins to rise against 6.5 N of viscous drag. If its buoyancy is 32 N, what is its acceleration?

Solution First, we have to find the net force on it. Its weight is $mg = (2.0\text{ kg})(9.8\text{ m/s}^2) = 19.6\text{ N}$. Since it is rising, the viscous drag is another downward force, so the total downward force is $19.6\text{ N} + 6.5\text{ N} = 26.1\text{ N}$. Since the buoyancy pushing it upward is 32 N, the net upward force is $32\text{ N} - 26.1\text{ N} = 5.9\text{ N}$. Now that we know the net force, we can find the acceleration:

$$a = \frac{F}{m} = \frac{5.9\text{ N}}{2.0\text{ kg}} = 3.0\text{ m/s}^2$$

Acceleration must be calculated from the vector sum of all forces acting on the object.

Try This What is the acceleration of a 1,200-kg boat if its motor produces 8,500 N of forward thrust and the viscous drag is 6,200 N?

4-4. Momentum

Which does more damage in striking a tree, a Cadillac or a Toyota? If both hit at the same speed, the Cadillac, with its larger mass, will surely exert more force on the tree. But suppose the Toyota is going much faster than the Cadillac. Now is there any way to figure it out?

Well, maybe we can. We might start by rewriting Equation 4-2 like this:

$$F = m \, \frac{\Delta v}{\Delta t}$$

Now multiply both sides by Δt:

$$F \, \Delta t = m \, \Delta v$$

and we recognize the left-hand side of this equation as the impulse that stops the car. Remember, as you learned in Chapter 2, that the force the tree exerts on the car has to be the same in magnitude as the force the car exerts on the tree. Both are damaged in the collision. And the amount of damage depends on the product of the mass of the car and the speed at which it was going before the tree abruptly brought it to rest.

The quantity *mv*—mass times velocity—is called *momentum*. In bringing the car to rest, the tree suddenly reduced the momentum of the car to zero. And the amount by which the momentum changed is equal to the impulse applied to the car:

$$\mathbf{F} \, \Delta t = \Delta(m\mathbf{v}) \qquad \text{(Equation 4-4)}$$

Or: The impulse applied to anything is equal to the amount by which its momentum changes.

This equation says nothing that we have not already used in Equation 4-2, but it is often more convenient. In some problems, it saves us the trouble of calculating the acceleration as a separate step. See Sample Problems 4-10 and 4-11.

Sample Problem

4-10 What braking force is needed to bring a 2,200-kg car going 18 m/s to rest in 6.0 s?

Solution Using Equation 4-4, we have

$$F(6.0 \text{ s}) = (2,200 \text{ kg})(18 \text{ m/s}) - 0$$

which says that the momentum is changing from *mv* to 0. This gives $F = 6,600$ N.

Sample Problem

4-11 A 650-kg rocket is to be speeded up from 440 m/s to 520 m/s in outer space. If the thrust of the engine is 1,200 N, for how long must the engine be fired?

Solution The change in the momentum of the rocket is (650 kg) × (520 m/s) − (650 kg) (440 m/s) = 52,000 kg · m/s. This must be equal to the impulse, so

$$(1{,}200 \text{ N}) \, (\Delta t) = 52{,}000 \text{ kg} \cdot \text{m/s}$$
$$\Delta t = 43 \text{ s}$$

There are some situations in which it is either difficult or impossible to use Equation 4-2, but Equation 4-4 comes to the rescue. This happens when the mass of the object is changing. In that case, even with a constant force, the acceleration is changing, and the problem becomes complex. Here is a case in point: A rocket ship traveling in outer space speeds up by turning on its engines. How much does its velocity change? You would have to know the thrust of the engine (F) and the period of time for which the engine burned (Δt). But if the fuel consumed is any substantial part of the mass of the rocket, the mass of the rocket is also changing. The impulse gives you the change in the momentum of the rocket ship, and you can use this to determine the final speed only if you know how much fuel was consumed. See Sample Problem 4-12.

Sample Problem

4-12 A rocket ship in outer space has a mass of 650 kg, including 120 kg of fuel, and it is going 140 m/s. Burning all the fuel produces a thrust of 1,200 N for 25 s. How fast is the ship then going?

Solution   The impulse applied to the ship is

$$P = F \, \Delta t = (1{,}200 \text{ N}) \, (25 \text{ s}) = 30{,}000 \text{ N} \cdot \text{s}$$

and this must equal the increase in the momentum of the ship. Its original momentum is

$$mv = (650 \text{ kg}) \, (140 \text{ m/s}) = 91{,}000 \text{ kg} \cdot \text{m/s}$$

Note that, since $N = \text{kg} \cdot \text{m/s}^2$, the impulse and the momentum are expressed in the same units and can be added. The final momentum is the initial momentum plus the impulse, or

$$mv_{\text{final}} = 91{,}000 \text{ N} \cdot \text{s} + 30{,}000 \text{ N} \cdot \text{s}$$

Since the fuel has been burned up, the final mass is only 530 kg, so

$$v_{\text{final}}(530 \text{ kg}) = 121{,}000 \text{ kg} \cdot \text{m/s}$$
$$v_{\text{final}} = 228 \text{ m/s}$$

A study of Equation 4-4 will explain why trapeze artists keep a net under them as they work. If they fall, they will have a certain definite momentum when they hit the ground. They must lose all this momentum; the impulse applied to their bodies will be $F \, \Delta t$. The purpose of the net is to increase the value of Δt, the time it takes them to come to rest. The hard, unyielding ground brings them to rest within a fraction of a second. The net spreads out the impulse over a much longer period of time, so that the force is proportionally smaller.

Core Concept

Impulse equals change in momentum:

$$\mathbf{F} \, \Delta t = \Delta(m\mathbf{v})$$

Try this   A pitcher throws a ball whose mass is 0.30 kg, bringing it from rest to a speed of 35 m/s. If the motion of his arm while holding the ball lasts 0.50 s, find (a) the momentum of the ball; (b) the impulse the pitcher applied to it; and (c) the force the pitcher used in throwing it.

4-5. The Law of Momentum

The greatest value of the concept of momentum is in the calculation of what happens when two or more objects interact. It turns out that we can often find out the results of a collision, for example, without knowing anything at all about the forces involved or how long they persist.

The explanation is this: Think of two objects, A and B, colliding. Object A exerts a force F on object B, and B exerts a force $-F$ on A. The negative sign indicates that the two forces are in opposite directions. They are equal to each other. Since the duration of the impact, Δt, is the same for both, the impulse that B exerts on A is the negative of the impulse A exerts on B.

Now, impulse is equal to change in momentum. It follows that, in the collision, the change in the momentum of A is the negative of the change in the momentum of B. The sum of the two changes is thus zero. In other words, during the collision, the total momentum does not change! The increase in the momentum of one object is exactly equal to the decrease in the momentum of the other. This gives us the very fundamental and important law of nature called the *law of conservation of momentum:* In an isolated system, the total momentum does not change.

You should think of using this rule whenever you have to deal with an interaction between two objects. For example, a little girl is standing on a wagon, and jumps off to the rear of it. To do so, she has to kick the wagon so that it moves forward. Before she jumped, the total momentum of the system was zero, so (if friction is small enough to ignore) it must continue to be zero afterward. Her momentum backward must equal the wagon's momentum forward. Sample Problem 4-13 shows how this idea can be used.

Sample Problem

4-13 The 35-kg girl is standing on a 20-kg wagon and jumps off, giving the wagon a kick that sends it off at 3.8 m/s. How fast is the girl moving?

Solution Total momentum is initially zero, and must remain zero. Therefore, the momentum acquired by the wagon in one direction equals the momentum acquired by the girl in the other:

$$(mv)_{girl} = (mv)_{wagon}$$
$$(35 \text{ kg}) (v_{girl}) = (20 \text{ kg}) (3.8 \text{ m/s})$$
$$v_{girl} = 2.2 \text{ m/s}$$

Now suppose the girl is running and jumps into the wagon. If you know how fast she is going, you can figure out how fast she and the wagon will be traveling together when she is in it. Before she hit the wagon, her momentum was the total momentum of the system, and that value does not change. See Sample Problem 4-14.

Sample Problem

4-14 The same 35-kg girl is now running along at 5.2 m/s and jumps into the 20-kg wagon. How fast is the wagon moving with the girl in it?

Solution At first the girl has all the momentum, and it must be equal to the total momentum of the girl and the wagon after she lands in it. So

$$(mv)_{\text{before}} = (mv)_{\text{after}}$$
$$(35 \text{ kg}) (5.2 \text{ m/s}) = (35 \text{ kg} + 20 \text{ kg})v_{\text{after}}$$
$$3.3 \text{ m/s} = v_{\text{after}}$$

In dealing with problems of this sort, be careful to take the direction of the momentum into account; it is a vector. In one dimension, it is enough to call one direction negative and the other positive. Thus, if a ball hits a wall perpendicularly, going 20 m/s, and bounces off at 15 m/s, the change in its velocity is 20 m/s − (−15 m/s) = 35 m/s. See Sample Problem 4-15 for an example.

Sample Problem

4-15 A 1.2-kg basketball traveling at 7.5 m/s hits the back of a 12-kg wagon and bounces off at 3.8 m/s, sending the wagon off in the original direction of travel of the ball. How fast is the wagon going?

Solution The momentum of the ball before it hit has to equal the sum of the momenta of the two objects after the collision. Let's call the original direction of travel of the ball positive. Then the equation becomes

$$(1.2 \text{ kg}) (7.5 \text{ m/s}) = (1.2 \text{ kg}) (-3.8 \text{ m/s}) + (12 \text{ kg}) (v)$$
$$1.1 \text{ m/s} = v$$

Core Concept

In a closed system, the total momentum does not change.

Try this A 2,000-kg limousine going east at 22 m/s strikes a 1,200-kg sports car going west at 30 m/s. How fast are the two cars going together after the collision?

4-6. Going in Circles Again

As we saw in Chapter 1, a change in the direction of motion is an acceleration, even if the speed remains constant. And when an object is traveling in a circular path, its direction is constantly changing. Its acceleration is centripetal—toward the center of the circle. The only way to produce a centripetal acceleration is to apply a centripetal force.

Consider, for example, a ball attached to the end of a string and whirled around overhead in a horizontal circle, as shown in Figure 4-2. If you let go of the string the ball will keep going in the direction in which it is traveling; it will fly off on a path that is tangent to the circle. To keep it going in the circular path, the string must constantly exert a force on the ball. The force is at all times perpendicular to the velocity, so it changes only the direction of the velocity, not the speed.

This, then, is the condition necessary for an object to travel at constant speed in a circular path: There must be a force acting on it, a *centripetal force* that is kept perpendicular to the velocity as the direction of the velocity changes. The force will be directed toward the center of the circle. Let's look at some examples.

When a clothes dryer spins around, the walls of the tub exert a centripetal force on the clothes, making them travel in a circular path. The

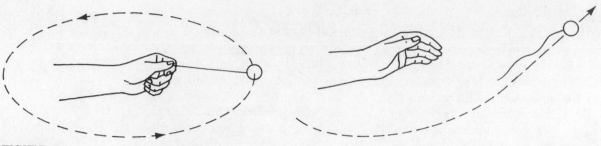

FIGURE 4-2

water can pass through the holes in the tub, so there is no effective centripetal force on it, and it flies off through the holes.

When a car travels in a circular path, the centripetal force is supplied by the friction of the road against the front tires. If the road is banked, the elastic recoil of the road helps by pushing the car toward the center of the circle.

When a satellite is in circular orbit above the earth, the earth's gravity is constantly pulling it toward the center of the earth, perpendicular to the satellite's velocity. Gravity is, in this case, a centripetal force.

How much is the centripetal force? Equation 4-2 tells us that the force that produces an acceleration is ma, and this law is not repealed when the force is centripetal. And we learned, in Equation 1-11b, that centripetal acceleration is equal to v^2/r—the square of its velocity divided by the radius of the circle in which it is traveling. Using these two relationships, we find that centripetal force is given by

$$F_c = \frac{mv^2}{r}$$

(Equation 4-6)

See Sample Problems 4-16, 4-17, and 4-18.

Sample Problem

4-16 The ball at the end of the string of Figure 4-2 has a mass of 150 g. It is traveling at 4.0 m/s and the string is 65 cm long. How much is the tension in the string?

Solution If we are to get the answer in newtons, we have to solve the problem in SI units; 150 g = 0.150 kg; 65 cm = 0.65 m. First, we need the acceleration of the ball, from Equation 1-11b:

$$a_c = \frac{v^2}{r} = \frac{(4.0 \text{ m/s})^2}{0.65 \text{ m}} = 24.6 \text{ m/s}^2$$

To get the force, use Equation 4-2:

$$F = ma = (0.150 \text{ kg})(24.6 \text{ m/s}^2) = 3.7 \text{ N}$$

Sample Problem

4-17 In outer space, a 1,200-kg spacecraft is traveling at 60 m/s. In order to change its direction, a rocket is fired off to one side, in a direction perpendicular to the velocity. If the thrust of the rocket is 3,820 N, what is the radius of the circle in which the spacecraft travels?

Solution From Equation 4-6,

$$r = \frac{mv^2}{F} = \frac{(1,200 \text{ kg})(60 \text{ m/s})^2}{3,820 \text{ N}} = 1,130 \text{ m}$$

Sample Problem

4-18 A 1.5-kg rock is tied to a string 40 cm long and spun around in a horizontal circle. If the string will break when the tension in it exceeds 250 N, what is the greatest speed the rock can have without breaking the string?

Solution From Equation 4-6,

$$v = \sqrt{\frac{Fr}{m}} = \sqrt{\frac{(250 \text{ N}) (0.40 \text{ m})}{1.5 \text{ kg}}} = 8.2 \text{ m/s}$$

Core Concept

The force that keeps an object going in a circular path at constant speed is equal to the mass of the object times its centripetal acceleration:

$$F_c = mv^2/r$$

Try This A bicycle and rider, with a combined mass of 55 kg, are traveling at 6.0 m/s. If they are going in a circular path whose radius is 30 m, how much is the force acting on the tires of the bicycle?

4-7. The Moon in Orbit

We saw in Chapter 1 that, when an artificial satellite is in orbit around the earth, it is in a state of free fall. Since its path is a circle, the force on it must be centripetal; that is, perpendicular to its velocity. That force is gravity.

The moon is also in free fall, in orbit around the earth. Let's find its acceleration. The radius of its orbit is

$$239,000 \text{ mi} \times \frac{1,600 \text{ m}}{\text{mi}} = 3.82 \times 10^8 \text{ m}$$

and the period of its revolution is

$$27.3 \text{ days} \times \frac{24 \text{ hr}}{\text{day}} \times \frac{60 \text{ min}}{\text{hr}} \times \frac{60 \text{ s}}{\text{min}} = 2.36 \times 10^6 \text{ s}$$

We can find out how fast it is going because we know that it travels the circumference of its orbit ($2\pi r$) in one period of revolution. Therefore its speed is

$$v = \frac{2\pi r}{T} = \frac{2\pi \times 3.82 \times 10^8 \text{ m}}{2.36 \times 10^6 \text{ s}} = 1.017 \times 10^3 \text{ m/s}$$

Now we can get its acceleration:

$$a = \frac{v^2}{r} = \frac{(1.017 \times 10^3 \text{ m/s})^2}{3.82 \times 10^8 \text{ m}} = 2.7 \times 10^{-3} \text{ m/s}^2$$

The moon's acceleration is extremely small, thousands of times less than the acceleration of an object falling freely near the surface of the earth. The earth's gravity pulls on the moon, but far more weakly than it pulls on us. This is only one of the pieces of evidence indicating that the force of gravity gets smaller at locations farther from the earth.

Using the actual known values, we can find out what sort of rule this weakening of gravity obeys. First, just how much weaker is gravity at the distance of the moon than it is here on earth? The ratio of the two values of acceleration due to gravity is

$$\frac{\text{acceleration of moon}}{g \text{ at surface of earth}} = \frac{2.7 \times 10^{-3} \text{ m/s}^2}{9.8 \text{ m/s}^2} = \frac{1}{3,600}$$

When you or I drop a rock, its acceleration is just about 3,600 times as great as the acceleration of the moon in its orbit.

The acceleration due to gravity is weaker because the moon is farther from the center of the earth than we are. The ratio of the two distances is this:

$$\frac{\text{moon to center of earth}}{\text{us to center of earth}} = \frac{243,000 \text{ mi}}{4,000 \text{ mi}} = 60$$

The moon is 60 times as far away, and experiences 1/3,600 of the acceleration due to gravity. The ratio of the two values of g (60:1) is the *inverse square* of the ratio of the two distances (1:3,600). The rule is general, and can be written this way:

$$g \propto \frac{1}{r^2} \qquad \text{(Equation 4-7)}$$

where g is the acceleration due to gravity at some point in space and r is the distance from that point to the center of the earth. Remember that the acceleration of an object in free fall depends *only* on its location; when a meteor flashes past the moon, its acceleration toward the earth is the same as the moon's, no matter how big it is, which way it is going, where it came from, or what its speed is.

Once we know this, we can calculate the acceleration of any object falling freely in the space around the earth. Sample Problem 4-19 shows how this is done.

Sample Problem

4-19 What is the acceleration of a piece of space junk when it is at an altitude of 22,000 km?

Solution Note that the problem does not specify which way the object is traveling. It will have the same acceleration whether it is going up, down, or sideways in orbit, so long as it is in free fall.

Since the r in Equation 4-7 is distance from the center of the earth, the first thing we need is the correct value of r for the object. This is its altitude plus the radius of the earth: 22,000 km + 6,400 km = 28,400 km.

Equation 4-7 can be written as

$$\frac{g_1}{g_2} = \left(\frac{r_2}{r_1}\right)^2$$

where g_1 and g_2 are values of acceleration due to gravity and r_1 and r_2 are corresponding radial distances from the center of the earth. Since we know the value of g at the surface (6,400 km from the center), we may write

$$\frac{9.8 \text{ m/s}^2}{g_2} = \left(\frac{28,400 \text{ km}}{6,400 \text{ km}}\right)^2$$

from which $g_2 = 0.50 \text{ m/s}^2$

The earth is an ordinary planet, with many of the same properties as any other planet or heavenly body. If Equation 4-7 works in the space around the earth, it should also work in the space around any other heavenly body. And it surely does. All the planets of the solar system, for example, follow the rule. Their accelerations are inversely proportional to the square of their distances from the sun, for it is the gravity of the sun that keeps them in orbit.

Further, if some object is in the space near a celestial object, the effect of gravity will be to exert a force on that object—the force we call the object's weight. Since that force is always mg, the force of gravity on any object also drops off inversely as the square of the distance from the body that produces the gravity. See Sample Problem 4-20.

Sample Problem

4-20 What is the weight of a 66-kg meteor when it is 12,000 km above the surface of the earth?

Solution Its weight is mg, so we have to find the value of g at that position. Use the inverse square rule as before:

$$\frac{g_1}{g_2} = \left(\frac{r_2}{r_1}\right)^2$$

Remembering that the distances must be measured from the center of the earth, we have

$$\frac{9.8 \text{ m/s}^2}{g_2} = \left(\frac{18,400 \text{ km}}{6,400 \text{ km}}\right)^2$$

$$\frac{9.8 \text{ m/s}^2}{g_2} = 8.27$$

$$g_2 = 1.19 \text{ m/s}^2$$

Then $w = mg = (66 \text{ kg})(1.19 \text{ m/s}^2) = 79$ N.

Core Concept

The force of gravity and the acceleration it produces drop off inversely as the square of the distance from the center of the star or planet that produces the force:

$$g \propto 1/r^2$$

Try This Mars has a radius of 3.4×10^6 m, and the acceleration due to gravity at its surface is 3.3 m/s^2. If a 300-kg meteor passes at a distance of 8.8×10^6 m from the center of Mars, find (a) the acceleration of the meteor; and (b) its weight at that position.

4-8. Gravity is Everywhere

What determines the amount of gravitational force the earth exerts on the moon, or on anything else? From the physics you already know, we can figure this out.

We saw in Chapter 2 that the weight of anything, which is the force of gravity acting on it, is equal to mg. The force of the earth's gravity acting on objects on earth, then, is proportional to the masses of the objects:

$$F_{\text{grav}} \propto m_o$$

We also know that every force is one half of an interaction pair; if the earth is pulling on any object, the object must be pulling on the earth with an equal force in the opposite direction. This force the moon exerts on the earth, for example, produces a detectable wobble in the motion of the earth. The only reason it is so small is that the earth has such a large mass that the force cannot produce much acceleration.

How much force does an object on earth exert on the earth? If the force the earth exerts on the object is proportional to the mass of the object, it is reasonable to suppose that the force the object exerts on the earth is proportional to the mass of the earth:

$$F_{grav} \propto m_e$$

Since this force is the same in both directions, it must be proportional to *both* the mass of the moon and the mass of the earth:

$$F_{grav} \propto (m_o)(m_e)$$

We also know that this force varies inversely as the square of the distance from the earth to the object, so

$$F_{grav} \propto \frac{m_o\, m_e}{r^2}$$

We must assume that the earth and the moon obey the same laws of nature as all other objects in the universe. This relationship, as applied to everything, is called the *law of universal gravitation*. It states that any two objects in the universe attract each other with a force proportional to the product of their masses and inversely proportional to the square of the distance between them. Newton was the first to prove that this law could explain the motions of objects in the solar system.

By inserting a constant, whose value depends only on the choice of units, we can write the law algebraically this way:

$$F_{grav} = \frac{Gm_1 m_2}{r^2} \qquad \textbf{(Equation 4-8)}$$

G is the same for any pair of objects at all; it is a universal constant whose value in the SI is

$$G = 6.67 \times 10^{-11}\ \mathrm{N \cdot m^2/kg^2}$$

This is an extremely small number. What it tells us is that the gravitational attraction between ordinary objects is usually too small to be detected. Gravity becomes a significant force only when one of the objects is very large—like the earth, for example. Sample Problem 4-21 shows how this law can be applied in determining the force between two ordinary objects.

Sample Problem

4-21 What is the gravitational attraction between a 70-kg boy and a 60-kg girl who are 3 m apart?

Solution Apply Equation 4-8, the law of universal gravitation:

$$F_{grav} = \frac{Gm_1 m^2}{r^2}$$

$$F_{grav} = \frac{(6.67 \times 10^{-11}\ \mathrm{N \cdot m^2/kg^2})(70\ \mathrm{kg})(60\ \mathrm{kg})}{(3\ \mathrm{m})^2}$$

$$F_{grav} = 3 \times 10^{-8}\ \mathrm{N}$$

which is insufficient to move two fruit flies toward each other.

All objects attract each other with a force proportional to the masses of both and inversely proportional to the distance between them:

$$F_{grav} = \frac{Gm_1m_2}{r^2}$$

Try This What is the force of attraction between two asteroids 4.0×10^5 m apart if their masses are 5,700 kg and 14,000 kg, respectively?

4-9. The Mass of the Earth

If you examine Equation 4-8, you will find something very interesting about it. If we know the value of G, we can determine the mass of the earth! All the other quantities in the equation are known.

The value of G was determined over 150 years ago. It was done by means of an extremely sensitive force-measuring device which could measure the gravitational attraction between a lead sphere and a brass sphere of ordinary size. Knowing the two masses, the force, and the distance between the spheres, it was possible to calculate G.

Now let's find the mass of the earth. First, divide both sides of Equation 4-8 by m_1, the mass of any object near the earth. We are trying to find m_2, the mass of the earth. After we have divided, the left side of the equation contains only the expression

$$\frac{F_{grav}}{m_1}$$

which is nothing but w/m for an object near the earth. As we saw before, this quantity is g, the acceleration due to gravity. Then Equation 4-8 becomes

$$g = \frac{Gm_2}{r^2} \quad \text{or} \quad m_2 = \frac{gr^2}{G}$$

Now, we know that, at the surface of the earth, $g = 9.8$ m/s^2. Since r is the distance from the center of the earth, the value of r at the surface is the radius of the earth, 6.4×10^6 m. Now all we have to do is substitute our known values in the expression for m_2:

$$m_2 = \frac{(9.8 \text{ m/s}^2)(6.4 \times 10^6 \text{ m})^2}{6.67 \times 10^{-11} \text{ N} \cdot \text{m}^2/\text{kg}^2}$$

Making the appropriate substitution of N = kg $\cdot$ m/s^2, the answer is 6.0×10^{24} kg.

Many constants can be found by use of the universal gravitation law.

Try This Using the constants given at the end of section 4-7, determine the mass of Mars.

4-10. Weightlessness

When astronauts are in an orbiting spacecraft, gravity seems to have disappeared. They, and everything in the craft, float freely around the cabin. Why?

You might notice the same phenomenon if you are ever unlucky enough to be trapped inside a falling elevator. Of course, it would have to be a frictionless elevator so that it would fall freely, accelerating at 9.8 m/s^2. Then, if you should drop a pencil, it would stay just exactly where you left it, floating freely in space.

If someone is watching this whole event from outside the elevator, he would see what is happening somewhat differently. He would say, "Your pencil is falling. It is obvious, however, that it has the same acceleration as the elevator, so it never gets any closer to the floor of the elevator. You have that acceleration also, just like everything else in the elevator. You only *think* gravity has disappeared."

What will a scale read if you step on it while the elevator is falling? If you step on a scale in your own bathroom, you are in equilibrium. The scale pushes up on you with a force equal to your weight, and you push down on the scale with an equal force (action equals reaction). In the falling elevator, you fall with an acceleration g because the only force acting on you is gravity. There is no upward force at all. The scale reads zero.

This condition of weightlessness exists for all measurements made inside the elevator. It occurs because the elevator is in free fall. Remember, however, that when a satellite is in orbit, it also is in free fall, accelerated centripetally by the force of gravity. And as long as the spacecraft is in free fall, everything in it has the same acceleration, and everything appears weightless to anyone inside.

Acceleration of a vehicle you are in by some force other than gravity produces a gravity-like effect. When you accelerate your car forward, you feel yourself pushed back; when you turn to the right, you feel yourself pushed to the left. If your car turns in a circle, you feel yourself pushed to the outside of the circle, as in Figure 4-3, and you say that you feel a centrifugal force. Anyone standing on the road outside looks at you and says, "Nonsense. You only think there is a force pushing you outward. Actually, your body is trying to travel in a straight line while the car is turning in a circle."

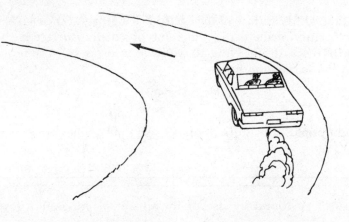

FIGURE 4-3

Core Concept

Acceleration of a vehicle changes the apparent gravity inside it, and when the vehicle is in free fall there are no gravitational effects inside.

Try This You are in a spacecraft in outer space somewhere, and everything inside it is weightless. Suddenly the things around you start to fall. What could you conclude?

We have, by now, sent a great many artificial satellites into orbit. There are thousands of bits of expensive hardware, and lots of the junk they released, orbiting the earth.

To put a satellite into orbit, you cannot simply boost it up there and drop it. It has to be lifted to a precisely predetermined altitude, pointed in exactly the right direction, and then given a specific speed. If you want the orbit to be circular, there is only one speed that will work at any given altitude.

The reason is that the circular orbit is produced by the centripetal force, which is gravity in this case. The force of gravity acting on an object is its weight, mg. To produce a circular orbit, this force, according to Equation 4-6, must be equal to a specific value, mv^2/r. Consequently,

$$mg = \frac{mv^2}{r} \qquad \text{or} \qquad g = \frac{v^2}{r}$$

Note that the mass of the satellite does not matter.

We have no control over the value of g; it depends strictly on r, the distance from the center of the earth. The equation tells us therefore that, at any given r, there is only one value of v that will produce a circular orbit.

Figure 4-4 illustrates how a satellite is typically launched. There are three parts when it takes off from the earth. Stage 1 is an enormous, powerful booster rocket that lifts the system several miles and gets it going at a fast speed. Then stage 1 separates and falls to earth. Stage 2 is a much smaller rocket that completes the lift to the proper altitude and then starts to fall. When the rocket is at the peak of its trajectory, traveling tangentially (horizontally), stage 2 separates, and stage 3 fires to give the satellite exactly the required speed. The satellite will have some smaller rockets of its own, which can be fired on radio command in case things do not go exactly right.

What speed must the satellite have in order to remain in a circular orbit? This can be found by first determining the value of the acceleration due to gravity at the desired altitude, using Equation 4-7. Sample Problems 4-22 and 4-23 show how.

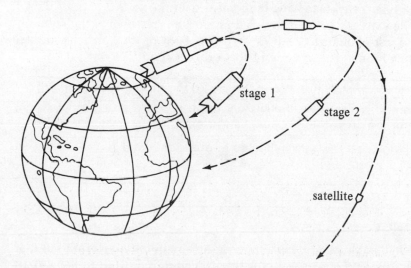

stage 1

stage 2

satellite

FIGURE 4-4

Sample Problem

4-22 How fast must a satellite be traveling to be in circular orbit at an altitude of 12,000 km?

Solution We need g and r. The value of r is the radius of the orbit, which is the altitude plus the radius of the earth: 12,000 km + 6,400 km = 18,400 km. To find g, use Equation 4-7:

$$\frac{g}{9.8 \text{ m/s}^2} = \left(\frac{6,400 \text{ km}}{18,400 \text{ km}}\right)^2$$

$$g = 1.19 \text{ m/s}^2$$

This is the acceleration of the satellite, which, from Equation 1-11b, must equal v^2/r. Therefore

$$v = \sqrt{ar} = \sqrt{\left(1.19 \frac{\text{m}}{\text{s}^2}\right)(1.84 \times 10^7 \text{ m})}$$

$$v = 4,700 \text{ m/s}$$

Sample Problem

4-23 The spacecraft Columbia travels in a low orbit, not far above the surface of the earth. How fast does it go?

Solution Near the surface of the earth, $g = 9.8$ m/s^2 and $r = 6.4 \times 10^6$ m. Then

$$v = \sqrt{ar} = \sqrt{\left(9.8 \frac{\text{m}}{\text{s}^2}\right)(6.4 \times 10^6 \text{ m})} = 7,900 \text{ m/s}$$

If the satellite is traveling too fast when it is released, or too slow, the orbit will be an ellipse rather than a circle. If it goes much too slow, the ellipse will intersect the surface of the earth, and the satellite will fall. If it is going more than twice the correct orbiting speed, the ellipse will open out into a hyperbola, and the satellite will escape into outer space, never to return.

Strange as it sounds, space is getting crowded. At least, one orbit is getting crowded because it is an especially useful one. This is the *synchronous* orbit, which is used by communications satellites. If you want to use a satellite to relay radio signals from the United States to Europe, you will want your relay satellite to hang motionless over the middle of the Atlantic Ocean. The way to do that is to put it into orbit directly over the equator, going just fast enough to keep up with the spin of the earth. Its orbital period is 24 hours. The altitude of this orbit is about 5.6 earth radii.

Core Concept

To put an artificial satellite into circular orbit, its speed must be such that the force of gravity, at its altitude, can bring it into a circular path.

Try This How fast must a satellite go in order to be in circular orbit at an altitude of 2 earth radii?

4-12. Making Things Spin

About $4\frac{1}{2}$ billion years ago, the earth condensed out of a swirling mass of rocks and dust and started to spin. It has been spinning ever since. With nothing to slow it down, no torque applied, it is in a state of *rotational equilibrium*, spinning with no change in its rotational speed. (This is not quite true, for it has slowed down considerably, mainly because of the impact of meteorites.)

The point is that there is a simple analogy between rotational and translational motion. It goes like this:

> An applied force changes momentum; with no force applied, the object is in equilibrium and its momentum does not change; in interactions, momentum is conserved.

> An applied torque changes *angular momentum*; with no torque applied, the object is in rotational equilibrium and its angular momentum does not change; in interactions, angular momentum is conserved.

For an object whose mass is concentrated in a small space, there is a deceptively simple expression for angular momentum:

$$\text{angular momentum} = mvr \qquad \text{(Equation 4-12)}$$

This is for an object traveling in a circular path; m is its mass, v is its speed, and r is the radius of the circle in which it travels.

Figure 4-5 illustrates a situation in which a force is applied, but no torque. Speed changes with no change in angular momentum. A string, held at the bottom, passes through a glass tube and is attached at the other end to a ball. The ball spins in a horizontal circle. Now what happens if you pull the string down so that the radius of the ball's path gets smaller? No torque is being applied because the force is along a radius, and it takes a tangential force to produce torque. Nevertheless, the ball speeds up! Angular momentum is conserved, so as r gets smaller, v gets larger. See Sample Problem 4-24 for the calculation.

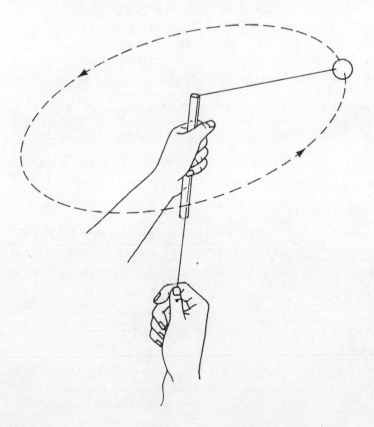

FIGURE 4-5

Sample Problem

4-24 If the ball of Figure 4-5 is going at 2.5 m/s when the radius of its path is 40 cm, how fast will it be going when the string is pulled in enough to reduce the radius to 15 cm?

Solution With no torque applied, angular momentum is conserved. Since there is no change in mass, $v_1 r_1 = v_2 r_2$, so (2.5 m/s)(40 cm) = v_2(15 cm) and $v_2 = 6.7$ m/s.

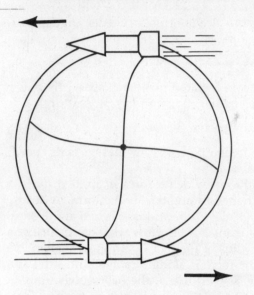

FIGURE 4-6

Figure 4-6 illustrates a situation in which the angular momentum *does* change. It is a rocket-driven pinwheel, such as you might see in a Fourth of July celebration. Each rocket applies a tangential force. There is no net force because the rockets point in opposite directions. Since they are not in line with each other, there is a net torque, and the wheel spins.

Unfortunately, calculations dealing with angular momentum are rarely this simple. The reason is that most spinning objects are not point masses; they have their masses distributed in space. A bowling ball consists of a great many atoms; if you set it spinning on its axis, every atom in it has its own mass, its own radial distance from the center, and its own velocity. Atoms on the axis have no angular momentum at all; for them, $r = 0$. The atoms with the greatest angular momentum are those farthest from the center and those with largest mass. These calculations can be done, but they require the methods of integral calculus.

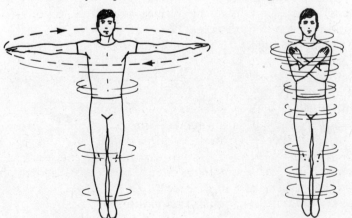

FIGURE 4-7

A figure skater can make use of the principle of conservation of angular momentum without knowing any mathematics at all. The skater in Figure 4-7 sets himself spinning, with his arms outstretched. Then he brings his arms in close to the center of his body. This reduces the *r* values for many of his atoms. Angular momentum is conserved only if the *v*'s increase, and he spins much faster. To come out of the spin, he stretches his arms out again, slowing the spin enough to go gliding off gracefully.

Core Concept

In any system, angular momentum changes only if a torque is applied.

Try This A 40-kg boy is on a merry-go-round that spins at one revolution every 6.0 s. If he is 3.5 m from the center of rotation, what is his angular momentum?

Summary Quiz

For each of the following, fill in the missing word or phrase:

1. Impulse is the product of force and _____ .
2. The change in the velocity of an object is proportional to the _____ applied to it.
3. The ratio of net force applied to an object to the acceleration it produces is the _____ of the object.
4. A kg·m/s^2 is called a _____ .
5. If there are several forces on an object, its acceleration depends on its mass and the _____ force.
6. Momentum is the product of mass and _____ .
7. The change in the momentum of an object is equal to the _____ applied to it.
8. In any interaction between two or more isolated objects, the total _____ does not change.
9. An object will move in a circular path at constant speed if a force is applied to it in a direction that is kept _____ to its velocity.
10. The force that accelerates an object into a circular path is called a _____ force.
11. The force of _____ is the centripetal force on the moon.
12. The force of gravity between two objects is directly proportional to the product of their _____ .
13. The force of gravity between two objects is inversely proportional to the square of the _____ .
14. The acceleration due to gravity at the surface of a planet depends on the _____ and the _____ of the planet.
15. There is no perceptible gravity within an orbiting spacecraft because the craft is in a state of _____ .
16. To place a satellite in circular orbit, it must be lifted to the right height and given the correct _____ .
17. The acceleration of a satellite in orbit depends only on its _____ .
18. The centripetal force acting on a satellite is produced by _____ .
19. The period of a satellite in synchronous orbit is _____ .
20. A change in angular momentum is produced by applying a _____ .
21. Angular momentum increases, with no change in speed, when an object is moved _____ its center of rotation.
22. In any isolated system, total angular momentum is _____ .

Problems

1. If a jet engine provides a thrust of 45,000 N, how long must it fire to produce a million N · s of impulse?
2. A rocket-driven sled speeds up from 40 m/s to 55 m/s in 5.0 s, using an engine that produces 3,500 N of thrust. How much thrust would be needed to get the same increase in speed in 2.0 s?
3. What force is needed to speed up a frictionless 60-kg cart from 4.0 m/s to 6.5 m/s in 3.0 s?
4. A 1,400-kg car strikes a telephone pole and comes to rest, driving the front bumper 40 cm into the engine. This means that it continued to travel 40 cm while the retarding force was being applied. If it was going 25 m/s before it hit, find (a) its average acceleration while coming to rest; and (b) the average force that produced the damage.
5. What force must the brakes and tires apply to a 2,800-kg truck going 30 m/s to bring it to rest in 8.0 s?
6. A 680-kg rocket is to be lifted off the surface of the moon, where $g = 1.67$ m/s^2. What force is needed to give it an upward acceleration of 2.0 m/s^2?
7. How much is the frictional force acting on a 75-kg cart if a push of 220 N gives it an acceleration of 2.0 m/s^2?
8. A 35-kg girl on roller skates, standing still, throws a 6-kg medicine ball forward at 3.5 m/s. How much is her recoil velocity (the backward speed she acquires as a result of the throw)?
9. What is the recoil velocity of a 7.5-kg rifle if it fires a 8.0-g bullet with a muzzle velocity of 640 m/s?
10. A 750-kg rocket, at rest in outer space, propels itself forward by ejecting 45 kg of hot gas, which leaves the nozzle at 85 m/s. Find (a) the change in the momentum of the fuel; and (b) the final speed of the rocket ship.
11. A rocket engine ejects 60 kg of hot gas at a speed of 95 m/s in 20 s. Find (a) the change in the momentum of the hot gas; and (b) the thrust of the engine.
12. Two hockey players make a head-on collision and cling to each other. One has a mass of 62 kg and is going 3.5 m/s; the other has a mass of 53 kg and is going 5.0 m/s. How fast are they going together after the collision?
13. What is the radius of the circle in which a 240-kg motorcycle and rider are traveling if they are going 35 m/s and the force exerted on the tires perpendicular to the velocity is 6,200 N?
14. A 45-kg woman is on a rotating platform in an amusement park, 2.5 m from the center of rotation. If she is going 4.5 m/s, what frictional force between her body and the platform is needed to keep her from flying off at a tangent?
15. What is the acceleration of a meteor passing by the earth at an altitude of 55,000 km?
16. How much is the gravitational force that keeps an artificial satellite of mass 3,500 kg in orbit around the earth at an altitude of 4,200 km?
17. Determine the gravitational attraction between two asteroids separated by 22,000 m if their masses are 450,000 kg and 700,000 kg, respectively.

5

Making Work Easier: Machines

5-1. The Pulley Principle

A piano mover, unable to fit the instrument into the staircase, decides to raise it outside the building to a window. He attaches it to a set of ropes and wheels which, somehow, make it possible for him to lift it with a force considerably smaller than the weight of the piano. How does it work?

Consider first the heavy block in Figure 5-1, suspended from two ropes. The upward force on the block is the tension in the ropes, and the sum of the two tensions must equal the weight of the block. If the whole system is symmetrical, each rope is under tension equal to half the weight of the block.

FIGURE 5-1

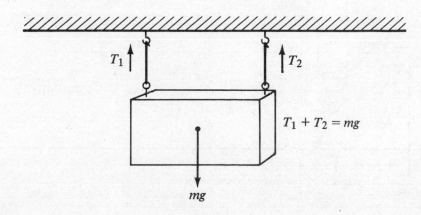

$$T_1 + T_2 = mg$$

$$mg$$

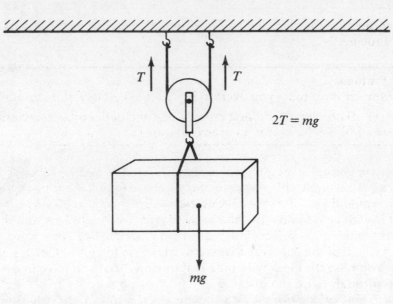

FIGURE 5-2

$2T = mg$

mg

Now look at Figure 5-2, where the block has been attached to a wheel. There is now only one rope, which passes over the wheel. The tension in the rope is the same throughout; if it were different on one side than on the other, the wheel would turn until the tension on the two sides equalized. The tension in the rope is still only half the weight of the block, since it exerts *two* upward forces on the block, as in Sample Problem 5-1.

Sample Problem

5-1 If the block of Figure 5-2 has a mass of 120 kg, how much is the tension in the rope?

Solution   The weight of the block is $mg = (120 \text{ kg})(9.8 \text{ m/s}^2) = 1,176$ N. This is divided equally between the two sides of the rope, so the tension in the rope is 588 N.

Now we have a system that helps in lifting things. Just fasten one end of the rope to a fixed support and pull on the other end, as in Figure 5-3.

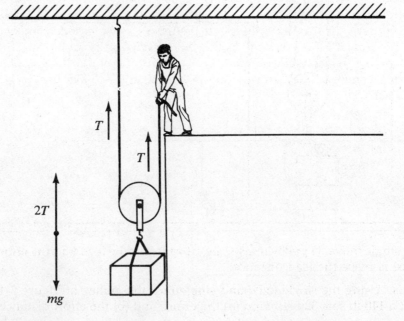

FIGURE 5-3

T

T

$2T$

mg

Now you can raise the block with a force equal to only half its weight, as in Sample Problem 5-2.

Sample Problem

5-2 How much force must you exert to lift the 120-kg block in Figure 5-3?

Solution   Since you are pulling on only one end of the rope, which has a tension of 588 N, that is the force you need.

Are you getting something for nothing? Well, yes and no. True, you can now lift the weight with less force, but you have to pull the rope farther than you would if you lifted the block directly. Every time you pull 10 feet of rope through your hands, the block rises 5 feet. You might look at it this way: If the block rises 5 feet, *both* sides of the supporting rope have to shorten 5 feet, and the only way to accomplish this is to pull 10 feet of rope through. You raise the block with only half the force, but you have to exert the force through twice the distance.

You might prefer to pull in a downward direction rather than upward, and you can manage this by attaching a fixed wheel to the support and passing the rope around it, as in Figure 5-4. The tension in the rope is still only half the weight of the block; the fixed pulley does nothing but change the direction of the force you exert.

Let's adopt some vocabulary. The weight of the object being lifted we will call the *load*, and the distance it rises is the *load distance*. The force you exert on the rope is the *effort*, and the distance through which you exert that effort is the *effort distance*.

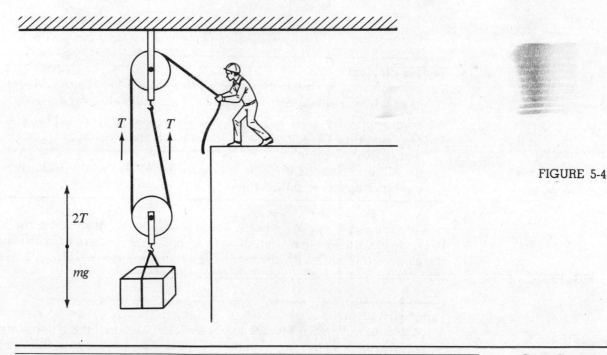

FIGURE 5-4

Core Concept

With a single movable pulley in use, the effort is half the load and the effort distance is twice the load distance.

Try This   Using the single fixed and single movable pulley of Figure 5-4, you lift a 140-lb sofa 32 feet. Find (a) the effort; and (b) the effort distance.

5-2. More Pulleys

There are ways to string up a system of pulleys that will reduce the effort still further. Figure 5-5 shows how the same two pulleys can be connected to a rope in such a way as to divide the load among three strands instead of two. This is done by fastening one end of the rope to the load instead of to the fixed support. Unfortunately, when you do this, you have to shorten all three strands when you raise the object, and the effort distance becomes three times the load distance. See Sample Problem 5-3.

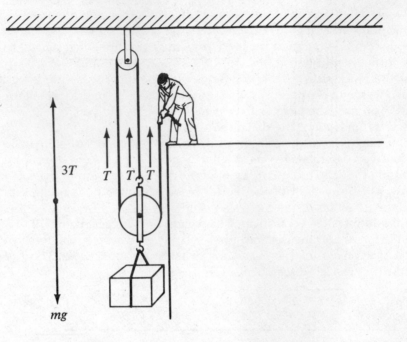

FIGURE 5-5

Sample Problem

5-3 Using the pulley system of Figure 5-5, the load is 480 N and it is to be lifted 6.0 m. Find (a) the effort; and (b) the effort distance.

Solution   (a) The load is divided among three supporting strands, so each supports 480 N/3 = 160 N; since that is the tension in the rope, that is also the effort.
(b) All three strands must be shortened by 6.0 m, so the amount of rope that has to be pulled through is 18 m.

By using more pulley wheels, you can reduce the effort still further. In Figure 5-6, the load has been divided among six strands of the rope, so the effort is only one-sixth the load. The effort distance, of course, is six times as far as the load rises. See Sample Problem 5-4.

Sample Problem

5-4 The pulley of Figure 5-6 probably has a lot of friction. Still, if you want to raise the load by 5 m, how much rope do you have to pull through?

Solution   Friction does not affect the geometry of the system; you still have to shorten all six supporting strands, so you must pull 30 m of rope. Note that the strand you are holding does not count, because it is not connected to the load.

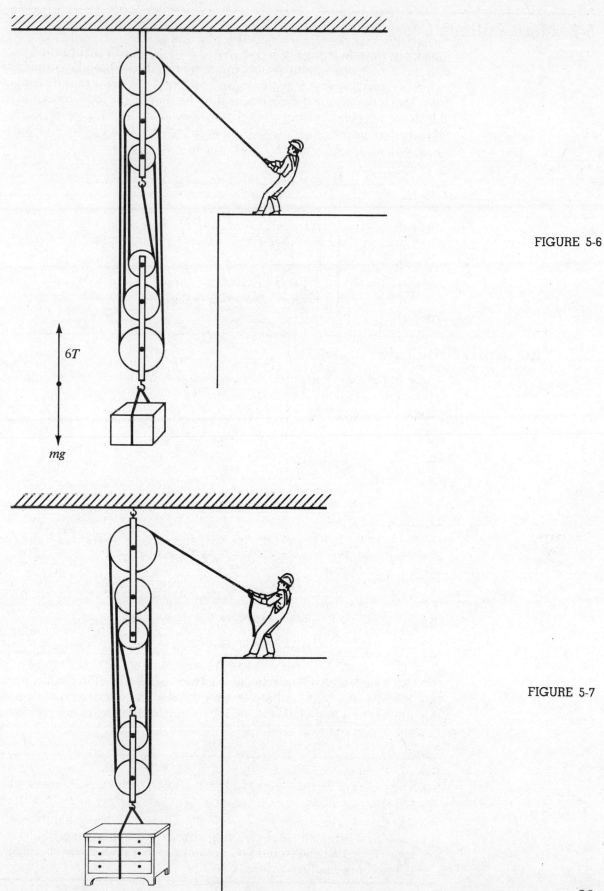

$6T$

mg

FIGURE 5-6

FIGURE 5-7

Unfortunately, there is a limit to how much you can reduce the effort. The analysis we did so far neglects a few things, such as friction and the weight of the movable pulleys themselves. Every time you add a pulley, you increase the friction in the system; if it is a movable pulley—the only kind that produces a reduction in force—you have to lift it along with the load. The effort in any real system is always larger than the ideal effort we calculated by dividing up the load. If there are a lot of pulleys, it may be considerably larger. And friction, while it increases the force you must exert, has no effect on the distance you have to pull that rope.

Core Concept

Effort distance is load distance times the number of supporting strands; effort is larger than load divided by the number of strands.

Try This The pulley system of Figure 5-7 is being used to lift a 80-kg bureau 22 m. Find (a) the distance the rope must be pulled; and (b) the minimum force that must be exerted.

5-3. The Work Principle

While pulleys are useful, they do not give you something for nothing. Ignoring the problem of friction, the input and output forces are in inverse ratio to the respective distances:

$$\frac{\text{effort}}{\text{load}} = \frac{\text{load distance}}{\text{effort distance}}$$

or, to put it another way,

$$(\text{effort})(\text{effort distance}) = (\text{load})(\text{load distance})$$

The frictionless pulley, then, does not alter the product of force and distance; it is the same for the mover who pulls on the rope as it is for the piano. This product occurs repeatedly in physical situations, so it is given a special name. Force times distance is called *work*.

Work is done whenever a force moves something through a distance. If you stand still holding a boulder over your head, you might get tired, but—in the physical sense—you are doing no work.

There is another limitation on the definition of work. Only the force in the direction of motion counts. For example, look at Figure 5-8—the child on the sled. The tension in the rope is pulling the sled, but it is also lifting the sled. Only the component of the force that is acting in the direction the sled is going is doing work on the sled. Since the sled is moving horizontally, the horizontal component of the tension is the only part that

FIGURE 5-8

is doing the work of moving the sled. This component, according to Equation 1-4, is the force times the cosine of the angle that it makes with the direction of motion. The full definition of work is

$$W = \mathbf{F}\,\Delta\mathbf{s}\,\cos\theta \qquad \text{(Equation 5-3)}$$

For calculations of work, see Sample Problems 5-5, 5-6, and 5-7. Note that the unit in which work is expressed is the product of force and distance—the foot-pound (ft · lb) in the English system and the newton-meter (N · m) in SI. A newton-meter is called a joule (J).

Sample Problem

5-5 Using a force of 75 N, you push a crate across the floor a distance of 5.0 m. How much work do you do?

Solution If you have any sense at all, you will exert your force in the direction you want the crate to go, so the angle between force and distance moved is 0. Since $\cos 0 = 1$, $W = F\,\Delta s = (75\text{ N})(5.0\text{ m}) = 375\text{ N} \cdot \text{m} = 375\text{ J}$.

Sample Problem

5-6 In trying to pull a stump out of the ground, you attach a rope to it, which makes an angle of 30° with the vertical. You pull for 4.0 min with a force of 350 N, but you can't budge the stump. How much work do you do?

Solution Since $\Delta s = 0$, you do no work.

Sample Problem

5-7 If the sled of Figure 5-8 is pulled with a force of 160 N for 30 m, the rope making an angle of 25° with the horizontal, how much work is done?

Solution From Equation 5-3,

$$W = \mathbf{F}\,\Delta\mathbf{s}\,\cos\theta$$
$$W = (160\text{ N})(30\text{ m})(\cos 25°)$$
$$W = 4350\text{ N} \cdot \text{m} = 4,350\text{ J}$$

Core Concept

Work is the product of force, distance traveled, and the cosine of the angle between them:

$$W = \mathbf{F}\,\Delta\mathbf{s}\,\cos\theta$$

Try This A roofer puts 20 kg of tiles on his back and carries them up a ladder 5.0 m long. If the ladder makes an angle of 22° with the vertical, how much work does he do? (Remember, to lift something you must exert a vertical force!)

5-4. Efficiency

In all our pulleys, the force was always exerted in the direction of motion, so the angle θ was 0. Since $\cos 0 - 1$, we do not have to worry about the angle in dealing with pulleys.

The effort times the effort distance is called the *work input*, and the load times the load distance is the *work output*. While there are no real frictionless pulleys, we can use the ideal of a frictionless, weightless pulley in doing useful calculations. With such an ideal pulley, the work output is exactly the same as the work input. If you use this idea to calculate how hard you will have to pull on the rope, the best you can do, when you have finished the calculation, is to say that the effort will be *at least* the amount you figure. How much more it will be depends on the friction and on the weight of the wheels.

With a real pulley, you always have to pull a little harder than that calculated value. The work input is therefore always more than the work output. The ratio between the work output and the work input is called the *efficiency* of the machine:

$$\text{efficiency} = \frac{W_{out}}{W_{in}} \qquad \text{(Equation 5-4)}$$

Efficiency is usually expressed as a percent. It tells you what fraction of the work you put into a machine comes out as useful work at the other end. See Sample Problems 5-8, 5-9, and 5-10.

Sample Problem

5-8 To lift a 540-N load 12.0 m, you turn a crank with a force of 110 N, pushing it through a distance of 72 m. What is the efficiency of the system?

Solution The work output is (540 N)(12.0 m) = 6,480 J. The work input is (110 N)(72 m) = 7,920 J. The efficiency is

$$\frac{W_{out}}{W_{in}} = \frac{6,480 \text{ J}}{7,920 \text{ J}} = 0.818$$

or 82%.

Sample Problem

5-9 Using a machine that is 90% efficient, you have to lift a 1,200-N load 3.0 m. How much work do you have to do?

Solution The work output is (1,200 N)(3.0 m) = 3,600 J. Since 0.90 = W_{out}/W_{in},

$$W_{in} = \frac{W_{out}}{0.90} = \frac{3,600 \text{ J}}{0.90} = 4,000 \text{ J}$$

Sample Problem

5-10 In a three-strand pulley, a 130-kg load is lifted 25 m, using an effort of 540 N. What is the efficiency of the pulley?

Solution The load is the weight being lifted, or

$$w = mg = (130 \text{ kg})\left(9.8 \tfrac{m}{s^2}\right) = 1,274 \text{ N}$$

With a three-strand pulley, the effort distance is three times the load distance, or 75 m. Then, from Equation 5-4,

$$\text{efficiency} = \frac{W_{out}}{W_{in}}$$

$$\text{efficiency} = \frac{(1,274 \text{ N})(25 \text{ m})}{(540 \text{ N})(75 \text{ m})}$$

$$\text{efficiency} = 0.786 = 79\%$$

Every additional pulley added to a system reduces its efficiency by introducing more useless weight and more friction. Well-oiled wheels are the best method of increasing efficiency.

Core Concept

Efficiency is the fraction of the work input that emerges as useful work output.

Try This   How much work is done in lifting a load by means of a pulley system 80% efficient if the rope is pulled for 30 m with a force of 220 N?

5-5. Leverage

A pulley is a device for transforming work by changing the values of force and distance that compose it. There are many such devices. Collectively, they are known as *simple machines*.

As I look around in my study, I have no difficulty in picking out a dozen simple machines in the room: the light switch, a doorknob, the handle of a pencil sharpener, the telephone dial, a knife, a screwdriver, even the keys of the typewriter on which I am writing this   all are devices that can convert the work I put into them into usable form. Without thinking about it, everybody uses a hundred simple machines every day.

Many of these devices are *levers*. A lever consists of a rigid bar, pivoted at some point. An effort force applied to the bar at some point produces a different force on a load at some other position on the bar. The crowbar of Figure 5-9 is typical. The load is the weight of the rock being lifted. The effort is the force exerted by the person who is trying to move the rock.

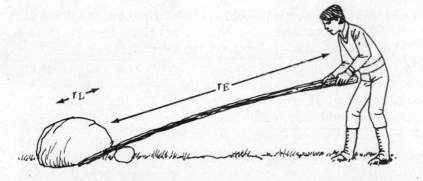

FIGURE 5-9

Usually, the purpose of a machine is to make it possible to exert a large load force with a smaller effort. The machine magnifies force. The amount of this magnification, the ratio of load to effort, is called the *mechanical advantage* of the machine. It can be defined algebraically as the ratio of the two forces:

$$MA = \frac{F_L}{F_E}$$ **(Equation 5-5a)**

In a pulley, the mechanical advantage is equal to the number of strands of rope supporting the load. For a lever, it can be found by considering the torques acting on the bar.

According to Equation 2-11, the torque around the pivot that is exerted by the worker is $F_E r_E$, where r_E (the *effort arm*) is the distance from the point where the effort is applied to the pivot. Similarly, the torque produced by the weight of the rock is $F_L r_L$, where r_L is the *load arm*. If the system is rotating in equilibrium, these two torques must have the same magnitude, so

$$F_E r_E = F_L r_L$$

From which we find that the mechanical advantage, F_L/F_E, is given by

$$MA_{lever} = \frac{r_E}{r_L}$$

(Equation 5-5b)

See Sample Problems 5-11 and 5-12.

Sample Problem

5-11 The crowbar of Figure 5-9 is 2.0 m long and is pivoted 30 cm from the rock. (a) What is its mechanical advantage; and (b) if the free end is pushed down with a force of 220 N, how much force is exerted on the rock?

Solution (a) The mechanical advantage of a lever is the ratio of effort arm (distance from effort to pivot) to load arm (distance from load to pivot):

$$\text{mechanical advantage} = \frac{1.7 \text{ m}}{0.3 \text{ m}}$$

$$\text{mechanical advantage} = 5.7$$

(b) Neglecting friction, the mechanical advantage is the ratio of load to effort. Therefore

$$\text{load} = 5.7 \times \text{effort}$$
$$\text{load} = 5.7 \times 220 \text{ N} = 1{,}250 \text{ N}$$

Sample Problem

5-12 A jack handle 65 cm long is pivoted at one end and pushed down at the other. If it exerts a force of 240 N at a point 12 cm from the pivot, find (a) its mechanical advantage; and (b) the effort force.

Solution The mechanical advantage of a lever is the ratio of effort arm to load arm: 65 cm/12 cm = 5.4. (b) Neglecting friction, the mechanical advantage of any machine is the ratio of load to effort, so 5.4 = 240 N/F_E, and F_E = 44 N.

For many kinds of levers, friction at the pivot is quite small, so efficiencies approach 100 percent, and the arm ratio is very near the force ratio. Usually, no correction is needed.

FIGURE 5-10

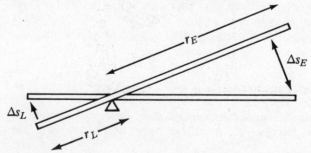

Does the law of work apply to levers? Surely. Look at Figure 5-10, which shows the movement of a lever. It is clear that the two triangles are similar, so

$$\frac{r_L}{r_E} = \frac{\Delta s_L}{\Delta s_E}$$

Substituting this ratio into Equation 5-5b will show that the work done, $F \Delta s$, is the same at the input and output ends.

Levers are classified according to the relative positions of the pivot, load, and effort. The three classes are represented by the tools shown in Figure 5-11. In the pliers (first class) the pivot is between the effort and the load. In the nutcracker (second class) it is the load that is between the other two. And in the sugar tongs (third class), the effort is in the middle.

FIGURE 5-11

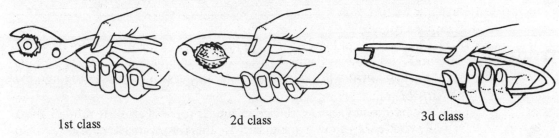

1st class   2d class   3d class

Note that in the third-class lever (the sugar tongs), the load arm is longer than the effort arm, so the mechanical advantage is less than 1. This lever magnifies distance at the expense of force. This can also be done with a first-class lever, as in the rather primitive device for drawing water in Figure 5-12. The analysis of this gadget is illustrated in Sample Problem 5-13.

Sample Problem

5-13 The man in Figure 5-12 is pushing on the end of a pole 4.5 m long, and he is 1.2 m from the pivot. The pail of water weighs 35 N. Find (a) the mechanical advantage; and (b) the force the man has to exert.

Solution (a) Mechanical advantage is the ratio of effort arm to load arm, or 1.2 m/3.3 m = 0.36.

(b) Mechanical advantage is load/effort, or 0.36 = 35 N/F_E; therefore, F_E = 97 N.

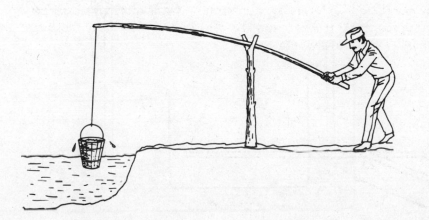

FIGURE 5-12

Mechanical advantage is the ratio of load to effort; in a lever, it is equal to the ratio of effort arm to load arm:

$$MA = \frac{F_L}{F_E}; \quad MA_{lever} = \frac{r_E}{r_L}$$

Try This The nutcracker of Figure 5-11 is 18 cm long, and the nut is 5 cm from the pivot. Find (a) the mechanical advantage of the nutcracker; (b) the effort needed to apply a force of 220 N on the nut; (c) the distance the handle has to move in order to drive the jaws 0.5 cm into the nut; (d) the work input; and (e) the work output.

5-6. Hydraulics

Liquids are nearly incompressible. This property makes them suitable as means of transforming work.

A hydraulic jack is a device in which force is applied to the oil in a small cylinder. As shown in Figure 5-13, this force causes some of the oil to be transferred to a larger cylinder. This forces the piston in the larger cylinder to rise, lifting a load.

This device takes advantage of the fact that oil, being nearly incompressible, transmits whatever pressure is applied to it. The pressure applied in the small cylinder appears unchanged in the big one, pushing up its piston.

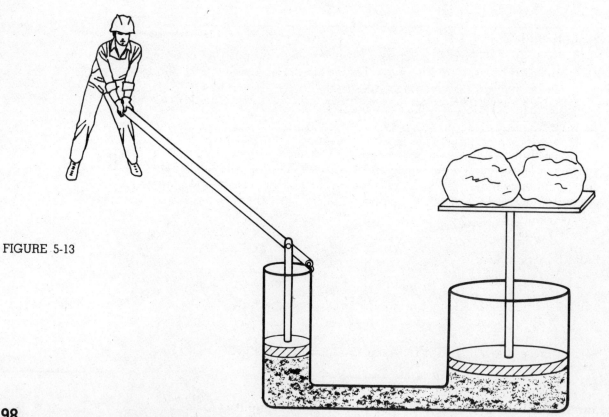

FIGURE 5-13

What is the mechanical advantage of a hydraulic jack? According to Equation 3-3, pressure is force per unit area, or F/A. Since the pressure is the same in both cylinders, we can write

$$\frac{F_E}{A_E} = \frac{F_L}{A_L}$$

where the A's stand for the areas of the two pistons. Solving this for the mechanical advantage, which is F_L/F_E (Equation 5-5a), we get

$$MA_{hydr} = \frac{A_L}{A_E} \qquad \text{(Equation 5-6a)}$$

This equation gives the mechanical advantage of the hydraulic part of the jack. The handle is a second-class lever, which has a mechanical advantage of its own. The handle multiplies the effort force; the hydraulic jack takes the output force of the handle and multiplies *that*. Thus, the mechanical advantage of the combination is the product of the MA's of the two parts. See Sample Problem 5-14.

Sample Problem

5-14 In the hydraulic jack of Figure 5-13, the handle is 80 cm long and it pushes on the piston rod 10 cm from its end. The area of the effort piston is 20 cm^2, and the area of the load piston is 230 cm^2. Find (a) the mechanical advantage of the jack; and (b) the effort needed to lift 1,650 kg.

Solution (a) The mechanical advantage of the lever is

$$\frac{\text{effort arm}}{\text{load arm}} = \frac{80 \text{ cm}}{10 \text{ cm}} = 8.0$$

and the mechanical advantage of the hydraulic press is

$$\frac{\text{area of load piston}}{\text{area of effort piston}} = \frac{230 \text{ cm}^2}{20 \text{ cm}^2} = 11.5$$

Since the lever multiplies force by 8.0 and the hydraulic press takes its output and multiplies it by 11.5, the total mechanical advantage is $11.5 \times 8.0 = 92$.

(b) The load is the weight of 1,650 kg, or (1,650 kg) (9.8 m/s^2) = 16,170 N. Neglecting friction, the effort is this quantity divided by 92, or 176 N.

Does the hydraulic jack increase your work? Of course not! Whatever volume of oil is pumped out of the effort cylinder goes into the load cylinder. In either cylinder, that volume is the product of the cross-section area of the cylinder and the distance the piston moves. Since these two volumes are the same,

$$A_E \, \Delta s_E = A_L \, \Delta s_L$$

From Equation 5-6a,

$$MA_{hydr} = \frac{A_L}{A_E} = \frac{\Delta s_E}{\Delta s_L}$$

while Equation 5-5a tells us that in any frictionless machine

$$MA = \frac{F_L}{F_E}$$

Setting these two values of MA equal to each other, we get

$$\frac{F_L}{F_E} = \frac{\Delta s_E}{\Delta s_L} \qquad \text{or} \qquad F_L \, \Delta s_L = F_E \, \Delta s_E$$

and lo and behold! the work output equals the work input.

Pulley, lever, or hydraulic jack—with no friction, the work output equals the work input. Let's assume that this works for all other machines as well; it does! Then we can get the ideal mechanical advantage of *any* machine—that is, its MA if it is assumed frictionless—just by studying the geometry of the machine. A look at the previous equation will then justify this statement:

$$\text{ideal MA} = \frac{\Delta s_E}{\Delta s_L} \qquad \text{(Equation 5-6b)}$$

In even the most complex machine, we can get its ideal mechanical advantage just by figuring out how far the load moves when we move the effort a known distance. See Sample Problem 5-15.

Sample Problem

5-15 A set of pulleys operates a lever that pushes on a hydraulic press that operates another lever. To make the last lever arm move 12 cm, it is necessary to pull 15 m of rope through the pulley. The last lever is lifting a 22,000 N load, and the force being applied to the pulley rope is 280 N. Find (a) the ideal mechanical advantage of the system; and (b) its efficiency.

Solution (a) Ideal MA $= \dfrac{\Delta s_E}{\Delta s_L} = \dfrac{15 \text{ m}}{0.12 \text{ m}} = 125$

(b) efficiency $= \dfrac{W_{\text{out}}}{W_{\text{in}}} = \dfrac{(22{,}000 \text{ N})(0.12 \text{ m})}{(280 \text{ N})(15 \text{ m})} = 0.63$, or 63%

Core Concept

Ideal mechanical advantage is the ratio between effort distance and load distance; for a hydraulic jack, it is equal to the ratio of the area of the load piston to that of the effort piston:

$$\text{ideal MA} = \frac{\Delta s_E}{\Delta s_L}; \qquad \text{MA}_{\text{hydr}} = \frac{A_L}{A_E}$$

Try This A hydraulic press has an output piston of area 350 cm^2 and an input piston of 25 cm^2. It is 90% efficient and is being used to raise a 400-kg load. Find (a) its ideal mechanical advantage; (b) the effort that would be required if there were no frictional forces; and (c) the actual effort needed.

5-7. A Loading Ramp

A ramp is a device commonly used to aid in lifting. To raise a heavy load a couple of feet onto a platform, it is common practice to place it on a dolly and wheel it up an inclined plane.

The work output of an inclined plane is the work that would have to be done to lift the load directly: the weight of the load times the vertical distance it goes. The work input is the actual force exerted in pushing the dolly up the ramp times the length of the ramp.

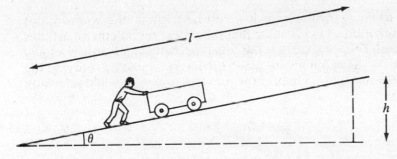

FIGURE 5-14

As you saw in Chapter 2, the dolly can be pushed up the ramp with a force $F_E = w \sin \theta$. Since $F_L = w$, we have

$$\sin \theta = \frac{F_E}{F_L}$$

A glance at the geometry of the plane (Figure 5-14) shows that

$$\sin \theta = \frac{h}{l}$$

where h is the height of the ramp and l is its length. This gives three expressions for the ideal mechanical advantage of an inclined plane:

$$\text{ideal MA} = \frac{F_L}{F_E} = \frac{l}{h} = \frac{1}{\sin \theta}$$

The equations show that the work output (hF_L) equals the work input (lF_E). As usual, the effort must be increased whenever the efficiency is less than 100 percent. See Sample Problems 5-16 and 5-17.

Sample Problem

5-16 What is the ideal mechanical advantage of a ramp 5.0 m long being used to push a cart up to a platform 1.5 m high?

Solution   The ideal mechanical advantage is

$$\frac{\text{effort distance}}{\text{load distance}} = \frac{\text{length}}{\text{height}} = \frac{5.0 \text{ m}}{1.5 \text{ m}} = 3.3$$

Sample Problem

5-17 A ramp is inclined at 26°, and it takes a force of 180 N to push a cart weighing 350 N up it. What is the efficiency of the ramp?

Solution   The ideal mechanical advantage is 1/sin 26° = 2.3. With no friction, the force needed would be (350 N)/2.3 = 153 N. Since the force actually needed is 180 N, the efficiency is (153 N)/(180 N) = 85%.

Core Concept

The ideal mechanical advantage of an inclined plane is equal to its length divided by its height:

$$\text{ideal MA}_{\text{plane}} = \frac{l}{h} = \frac{1}{\sin \theta}$$

Try This   On a ramp 2.5 m long and 0.5 m high, a force of 420 N is being used to push a loaded dolly. If the efficiency of the ramp is 85%, how much does the dolly weigh?

5-8. A Vise

The vise of Figure 5-15 is a complex machine in which the handle acts as a lever operating a new kind of machine: a screw. How can we calculate the constants of this gadget?

FIGURE 5-15

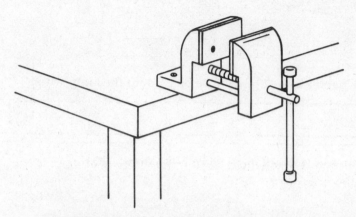

It would be very difficult to calculate the ratio of the force the jaws apply to the force on the handle. The best we can do is work with the distances, using Equation 5-6b.

A screw consists of a single continuous spiral wrapped around a cylinder. The distance between ridges is known as the *pitch* of the thread, as shown in Figure 5-16. Every time the screw makes one complete turn, the screw advances a distance equal to the pitch. In the vise, one complete turn is made when the end of the handle travels in a circle whose radius is the length of the handle (*l*). Therefore, when the effort moves a distance $2\pi l$ the load moves a distance equal to the pitch of the thread. Therefore, for a screw,

$$\text{ideal MA} = \frac{\Delta s_E}{\Delta s_L} = \frac{2\pi \ (\text{length of handle})}{\text{pitch of thread}}$$

However, if you use this expression to calculate the forces, you will get it all wrong. The vise is a high-friction device. It has to be, for it is the friction that keeps it from opening when you tighten it. A vise is a self-locking machine because its efficiency is considerably under 50 percent. To see what some typical values look like, see Sample Problem 5-18.

FIGURE 5-16

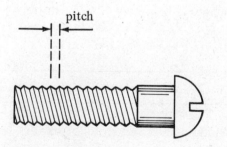

pitch

Sample Problem

5-18 A vise is made with a handle 10 cm long and threads with a pitch of 0.15 cm. If its efficiency is 30%, how much effort is needed to produce a force of 3,500 N at the jaws?

Solution The ideal mechanical advantage of the vise is

$$\frac{2\pi \,(10 \text{ cm})}{0.15 \text{ cm}} = 419$$

Therefore, the effort distance is 419 times the load distance. Then

$$\text{efficiency} = \frac{W_{out}}{W_{in}} = \frac{(\text{load})(\text{load distance})}{(\text{effort})(\text{effort distance})}$$

$$0.30 = \frac{(3,500 \text{ N})(\text{load distance})}{(\text{effort})(419)(\text{load distance})}$$

$$\text{effort} = \frac{3,500 \text{ N}}{419 \times 0.30} = 28 \text{ N}$$

Core Concept

A vise is a high-friction device whose ideal mechanical advantage is the circumference of the effort circle divided by the pitch of the thread:

$$\text{ideal MA}_{screw} = \frac{2\pi \, l}{\text{pitch}}$$

Try This What is the efficiency of a vise if its handle is 12 inches long, the screw has four threads to the inch, and it takes a force of 45 pounds on the handle to produce an output force of 1,500 pounds between the jaws?

5-9. Machines That Spin

What is the mechanical advantage of a winch, such as that shown in Figure 5-17? The principle is not much different from that of a lever. Since the crank and the shaft turn together, the torque exerted by the effort (the force on the handle) must be equal to the torque exerted by the load (the tension in the rope). Therefore, from Equation 2-11,

$$F_E r_E = F_L r_L$$

where the r's are the radius vectors of the effort and load, respectively. From this it follows that

$$\text{ideal MA}_{winch} = \frac{F_L}{F_E} = \frac{r_E}{r_L}$$

and Sample Problem 5-19 shows how this works out.

FIGURE 5-17

Sample Problem

5-19 What is the ideal mechanical advantage of the windlass of Figure 5-17 if the crank is 30 cm long and the shaft on which the rope is wound has a diameter of 12 cm?

Solution Ideal mechanical advantage is the ratio of radius of the crank to radius of the shaft:

$$\text{ideal MA} = \frac{30 \text{ cm}}{6 \text{ cm}} = 5$$

In mechanical devices, gears are commonly used to change torque. Consider the gears of Figure 5-18, for example. We assume that both gears are mounted on shafts of equal diameter, and that the small gear is driving the large one. What is the mechanical advantage of this combination?

FIGURE 5-18

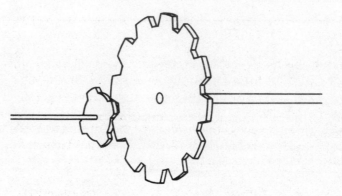

First of all, the teeth must have the same size and spacing on both gears in order for them to mesh properly. With 12 teeth in the large gear and only 4 in the small one, the small gear has to make three complete revolutions to make the big one turn once. The large, load gear moves only one-third as far as the smaller, effort gear. And the ratio of the two distances is the same as the ratio of the number of teeth in the two gears. Then, applying Equation 5-6b, we can say

$$\text{ideal MA}_{\text{gear}} = \frac{n_{\text{L}}}{n_{\text{E}}}$$

where the n's stand for numbers of teeth.

Gears convert torque. Where the teeth mesh, the force that each gear exerts on the other is the same. That means that the torque in the large gear is greater than that in the small one, for torque is Fr. Since the numbers of teeth is in the same ratio as the two radii, the torques are in the same ratio as the numbers of teeth—and that ratio is the mechanical advantage of the gears. See Sample Problem 5-20.

Sample Problem

5-20 A gear with 12 teeth, mounted on a shaft 6.0 cm in diameter, drives a gear with 54 teeth whose shaft is 2.0 cm in diameter. (a) What is the mechanical advantage of the gears? (b) If the torque on the large shaft is 50 N · m, how much is the torque on the small shaft?

Solution (a) The ideal mechanical advantage of the gears is the ratio of teeth in the driven gear to those in the driving gear: 54/12 = 4.5.

(b) The gears multiply the torque by 4.5, so the torque for the driven gear is $4.5 \times 50 \text{ N} \cdot \text{m} = 225 \text{ N} \cdot \text{m}$. The shaft diameters do not matter; each gear is firmly mounted on its own shaft, so the torque at the other end of the shaft is the same as the torque acting on the gear.

In a winch, mechanical advantage is the ratio of radii ($MA = r_L/r_E$); in a pair of gears, mechanical advantage is the ratio of numbers of teeth ($MA = n_L/n_E$).

Try This In the gears of Figure 5-18, the torque on the shaft of the small gear is $60 \text{ N} \cdot \text{m}$. Find (a) the mechanical advantage of the gear set; and (b) the torque on the other shaft.

5-10. Power

When the piano mover rigs his tackle, he has to consider many factors. For one, the more pulleys he puts in, the longer it will take him to get the job done. If he has to pull more rope—using less force, to be sure—he will have to keep pulling for a longer time.

There is a definite limit to the amount of work the mover can do in a given time. The rate at which he does work is called his *power*. Power is work done per unit time, defined by the equation

$$P = \frac{W}{\Delta t}$$

(Equation 5-10)

The English unit of power is the foot-pound per second, and it takes 550 of them to make one horsepower. The SI unit is the joule per second, or *watt* (W). A horsepower is 746 watts. The watt is a very small unit, and the kilowatt (=1,000 W) is commonly used. See Sample Problems 5-21 and 5-22.

Sample Problem

5-21 Using a five-strand pulley, a mover pulls on the rope with a force of 140 N, moving 22 m of rope in 2.0 min. (a) What is the power input to the pulley system? (b) Neglecting friction, what is the power output?

Solution (a) Power is work done per unit time, or

$$P = \frac{(140 \text{ N})(22 \text{ m})}{120 \text{ s}} = 26 \text{ J/s} = 26 \text{ W}$$

(b) With no friction, the power output is the same as the power input, since the work output takes the same length of time as the work input.

Sample Problem

5-22 An electrically operated hoist must lift up to 500 kg of material a distance of 75 m, to the top of a building under construction. If it is run by a motor that produces 20 kW of power, how long must it take to do the job?

Solution The work that has to be done is

$$F \, \Delta s = (500 \text{ kg})\left(9.8 \frac{\text{m}}{\text{s}^2}\right)(75 \text{ m}) = 367,500 \text{ J}$$

Then, from Equation 5-10,

$$\Delta t = \frac{W}{P} = \frac{367{,}500 \text{ J}}{20{,}000 \text{ W}} = 18 \text{ s}$$

In all the machines we have discussed so far, work comes out the load end as it goes in at the effort end. Thus, the power output of any machine is equal to the power input. Machines do not increase your power. A pulley or a windlass will spread the work out over a longer period of time, so that you can do it with the power available in your muscles and without straining for a force larger than convenient. For an example, see Sample Problem 5-23.

Sample Problem

5-23 Using a pulley system with an efficiency of 80%, a man must lift a 120-kg desk 22 m. If he can work at the rate of 450 W, how long will it take him?

Solution The work output is

$$(120 \text{ kg})\left(9.8 \frac{\text{m}}{\text{s}^2}\right)(22 \text{ m}) = 25{,}900 \text{ J}$$

However, because of efficiency loss, he must do more than that:

$$W_{\text{in}} = \frac{W_{\text{out}}}{\text{eff}} = \frac{25{,}900 \text{ J}}{0.80} = 32{,}400 \text{ J}$$

From Equation 5-10,

$$\Delta t = \frac{W}{P} = \frac{32{,}400 \text{ J}}{450 \text{ J/s}} = 72 \text{ s}$$

Core Concept

Power is work done per unit time:

$$P = \frac{W}{\Delta t}$$

Try This What power is being produced by a 40-kg child who runs upstairs, going a vertical distance of 15 m in 25 s?

5-11. The Drop Hammer

A drop hammer is a device used for driving pilings into soft ground or mud. It consists of a massive chunk of metal that is lifted by an engine of some kind and then dropped onto the piling. The effort is a comparatively small force, the weight of the hammer, and it may be lifted a considerable distance. At the load end, where it hits the piling, it exerts a much larger force over a small distance.

There is no way to calculate a definite mechanical advantage for a drop hammer, since the distance the piling is driven into the ground depends on how sharp it is and on how soft the mud is. Nevertheless, it is possible to show that the work output equals the work input. Here's how.

The work input is the work done in lifting the hammer a vertical distance Δh. Since the force needed is mg, the weight of the hammer, the work input is

$$W_{in} = mg \, \Delta h$$

When the hammer is dropped, it accelerates from rest, with acceleration g, until it reaches a speed v at the instant it touches the piling. Then, from Equation 1-7c,

$$v = \sqrt{2g \, \Delta h}$$

Now the hammer, going at speed v, drives the piling into the ground a distance s. It comes to rest with acceleration a. Again, from Equation 1-7c,

$$v = \sqrt{2as}$$

The v's in these two equations are the same, since both represent the speed of the hammer at the instant it first touches the piling. Therefore,

$$\sqrt{2as} = \sqrt{2g \, \Delta h}$$

and

$$as = g \, \Delta h$$

Multiplying both sides of this equation by the mass of the hammer gives

$$mas = mg \, \Delta h$$

and, since $mg \, \Delta h = W_{in}$,

$$W_{in} = mas$$

Equation 4-2 tells us that the force that slows up the hammer is equal to ma, so

$$W_{in} = Fs$$

F is the force on the hammer, which must be equal to the force the hammer exerts on the piling. Also, s is the distance the piling moves, since the hammer and the piling move together. Therefore the work done on the piling, the work output of the hammer, is Fs, and

$$W_{in} = W_{out}$$

Core Concept

The work output of a drop hammer is equal to the work input.

Try This The mass of the drop hammer is 220 kg and it is lifted 5.0 m before being dropped. If it drives the piling 0.50 m into the ground, what is the average force it exerts on the piling?

5-12. Work in Storage

There is a great difference between the drop hammer and any of the other machines we have studied. In a drop hammer, the work does not come out while it is going in. If the hammer is at its high position when the quitting whistle blows, it can be dropped the next morning. The work can be stored in the hammer.

Can you tell, just by looking at the hammer, whether or not there is any work stored in it? Easily. Just see whether the hammer is at the top or the bottom of its track. If it is at the top, it is in a *high-energy state*. At the bottom, it is in a low-energy state.

It is not easy to give a quick definition of exactly what we mean by energy, and we will have to take it a little at a time. Let's start this way: Whenever work is done on a system, its energy increases by an amount equal to the work done. If the system does work, its energy decreases by an amount equal to the work done. This tells us that we measure energy in the same units we use for work: joules or foot-pounds.

Doing work can make all kinds of changes in a system, and all of them represent increases in the energy of the system. Pushing something to a higher elevation is one way of increasing its energy. The energy an object has because of its elevation is called *gravitational potential energy*. The gravitational potential energy of the pile driver is equal to the work done in lifting it from its zero level to some height h. Thus,

$$E_{grav} = mgh \qquad \text{(Equation 5-12)}$$

Sample Problems 5-24 and 5-25 are calculations of E_{grav}.

Sample Problem

5-24 What is the gravitational potential energy of a drop hammer whose mass is 850 kg when the hammer is at a height of 5.0 m?

Solution   From Equation 5-12,

$$E_{grav} = mgh = (850 \text{ kg})\left(9.8 \tfrac{m}{s^2}\right)(5.0 \text{ m})$$

$$E_{grav} = 41,650 \text{ J}$$

Note the units: $J = N \cdot m$ and $N = kg \cdot m/s^2$, so $J = kg \cdot m^2/s^2$.

Sample Problem

5-25 If you do 1,250 J of work in carrying a load of books up a flight of stairs, covering a vertical distance of 12.0 m, what is the mass of the books?

Solution   From Equation 5-12,

$$m = \frac{E_{grav}}{gh} = \frac{1,250 \text{ J}}{(9.8 \text{ m/s}^2)(12.0 \text{ m})} = 11 \text{ kg}$$

Suppose you push the object up an inclined plane instead of lifting it directly. Then how much is its gravitational potential energy? We have seen that the work done in moving something up an inclined plane is exactly the same as the work done in lifting it directly to the same height. Gravitational energy depends only on *vertical* displacement. That is what the h stands for.

Carrying something horizontally takes no work at all, no matter how tired it makes you. This can be seen from the original definition of work (Equation 5-3):

$$W = \mathbf{F} \, \Delta \mathbf{s} \cos \theta$$

Carrying horizontally, you are exerting a force that is straight up, supporting the weight. The displacement is horizontal, at a right angle to the force. And cos 90° = 0. No work. If you have to move a load horizontally, do not hire someone to carry it. Put it on a frictionless cart and give it the gentlest shove. It will keep going forever.

Doing work on a system increases the energy of the system; if the object is lifted, its gravitational potential energy is *mgh*.

Try This If a 50-kg woman walks up a set of ramps to the third floor of a building, a vertical distance of 11 m, how much is her gravitational potential energy? Consider it to be zero at ground level.

Summary Quiz

For each of the following, fill in the missing word or phrase:

1. With an object suspended from a single movable pulley, the effort is _____ of the load.
2. With an object suspended from a single movable pulley, the effort distance is _____ the load distance.
3. In any pulley system, the number of strands of rope supporting the load multiplied by the load distance will be equal to the _____ .
4. Force multiplied by the distance the force moves something is called _____ .
5. In an ideal pulley, _____ input equals _____ output.
6. In an ideal pulley, the tension in the rope times the number of strands is equal to the _____ .
7. In a real pulley, the work output is always _____ than the work input.
8. The ratio of work output to work input is called _____ .
9. In a lever, the ratio of effort arm to load arm is the _____ of the lever.
10. The mechanical advantage of a hydraulic lift is equal to the ratio of the _____ of the two pistons.
11. The ideal mechanical advantage of any machine is equal to the ratio of effort _____ to load _____ .
12. The ideal mechanical advantage of an inclined plane is the ratio of _____ to _____ .
13. A vise is a self-locking machine because of its high _____ .
14. The mechanical advantage of a pair of gears is equal to the ratio of the _____ of the two gears.
15. A watt is equal to a _____ per second.
16. The rate at which work is done is called _____ .
17. The work input to a drop hammer is the distance it is lifted times its _____ .
18. When a drop hammer is at the top of its track, it is in a state of high _____ .
19. Gravitational potential energy is the product of _____ and _____ .
20. Doing work on a system always increases its _____ .

Problems

1. Using a system of one fixed and one movable pulley, like that of Figure 5-4, it takes a force of 450 N applied to the rope to lift a 80-kg load 12 m. Find (a) the ideal mechanical advantage of the pulley; (b) the amount of rope that has to be pulled through the pulley; (c) the work output; (d) the work input; and (e) the efficiency.
2. What is the ideal mechanical advantage of the pulley system shown in Figure 5-19?
3. If the efficiency of the pulley of Figure 5-19 is 80%, how much effort is required to lift a 750-lb load?

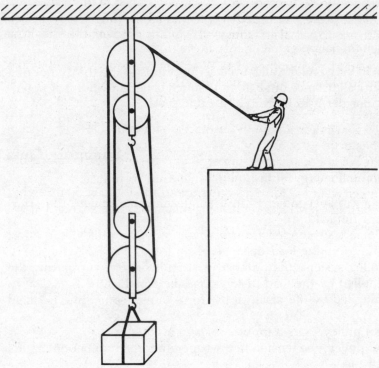

FIGURE 5-19

4. You lift a 25-kg crate vertically 1.2 m and then carry it horizontally 3.5 m to deposit it on a shelf. How much work do you do?

5. How much work do you do in pushing a lawn mower 30 ft, using a force of 25 lb, if the shaft makes an angle of 30° with the ground?

6. You turn a crank connected to a set of gears that lifts a 1,400-N weight 2.5 m. If the total distance you turn the crank is 50 m, how much is the *least* effort you will have to exert?

7. The crank arm of a bicycle is 14 cm long and turns a gear with 26 teeth. This gear is connected by a sprocket chain to a gear of 10 teeth, which turns a wheel with a radius of 33 cm. Find (a) the ideal mechanical advantage of the gear system; (b) how many times the wheel goes around for each turn of the crank; and (c) the mechanical advantage of the bicycle.

8. A deep-sea fishing rod 3.2 m long is pivoted at its butt end and held 60 cm from the butt. How much force must be exerted to lift a 22-kg fish out of the water?

9. What is the mechanical advantage of a Stillson wrench if it is being used to turn a pipe 8.0 cm in diameter and the handle is 25 cm long?

10. What is the mechanical advantage of a pry bar 2.0 m long being used to pry up a nail if the bar is pivoted 12 cm from the nail?

11. A hydraulic jack has an effort piston with an area of 10 cm^2 and a load piston of 120 cm^2. Neglecting friction, what effort is needed to lift 330 kg?

12. A cart loaded to 240 kg is being pushed up a ramp 3.2 m long, to raise the cart 1.0 m. If the efficiency of the whole system is 75%, find (a) the work output; (b) the work input; and (c) the effort.

13. A torque of 50 N · m is applied to a gear with 30 teeth, which drives another gear having 12 teeth. What is the torque in the second gear?

14. A 40-kg load is to be carried up a vertical distance of 10 m in 8.0 s. (a) How much power is needed? (b) If this job is done with a pulley system with a mechanical advantage of 5 and an efficiency of 70%, how much power is needed?

15. A motor operating a pump has 240 W of power. How long will it take the system to raise 100 kg of water 12 m?

16. A force of 300 N pushes a cart 60 m. How much does the energy of the cart increase?

17. How much gravitational potential energy is stored in a 12-kg medicine ball that has been lifted 2.0 m?

18. What is the weight of a drop hammer if it gives up 120,000 J of gravitational potential energy in dropping 10 m?

19. A 60-kg hiker walks 12 km and ends up at an altitude 250 m higher than his starting point. How much has his gravitational potential energy increased?

CHAPTER
6

Energy: They Don't Make It Any More

Looking for some way to define that mysterious quantity called energy, we said that, whenever work is done on a system, its energy increases. Doing work changes the state of the system to a condition of higher energy. Any change made by doing work on a system can therefore be used as a measure of the energy of the system. Lifting something to a higher level is one example, but there are many others.

If you do a lot of work in sandpapering a plank, for example, you must be adding energy to the system. So you look around to see how that energy can be measured. The plank is smoother—so perhaps smoothness is a measure of energy. The plank is also hotter, and you might well wonder whether there is some way of using the temperature as a measure of the work that has been done on the plank.

If you push a cart—doing work on it—you increase its energy, by definition. If that cart were on a horizontal, frictionless surface, it would keep going faster as long as you kept pushing it. It would surely seem that its velocity is some sort of measure of its energy.

And suppose you grab the ends of a spring and stretch it. Surely this takes a force, moving through a distance. So work is being done. Unquestionably, then, the length of the spring at any given moment is a measure of the energy in it.

113

Suppose you have to pump up a bicycle tire, doing work by pushing the air into a smaller space at higher pressure. Isn't this acceptable evidence that the pressure of a gas can be used as a measure of energy if only we know how to go about it?

Now take a look at the two bar magnets in Figure 6-1. At (a), they are shown close together; at (b) they are farther apart. The two N poles are facing each other. Knowing that two N poles repel each other, can you tell whether (a) or (b) is the high-energy state? All you have to do is ask yourself whether you would have to do work on the system to change it from state (a) to state (b), or vice versa. Since the magnets repel each other, work would have to be done to push them closer together, and (b) is the high-energy state. You can measure the energy of the system by the distance between the two N poles.

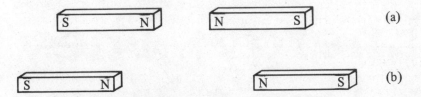

FIGURE 6-1

So we have many different ways of measuring energy. We classify the various forms of energy according to the way in which they can be measured. We already know that gravitational potential energy is measured by the height of something. If we measure energy by means of the velocity at which the thing is going, we call it *kinetic* energy. Energy measured in terms of temperature is called *thermal* energy. Energy stored in a stretched spring—or in any other deformed solid—is called *deformational* energy. And so on.

Core Concept

Energy can be measured in terms of many different kinds of physical quantities, and each has its own name.

Try This   It is known that a positive and a negative electric charge will attract each other. If such a pair of charges move closer to each other, does their energy increase or decrease?

6-2. Energy of Motion

A cart moving on a horizontal, frictionless surface will keep going at constant speed forever. To speed it up, you have to push it through a distance—do work on it. Then it follows that its speed must be a measure of its energy. The energy an object has because of its speed is called *kinetic energy*.

How can you determine the kinetic energy of a moving object? It must be equal to the work that was done in making it move. If an object is at rest and is pushed through a distance s with a force F, the work done on it is equal to the kinetic energy it acquires, so, from Equation 5-3,

$$E_{kin} = Fs$$

The force produces an acceleration; since it is the only force acting, $F = ma$ (Equation 4-2), so

$$E_{kin} = mas$$

The acceleration is uniform and starts at rest, so Equation 1-7c tells us that

$$v = \sqrt{2as}$$

from which

$$as = \frac{1}{2}v^2$$

and substituting this into the equation above gives the standard formula for kinetic energy:

$$E_{kin} = \frac{1}{2}mv^2 \qquad \text{(Equation 6-2)}$$

Therefore, we can determine the kinetic energy of anything if we know only its mass and velocity. See Sample Problems 6-1, 6-2, and 6-3.

Sample Problem

6-1 What is the kinetic energy of a 6.0-kg medicine ball traveling at 4.5 m/s?

Solution Use Equation 6-2:

$$E_{kin} = \frac{1}{2}mv^2$$

$$E_{kin} = \frac{1}{2}(6.0 \text{ kg})\left(4.5\frac{\text{m}}{\text{s}}\right)^2$$

$$E_{kin} = 60.75\frac{\text{kg} \cdot \text{m}^2}{\text{s}^2}$$

$$E_{kin} = 61 \text{ J}$$

Sample Problem

6-2 In pushing a frictionless cart from rest on a horizontal surface, 2,400 J of work is done. If the mass of the cart is 65 kg, how fast is it then going?

Solution The work done is the amount of kinetic energy it gains, starting with none. From Equation 6-2,

$$v = \sqrt{\frac{2E_{kin}}{m}} = \sqrt{\frac{2(2,400 \text{ J})}{65 \text{ kg}}} = 8.6 \text{ m/s}$$

Sample Problem

6-3 What is the mass of a baseball that has 110 J of energy when it is going 22 m/s?

Solution From Equation 6-2,

$$m = \frac{2E_{kin}}{v^2} = \frac{2(110 \text{ J})}{(22 \text{ m/s})^2} = 0.45 \text{ kg}$$

Like all energy, kinetic energy is a scalar. It is true that v is a vector, but any vector squared is a scalar, equal to the square of the magnitude of the vector. Kinetic energy does not change when the direction of motion changes. Consider, for example, a ball attached to a string and whirled around in a horizontal circle. A (centripetal) force is constantly being exerted to keep it in the circle. Does this change its kinetic energy?

It does not. Remember the definition of work (Equation 5-3)?

$$W = \mathbf{F} \, \Delta \mathbf{s} \cos \theta$$

In this case, the force is radially inward, while the velocity is tangential. Since the force is perpendicular to the displacement, $\theta = 90°$. The cosine of $90° = 0$, so you are doing no work on the ball. Its energy does not change, even though the direction of its velocity is constantly changing.

Core Concept

Kinetic energy $= \frac{1}{2}mv^2$; it is the energy anything has because of its motion.

Try This   What is the kinetic energy of a 1,200-kg car going 22 m/s?

6-3. Energy Transformed

Let's take another look at that drop hammer. Work was done on it to lift it, thereby increasing its (gravitational potential) energy. Yet the hammer did no work until it actually struck the piling. Where was all that energy just before the hammer hit the piling?

Well, the hammer was moving fast at that moment. In fact, from Equation 1-7c, we can easily see that its speed was

$$v = \sqrt{2gh}$$

from which it is easy to show that

$$\frac{1}{2}v^2 = gh$$

and multiplying by m gives

$$\frac{1}{2}mv^2 = mgh$$

In other words, the kinetic energy it has just before it hits the piling is equal to the gravitational potential energy it lost while falling through a distance h. Its total energy has not changed. We might have expected this, since no one was doing any work on it while it fell.

If you calculate the kinetic energy gained and the gravitational potential energy lost when a cart rolls down a frictionless inclined plane, you will get the same result. In the absence of friction, the total energy does not change. This result will hold for a hill of any shape at all.

Once we know this, certain problems become solvable. Consider, for example, the roller coaster cart of Figure 6-2. It is pushed a short distance on a level track and then released, so that it rolls downhill, uphill, back down, and so on. We would like to know how fast it is going at point P. An

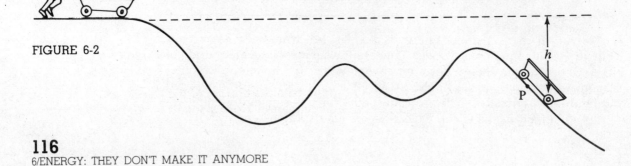

FIGURE 6-2

impossible problem, right? You have to know the slope of the hill at all points so you can calculate the acceleration everywhere, add the velocities, and so on.

Wrong. It's easy. The nice thing about energy is that it is a scalar, so you do not have to worry about directions. Also, once you know that its total value never changes, you can deal with the initial state and the final state and forget about everything that happens in between.

To take the simplest case first, let's consider a 50-kg cart starting at rest and rolling, frictionless, down a slope 20 m long and 10 m high. Its kinetic energy at the top is 0, and its gravitational potential energy, from Equation 5-12, is $mgh = (50 \text{ kg})(9.8 \text{ m/s}^2)(10 \text{ m}) = 4,900 \text{ J}$. The force pushing it downhill, as we saw in Chapter 5, is its weight times $h/l = (50 \text{ kg})(9.8 \text{ m/s}^2)(10 \text{ m}/20 \text{ m}) = 245 \text{ N}$. This produces an acceleration, according to Equation 4-2, of

$$a = \frac{F}{m} = \frac{245 \text{ N}}{50 \text{ kg}} = 4.9 \text{ m/s}^2$$

Now we can find out how fast it is going at the bottom, using Equation 1-7c:

$$v = \sqrt{2as} = \sqrt{2(4.9 \text{ m/s}^2)(20 \text{ m})} = 14 \text{ m/s}$$

So its kinetic energy at the bottom is

$$E_{kin} = \frac{1}{2}mv^2 = \frac{1}{2}(50 \text{ kg})(14 \text{ m/s})^2 = 4,900 \text{ J}$$

This is exactly equal to the gravitational potential energy it had before it started down the hill!

In a frictionless system like this, the amount of kinetic energy the object gains is equal to the potential energy it loses. It works just as well if the hill has curves in it. And it works if the cart is in motion before it starts down the hill. Using this principle, we can easily find out how fast the cart is going at any point on the roller coaster hill of Figure 6-2. Its kinetic energy at any point, such as P, is equal to the amount of kinetic energy it had originally, plus the amount of gravitational potential energy it lost while rolling downhill. For examples, see Sample Problems 6-4, 6-5, and 6-6.

Sample Problem

6-4 Starting at rest, the cart of Figure 6-2 glides frictionless to point P, which is 4.5 m below the top of the hill. How fast is it going at P?

Solution Considering its gravitational energy to have dropped to zero at P, its gravitational energy when it is at the top of the hill is $(mgh)_{start}$. Its kinetic energy starts at zero and increases to $(\frac{1}{2}mv^2)_{end}$. Since its total energy does not change,

$$(mgh)_{start} = \left(\frac{1}{2}mv^2\right)_{end}$$

The m's drop out, and the equation becomes

$$v = \sqrt{2gh} = \sqrt{2(9.8 \text{ m/s}^2)(4.5 \text{ m})} = 9.4 \text{ m/s}$$

Sample Problem

6-5 The same cart, going 9.4 m/s at point P, continues to the bottom of the path and climbs up to a point Q, 3.0 m higher than P. How fast is it then going?

Solution The total energy does not change, so

$$(E_{grav})_P + (E_{kin})_P = (E_{grav})_Q + (E_{kin})_Q$$

For convenience, we may consider P to be the level of zero gravitational energy. The equation then becomes

$$\left(\tfrac{1}{2}mv^2\right)_P = (mgh)_Q + \left(\tfrac{1}{2}mv^2\right)_Q$$

Again, the *m*'s drop out, and we have

$$\tfrac{1}{2}(9.4 \text{ m/s})^2 = (9.8 \text{ m/s}^2)(3.0 \text{ m}) + \tfrac{1}{2}v_Q^2$$

$$\tfrac{1}{2}v_Q^2 = 44.2 \text{ m}^2/\text{s}^2 - 29.4 \text{ m}^2/\text{s}^2$$

$$v_Q = \sqrt{2(14.8 \text{ J})} = 5.4 \text{ m/s}$$

Sample Problem

6-6 The cart has a mass of 650 kg and is pushed for 8.0 m with a force of 520 N and then released at the lip of the hill. How fast is it moving at point P, which is 3.0 m below the lip? (Neglect friction.)

Solution The cart has some kinetic energy when the pushing stops, equal to the work done in pushing it:

$$W = F \, \Delta s = (520 \text{ N})(8.0 \text{ m}) = 4{,}160 \text{ J}$$

When it is at point P, it has lost gravitational potential energy equal to

$$E_{grav} = mgh = (650 \text{ kg})\left(9.8 \tfrac{\text{m}}{\text{s}^2}\right)(3.0 \text{ m})$$

$$E_{grav} = 19{,}110 \text{ J}$$

All this was converted to kinetic energy, which is added to its original kinetic energy, giving a total of 23,270 J. Therefore,

$$E_{kin} = \tfrac{1}{2}mv^2$$

$$\tfrac{1}{2}(650 \text{ kg})v^2 = 23{,}270 \text{ J}$$

$$v = 8.5 \text{ m/s}$$

In practice, of course, it will be going slower than that because a real roller coaster is not frictionless. But you can be quite certain that it cannot be going faster than the speed you calculated.

Core Concept

When an object is moving in a frictionless system, the sum of its kinetic and potential energies does not change.

Try This A child in a wagon, combined mass 30 kg, is going 6.0 m/s on level ground. Then the wagon coasts up a hill. Find (a) its kinetic energy at the bottom of the hill; and (b) how high it rises on the hill before coming to rest. Neglect friction.

Suppose you had the job of figuring out how much work a rocket engine has to do to lift an artificial satellite into outer space. This work must be equal to the increase in its gravitational energy, so you look at the formula and start to calculate the value of *mgh*.

When you get to the *g*, you are in trouble. When the rocket takes off, it is in a place where $g = 9.8$ m/s². But it is rising into regions where *g* is getting smaller. What value should you use? It seems that the expression *mgh* is suitable only in cases where *h* is so small that *g* remains practically constant.

When dealing with outer-space distances, you need a different expression for gravitational potential energy. In arriving at such an expression, it is most convenient to adopt the convention that, when two objects are infinitely far apart, there is no gravitational potential energy. They do not influence each other, and it requires no work to move them around—at least as far as their effect on each other is concerned.

As the two objects come closer to each other, their gravitational potential energy decreases. This is consistent with the usual use of the *mgh* formula. But if the energy was zero at infinite separation, and less than that at closer distances, then it must always be negative at any real distance! The formula, in fact is this:

$$E_{grav} = \frac{-Gm_1m_2}{r} \qquad \text{(Equation 6-4)}$$

where *G* is the universal gravitation constant, the *m*'s are the masses of the two objects, and *r* is the distance between their centers. See Sample Problems 6-7 and 6-8.

Sample Problem

6-7 How much is the gravitational energy of a 720-kg meteorite when it is at an altitude of 2,200 km? (See Appendix 3 for constants.)

Solution   Apply Equation 6-4:

$$E_{grav} = \frac{-Gm_1m_2}{r}$$

$$E_{grav} = \frac{-(6.67 \times 10^{-11} \text{ N} \cdot \text{m}^2/\text{kg}^2)(6.0 \times 10^{24} \text{ kg})(720 \text{ kg})}{8.6 \times 10^6 \text{m}}$$

Note that the value of *r* must be measured in meters to the center of the earth. The answer comes out -3.4×10^{10} J.

Sample Problem

6-8 How much work must be done on a 5,800-kg satellite to lift it into orbit at an altitude of 2,600 km?

Solution   The work that has to be done is the difference between the satellite's gravitational potential energy when it is on earth and its energy when it is at the high altitude. First, calculate the energy when it is on earth; there, *r* is the radius of the earth, 6,400 km:

$$E_{grav} = \frac{-Gm_1m_2}{r}$$

$$E_{grav} = \frac{(6.67 \times 10^{-11} \text{ N} \cdot \text{m}^2/\text{kg}^2)(6.0 \times 10^{24} \text{ kg})(5,800 \text{ kg})}{6.4 \times 10^6 \text{ m}}$$

$$E_{grav} = -3.63 \times 10^{11} \text{ J}$$

When it is in orbit, r is its altitude plus the radius of the earth, or 9.0×10^6 m. Repeating the calculation with this value of r gives

$$E_{grav} = -2.58 \times 10^{11} \text{ J}$$

which is an increase of 1.05×10^{11} J.

Core Concept

For calculating gravitational potential energy over large distances, use a formula that gives a negative value, rising to zero at infinite separation:

$$E_{grav} = \frac{-Gm_1m_2}{r}$$

Try This If a 60-kg meteor falls to earth from a great distance, how much gravitational potential energy does it lose?

6-5. Escape from Earth

According to the formula for gravitational attraction, the force does not drop to zero at any finite distance. Yet it is possible to fire a rocket straight up, carrying a space probe, in such a way that it will coast away into outer space and never return.

The reason is that it is coasting continually into regions of ever-weakening gravity. Its acceleration is always toward the earth, slowing it down. But the acceleration is getting smaller and smaller as it moves away from the earth. If it is going fast enough when it is flung into space, it will still have some velocity left when the acceleration due to the earth's gravity is too small to make any difference.

To allow the space probe to escape, it has to be given a certain minimum speed when it is thrown upward. That speed is known as its *escape velocity*. Its value can be calculated by careful consideration of the energy changes of the space probe as it rises.

When anything rises in free fall, it is losing kinetic energy and gaining gravitational potential energy. The sum of the two does not change. If it is to escape, it needs enough total energy so that it will not come to rest until it is infinitely far away. At that point, its total energy is zero. Then its total energy is great enough so that, if you fire it straight up, it will keep going forever. The space probe (or any projectile) needs enough kinetic energy so that its total energy will be zero:

$$\frac{1}{2}mv^2 - \frac{Gm_1m_2}{r} = 0$$

where m and m_1 are both the mass of the projectile. Note that you can divide the equation through by the mass of the projectile and solve for v. Neglecting air drag, the equation becomes

$$v_{escape} = \sqrt{\frac{2Gm_2}{r}}$$

where m_2 is the mass of the earth and r is its radius. The mass of the projectile is not in the equation; all objects have the same escape velocity, which turns out to be about 11,000 m/s. This is about 20 times the speed of a rifle bullet. See Sample Problem 6-9 for the calculation.

Sample Problem

6-9 What is the escape velocity from the surface of the earth?

Solution The object needs enough kinetic energy so that the total energy will be at least zero. The text shows that this condition exists when

$$v > \sqrt{\frac{2Gm_2}{r}}$$

$$v > \sqrt{\frac{2(6.67 \times 10^{-11} \text{ N} \cdot \text{m}^2/\text{kg}^2)(6.0 \times 10^{24} \text{ kg})}{6.4 \times 10^6 \text{ m}}}$$

$$v > 11{,}200 \text{ m/s}$$

In practice, we often put space ships into orbit before sending them off into outer space. Their escape velocity is considerably smaller, since they are leaving from a point where r is larger.

Core Concept

Escape velocity, the same for all objects, is the velocity needed to bring the total energy to zero.

Try This What is the escape velocity from a point in space at an altitude equal to the radius of the earth?

6-6. What Really Happens?

All the energy conversions we have done so far suffer from a serious defect: They don't work. When we say "neglect friction," we are ignoring an aspect of the problem that might make our answer very different.

In some systems, the work done disappears entirely into friction, creating neither kinetic nor potential energy. Think of the sandpaper job again. In these cases, the only observable outcome of all that work is an increase in the temperature of the object. Things get hot when you rub them. Something is going on *inside* the material, something that makes it feel warmer and makes a thermometer reading rise. Whatever it is, it must be an increase in some kind of energy, because it results from doing work.

This energy inside things, the energy that makes them feel warm, is called *thermal energy*. It can be measured, by methods you will learn when you get to the next chapter. Careful measurements in many different kinds of mechanical systems have shown that, where there is friction, the amount of thermal energy produced is equal to the loss of kinetic and potential energy from the system. In other words, if you take the thermal energy into account, the total energy of the system does not change.

Using this principle, you can determine the amount of thermal energy produced in many situations. See, for example, Sample Problems 6-10 and 6-11.

Sample Problem

6-10 A boy in a wagon, combined mass 55 kg, coasts on level ground, slowing down from 4.5 m/s to 2.2 m/s. How much thermal energy is created?

Solution   Since the ground is level, there is no change in gravitational energy; all the lost kinetic energy is converted to thermal energy. Therefore,

$$E_{therm} = \left(\frac{1}{2}mv^2\right)_{start} - \left(\frac{1}{2}mv^2\right)_{end}$$

$$E_{therm} = \frac{1}{2}(55 \text{ kg})(4.5 \text{ m/s})^2 - \frac{1}{2}(55 \text{ kg})(2.2 \text{ m/s})^2$$

$$E_{therm} = 420 \text{ J}$$

Sample Problem

6-11 A 1,550-kg car going 20 m/s coasts uphill, coming to rest after rising a vertical distance of 11.0 m. How much thermal energy is produced?

Solution   The kinetic energy of the car at the bottom of the hill is

$$E_{kin} = \frac{1}{2}mv^2 = \frac{1}{2}(1{,}550 \text{ kg})(20 \text{ m/s})^2$$

$$E_{kin} = 3.10 \times 10^5 \text{ J}$$

As it rises up the hill, the amount of gravitational potential energy it gains is

$$E_{grav} = mgh = (1{,}550 \text{ kg})\left(9.8\tfrac{m}{s^2}\right)(11.0 \text{ m})$$

$$E_{grav} = 1.67 \times 10^5 \text{ J}$$

Since all the kinetic energy was lost as the car went up the hill, the rest of it must have been converted into thermal energy. This energy comes to 1.43×10^5 J.

The ability to calculate thermal energy, and to show that it accounts for the lost kinetic and potential energy, has led to what is perhaps the most firmly established and most fundamental law of physics. It is called the *law of conservation of energy*, or the *first law of thermodynamics:* In a closed system, the total energy remains constant.

As we shall see, this law has led to a greatly expanded concept of energy. Every time a system seemed to use up energy, or to create some out of nothing, physicists looked around to see whether there might not be some other form of energy involved, which they had neglected to take into account. So far, they have always found them.

Core Concept

In a closed system, the total amount of energy does not change.

Try This   A child in a wagon, total mass 30 kg, starts at rest and rolls downhill, going a vertical distance of 6.0 m. If the wagon was going 7.2 m/s at the bottom of the hill, how much thermal energy was created?

6-7. Energy of a Spin

Let a ball roll downhill and measure its speed at the bottom. If you assume that all of its lost gravitational potential energy is converted into kinetic energy, the value you get for its speed will be wrong. The ball is going too slowly at the bottom. There is some energy missing, even if you make a reasonable allowance for the creation of thermal energy.

When this happens, if we believe in the conservation of energy, we must look for the missing energy. In this case, we do not have to look very far. The ball is spinning when it gets to the bottom. Surely, some of the lost gravitational energy was used in making the ball spin. A spinning ball must have some *rotational kinetic energy*.

Calculation of rotational kinetic energy is difficult because it must take into account the shape of the object. Consider, as the simplest example, three kinds of objects: a thin ring, a disk, and a solid sphere. Assume that all three have the same mass and are spinning at the same rate. Do all have the same rotational kinetic energy?

Absolutely not. For the ring, we could probably figure out quite easily what the kinetic energy is, because all of its mass is at the same distance from the center and all of it is traveling at the same speed. But the mass of the disk is distributed from rim to center; points nearer the center are going more slowly than those at the rim and so have less kinetic energy. The center point has no kinetic energy at all. It follows that the disk, spinning with the same rotational speed as the ring, has less rotational kinetic energy. Calculations show that it has just half as much.

The sphere has even less, since it has more mass concentrated near the axis of rotation. Its kinetic energy is only two-fifths that of the ring. Calculations of rotational kinetic energy can be made provided the object has some sort of regular geometric shape. For the skater doing a spin, we would have to be satisfied with an approximation.

Kinetic energy can be stored conveniently only in rotational form, and the device that does it is a flywheel. Once set spinning, it can be persuaded to give up its energy gradually. A scheme has been attempted to run vehicles on this basis; you set a large, massive flywheel in the vehicle spinning. To run the vehicle, you connect the drive shaft to the flywheel through a clutch, and the kinetic energy of the flywheel is gradually used up in running the vehicle. For many reasons, this has not been practical.

Core Concept

Objects in rotation have rotational kinetic energy, which depends on their mass, their rotational speed, and the way their mass is distributed.

Try This Which has more rotational kinetic energy if both are spinning at the same rate, a solid wooden ball or a hollow steel ball with the same mass and diameter?

6-8. More Energies

Once it is recognized that an increase in temperature represents increased energy, we have to enlarge our definition. To make something warm (a teakettle, say), you do not have to do work on it. You can build a fire under it. The thermal energy of the fire can be transferred to the teakettle—with no work being done.

Well, then, if the fire contains energy, where did it come from? Somehow, it seems to be stored in the fuel—the gas, oil, coal, wood, whatever. The energy stored in the fuel is released in the form of thermal energy as a result of making a change in the chemical composition of the material. We combine the coal with oxygen and convert it to carbon dioxide, water vapor, soot, and ashes. The energy stored in the fuel is called *chemical potential energy*.

How did the energy get into the coal in the first place? Coal is the chemically altered remains of ancient plants. Plants grow by converting materials gathered from the air and the soil into more living tissue. In this process, the level of chemical potential energy of these materials is increased. This can be done only if there is some sort of energy available to be added to the materials. Plants use the energy of sunlight for this purpose. It comes in the form of light rays, and the energy can be called *radiant energy*.

Civilization depends absolutely on our ability to transform energy. Even a primitive farmer must do this; he feeds his oxen so that their muscles can convert the chemical energy stored in the fodder into the kinetic energy that pulls the plow. A tractor does the same thing, except that its "fodder" is gasoline. Engines of all kinds—starting with steam engines—convert the thermal energy of fuel into the kinetic energy that drives machinery.

At home, you drive machinery by plugging it into a wall outlet. The *electric energy* coming out of the wall is converted into mechanical or thermal energy in home appliances. A very large part of the fuel we burn turns generators that produce electric energy, which can be transmitted over long distance through wires.

It is important to note that all these transformations are quantitative. When a fire burns, every joule of thermal energy produced is accounted for by the disappearance of a joule of chemical potential energy. For each kind of energy, there is a formula that allows us to calculate it. The discovery of the correct formulas has been an important part of the growth of physics. So far, it has always been possible to find new energy terms to balance the equation, no matter what kind of process is involved.

Core Concept

Energy can be converted into many different forms, but its total quantity does not change.

Try This   What energy transformations take place when falling water is used to turn a turbine and make electricity?

6-9. Energy and Mass

The discovery of radioactivity, in 1892, presented a severe challenge to the law of conservation of energy. Here were lumps of rock that managed somehow to give out energy continuously, in some form that was not at all understood. The flow of energy continued indefinitely, even in total darkness.

The key to the rescue of the theory came with the publication of Albert Einstein's "special" theory of relativity (as distinguished from the "general" theory, which came later) in 1905. This bold and imaginative product of a brilliant young mind proposed a complete revision of all the concepts that lie at the very heart of physical thought—time, space, mass, energy, momentum. At first, the theory was often ridiculed, but experimental evidence has forced it into a position at the heart of the canon of physical law.

A foundation stone of the science of chemistry is the law of conservation of mass. This law says that no chemical reaction can change the total mass of material. Thus, if you burn a gallon of gasoline, the total mass of the gasoline and the oxygen that went into the reaction is the same as the mass of the exhaust gases that go out the tailpipe of your car. Through any reaction, energy is also conserved—the stored chemical energy is converted into the kinetic energy that runs your car plus a lot of useless thermal energy.

The most important corollary of the theory of relativity is that what we call mass and what we call energy are simply two different ways of measuring the same thing. Every kind of energy has mass, and all mass is a form of energy. The relationship between the two methods of measurement is given by the famous equation

$$E = mc^2 \qquad \text{(Equation 6-9)}$$

in which E is any kind of energy, m is the corresponding mass, and c is the speed of light, 3.00×10^8 m/s.

This theory revises the laws of conservation of mass and energy, uniting them into a single law. When you burn the gasoline, the energy it releases causes a corresponding reduction in its mass. The car in motion has more mass than it had at rest, the extra mass being its kinetic energy. The hot engine and the hot exhaust gases also have extra mass, which they lose as they cool off. Thus, when all the thermal energy has been dissipated and the car is once more at rest, the whole system has less mass than it had before the gasoline was burned. The chemist's law of conservation of mass is defective because it has always failed to take into account the mass of the energy converted.

How is it that, in a century of quantitative chemistry, no one detected this discrepancy? The reason is that c^2 is an extremely large number. The energy changes in a chemical reaction are not great enough to produce a mass discrepancy that is detectable. Even with the extraordinary accuracy of today's chemists, they can safely ignore any mass changes and continue business with the old rules. See Sample Problems 6-12, 6-13, and 6-14.

Sample Problem

6-12 A truck with a mass of 22,000 kg goes from rest to 27 m/s. How much does its mass increase?

Solution The increase in mass corresponds to the increase in kinetic energy, which is

$$E_{kin} = \frac{1}{2}mv^2 = \frac{1}{2}(22{,}000 \text{ kg})(27 \text{ m/s})^2 = 8.0 \times 10^6 \text{ J}$$

To get the amount of mass corresponding to this much energy, use Equation 6-9:

$$m = \frac{E}{c^2} = \frac{8.0 \times 10^6 \text{ J}}{(3.0 \times 10^8 \text{ m/s})^2} = 8.9 \times 10^{-11} \text{ kg}$$

Sample Problem

6-13. What is the energy equivalent of the mass of a drop of water that has a mass of about 10^{-5} kg?

Solution

$$E = mc^2 = (10^{-5} \text{ kg})(3.0 \times 10^8 \text{ m/s})^2 = 10^{12} \text{ J}$$

Sample Problem

6-14 In burning a gallon of gasoline, a half-billion joules (5.0×10^8 J) are converted into thermal energy. When all the exhaust products have cooled off, how much less is their mass than the mass of the original gasoline?

Solution From Equation 6-9,

$$m = \frac{E}{c^2} = \frac{5.0 \times 10^8 \text{ kg} \cdot \text{m}^2/\text{s}^2}{(3.0 \times 10^8 \text{ m/s})^2}$$

$$m = 6 \times 10^{-9} \text{ kg}$$

Core Concept

Mass and energy are two aspects of the same thing, and their measurement is related by the square of the speed of light:

$$E = mc^2$$

Try This If a 1,400-kg car is going 30 m/s, how much is the increase in its mass due to its kinetic energy?

6-10. The Energy of the Nucleus

An atom consists of an extremely dense nucleus (made of neutrons and protons) surrounded by a cloud of electrons. The protons have positive electric charges, and the electrons have negative charges. It is a basic rule of electricity that opposite charges attract each other. It is this attraction that keeps the electrons in the space around the nucleus. The outermost electrons are not held very strongly, and it is interchanges of these outer electrons with those of other atoms that produce chemical combinations.

The atomic nucleus is an extremely compact structure. The positively charged protons and the electrically neutral neutrons are tightly packed together. The laws of electricity tell us that the protons, all possessing the same charge, must repel each other. It must be a strong repulsion with the particles so close together. Then what keeps the nucleus from exploding?

There must be another force involved, a force of attraction that acts on protons and neutrons. It is far more powerful than the electric repulsion of the protons, but it acts only at the extremely short distances within the atomic nucleus. It is known as the *strong nuclear force*.

A *nuclear reaction* is different from a chemical reaction. Whereas a chemical reaction involves only the outermost electrons of the atom and is mediated only by the electric force, the nuclear reaction is a rearrangement of the neutrons and protons that make up the nucleus. New kinds of nuclei are created by splitting large nuclei or by combining small ones. These reactions involve the extremely powerful strong nuclear force. The energy changes in these nuclear transformations are therefore very large—large enough that they can be measured in terms of changes in the mass of the particles involved.

Einstein's equation has been tested and found to be correct in nuclear reactions. When large nuclei are split, or when small ones are combined, there is a detectable loss of mass and a corresponding enormous increase in other forms of energy. In an atomic explosion, this takes the form of radiation, thermal energy, and the kinetic energy of the blast. In a controlled nuclear reaction, it is only thermal energy that is produced, and this energy is used to run a conventional steam engine to make electricity.

A nuclear reaction involves a detectable mass change and an enormous release of energy because it is mediated by the strong nuclear force.

Try This How much energy is released in a nuclear explosion that converts 2 kg of uranium into 1,999.9 g of other substances?

6-11. Inevitable Losses

Energy is never used up; the total amount of it never changes. Then why do we have to keep drilling for oil?

The reason is that not all forms of energy are equally useful. Thermal energy in particular is useful only when it is at a very high temperature. Hot steam can turn a turbine that runs an electric generator. If you were to take that steam and mix it with cold water, you would still have the same amount of thermal energy you started with, but it would be spread thin, through a large mass of material, and the temperature would be low. You cannot turn a turbine with warm water.

A steam engine wastes a good deal of energy. It must be cooled, usually by running lots of cold water through it. The cooling water emerges warm and is discharged into a river or the ocean. The best steam engines are only about 48 percent efficient; 52 percent of the thermal energy given up by the steam leaves the engine in the form of useless warm water. According to the *second law of thermodynamics*, there is no way to convert thermal energy into other forms with 100 percent efficiency. In practice, you cannot come anywhere near 100 percent.

What happens to the other 48 percent of the energy, the useful part? It is used to drive an electric generator—which is also considerably less than 100 percent efficient. So another part of the total becomes useless, low-temperature thermal energy, because of the friction in the generator and the heat generated by the current flowing in its wires. What is left goes out into transformers and transmission lines, where some more is lost as heat. Finally it reaches your home.

The process continues at the point where the electric energy is used. You plug in a toaster, thereby converting some of the electricity to thermal energy—100 percent! If you run a mixer, it creates kinetic energy—and more thermal energy. Finally, all that kinetic energy does nothing but warm up the batter. In the end every bit of the energy that was in the steam in the form of usable, high-temperature thermal energy is dissipated into the environment as thermal energy once again. But this time, it is at low temperature—spread thin and useless.

And that is why we have to keep looking for oil. The first law of thermodynamics tells us that the total amount of energy does not change; the second law informs us that every time energy is converted from one form to another, part of it becomes useless, low-temperature thermal energy.

Every energy conversion results in degradation of part of it into a useless form.

Try This You drive your car up a hill and then come back and stop at the place where you started. Describe the ultimate fate of all the energy in the gasoline you burned.

Summary Quiz

Fill in the missing words or phrases:

1. The energy of a system is always increased when _____ is done on the system.
2. If two objects attract each other, their potential energy _____ when they are moved farther apart.
3. The energy of an object as measured by its motion is called _____ energy.
4. If the speed of an object doubles, its kinetic energy is multiplied by a factor of _____ .
5. Since energy does not depend on direction, energy is a _____ quantity.
6. As an object falls freely, it loses _____ energy and gains an equal amount of _____ energy.
7. Neglecting friction, when an object slides down a slope of any shape, its _____ energy remains constant.
8. When two objects are separated by a great astronomical distance, their gravitational energy is _____ .
9. On an astronomical scale, gravitational potential energy always has a _____ sign.
10. In order for a space probe to coast completely away from the earth and never return, the sum of its _____ and gravitational energies must be positive.
11. The speed at which an object's kinetic and gravitational potential energies barely add up to more than zero is called the object's _____ velocity.
12. Increase in temperature indicates an increase in _____ energy.
13. According to the first law of thermodynamics, the total amount of _____ in a closed system does not change.
14. Rotational kinetic energy of a sphere is _____ than that of a disk with the same mass, radius, and angular velocity.
15. A fire converts _____ energy to thermal energy.
16. The most convenient form for the transfer of energy is as _____ energy.
17. According to relativity theory, energy and _____ are different aspects of the same thing.
18. The mass corresponding to any energy can be found by dividing the energy by the square of _____ .
19. The nucleus of an atom is held together by the _____ force.
20. In contrast to a chemical reaction, a nuclear reaction produces a measurable change of _____ .
21. As energy is used, all of it eventually turns into useless _____ energy.

Problems

1. Using appropriate machinery, you lift a load by pulling 22 m of rope with a force of 150 N. How much does the energy of the whole system increase?
2. A frictionless 380-kg cart is pushed on a horizontal track for 15 m, using a force of 75 N. Find (a) the kinetic energy of the cart; and (b) its velocity.
3. A 20-kg boulder falls off a cliff and strikes the ground going 10.0 m/s. How much work does it do on the ground while coming to rest?

4. A pendulum, swinging back and forth, rises at the end of its swing to a position 15 cm higher than its lowest point. How fast is it going at the lowest point?

5. A 300-g toy rocket is fired straight upward, its engine providing a thrust of 6.5 N. The rocket engine burns out at an altitude of 25 m. Find (a) the work done by the rocket engine; (b) the gravitational potential energy of the rocket when it comes to rest at its highest position; and (c) the height above the ground at which the rocket comes to rest.

6. A 160-g golf ball is dropped from a height of 5.0 m and bounces back up, reaching a height of 4.2 m on the rebound. Find (a) the amount of thermal energy produced in the impact; and (b) the speed of the ball as it comes off the ground.

7. A force of 250 N is used to push a 30-kg wagon up a hill, a distance of 12 m, bringing it to a point 6.0 m higher than its starting point. If 10% of the work done is used in overcoming friction, find (a) the amount of work done; (b) the increase in the gravitational potential energy of the wagon; (c) the increase in the thermal energy of the system; (d) the increase in the kinetic energy of the wagon; and (e) the speed of the wagon at the top.

8. A rocket engine is to lift a 250-kg space probe so that it will coast all the way out of the solar system. (a) How much energy must the rocket supply? (b) If the thrust of the engine is 640,000 N, at what altitude will the engine stop firing?

9. A 60-kg meteor falls from outer space to the moon, which has a radius of 1.73×10^6 m and a mass of 7.48×10^{22} kg. How much work does it do on the surface of the moon when it strikes?

10. Using the data of Problem 9, determine the escape velocity from the surface of the moon.

11. The brakes are applied in a 1,200-kg car and it slows down from 22 m/s to 15 m/s. How much thermal energy is created in the brakes and tires?

12. How much is the gravitational potential energy of a 1,800-kg elevator when it is 15 m above its lowest level?

13. If 20,000 J of work is done in pumping water up to a height of 12 m, how much water is pumped?

14. A 1.5-kg rock traveling at 4.0 m/s strikes a stationary 2.5-kg lump of clay, and the two bodies then move together. Find (a) the speed of the combination after the impact; and (b) the amount of thermal energy produced in the collision. (Hint: Momentum is conserved and some of the kinetic energy is converted to thermal energy.)

15. It takes an average force of 60 N to pull the string of a bow a distance of 35 cm. How much energy is stored in the bow?

16. What mass of fuel is used in a nuclear power plant that produces nuclear energy at the rate of 10^9 W during a whole year?

17. How much energy is there in the mass of a grain of sugar, which has a mass of about 10^{-4} g?

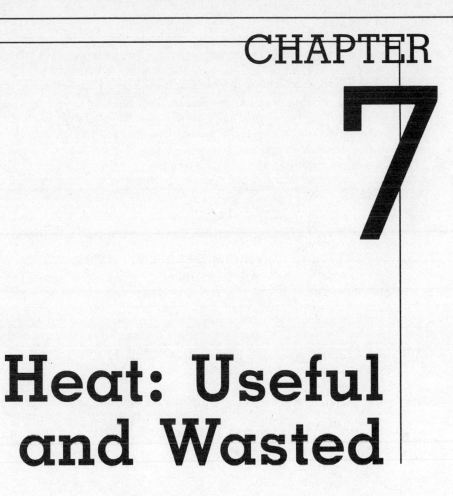

CHAPTER 7

Heat: Useful and Wasted

7-1. Heat and Temperature

As we saw in Chapter 6, we have good reason to believe that, when something gets warmer, its energy content goes up. We called that kind of energy *thermal energy*. With confidence in the law of conservation of energy, we think there ought to be some way to measure thermal energy and to show that the amount of it can account for the kinetic energy lost because of friction.

The first step is to quantify the statement that some things are warmer than others. We need a device that will measure the degree of warmness or coldness of something—a thermometer. Touching the thing with your hand can give you a rough idea of how hot it is, but your sense of temperature is highly unreliable, and you surely would not want to touch things that are very hot or very cold.

Fortunately, there are a lot of measureable properties of things that change when the objects heat up. A liquid or a solid expands when heated. The pressure in a gas increases. Electrical resistance changes. Any of these properties may be useful in defining what we mean by temperature.

What do we mean by the statement that two objects are at the same temperature? Rigorously defined, it means this: If the two objects are placed together, and neither one of them undergoes any change, then they are said to be at the same temperature.

If the objects are not at the same temperature, the hot one gets cooler and the cold one gets warmer, until both of them stop changing. Then, they are at the same temperature. No further change takes place, and the objects are said to be in a state of *thermal equilibrium*.

Since we start with the assumption than an object has more thermal energy in it if it is warmer, this means that the cool object has lost thermal energy and the warm one has gained some. Since energy is conserved, this implies that some energy has passed from the warm object to the cool one. Energy that flows from a warmer object to a cooler one because of the difference in temperature is called *heat*.

Core Concept

Heat is energy transferred because of a temperature difference; no heat is transferred spontaneously unless there is a temperature difference.

Try This To measure your body temperature, you insert a fever thermometer into your mouth. How do you know when thermal equilibrium has been reached?

7-2. Traveling Heat

For one object to warm another, actual contact is not necessary. The sun keeps us all warm.

There are three distinctly different mechanisms by which heat can be transferred. Each occurs under a particular set of conditions. The only thing they all have in common is the fundamental definition of heat: The transfer of energy takes place, in all cases, from something warmer to something colder.

If one end of a hot poker is placed in the fire and left there for any length of time, the other end gets hot. Heat is transferred through the rod, from the hot end to the cold end. This method of heat transfer is called *conduction*. It is a flow of energy through a material. Metals—particularly silver, copper, and aluminum—are good conductors of heat, while most nonmetals conduct poorly. The copper bottom of a frying pan conducts the heat in all directions, spreading it out evenly over the bottom. The handle is made of wood or plastic—poor conductors. These materials, when used for the purpose of preventing conduction of heat, are called *insulators.*

Water is a very poor conductor of heat. If you place a pan of water under the broiler of your stove, it is not difficult to get it boiling at the surface while it remains ice cold at the bottom. To heat the water uniformly throughout, you have to place the flame under the pot. This warms the water at the bottom, making it expand. Its density drops, and it therefore rises above the surrounding cool water. The result is a circulation that continuously brings colder water near the flame, until the entire potful gets to the boiling point. This process is called *convection*.

Heat transfer by convection occurs only under certain special conditions. You need a fluid, for one thing, so that the warm material can rise through the colder material. And the fluid must either be heated at the bottom or cooled at the top. The radiator in your room is placed near the floor, while the cooling element in your refrigerator is at the top of the box. We have a lot of devices for preventing transfer of heat by convection. They do this by preventing air from circulating. This is why we fill the wall spaces of our homes with foam, why we wrap ourselves in layers of wool, why we make our picnic refrigerator of styrofoam. Air is a very poor conductor, and if we can trap it, it will not be able to transfer heat by convection, either.

Heat comes to us from the sun through the vast vacuum of empty space. It travels in the form of rays, similar to light rays but invisible for the most part. Every warm object—including you—gives off rays of this sort. Cool objects absorb them, so the net transfer of heat, as always, is from warmer to colder objects. Heat being thus transferred by *radiation* travels in straight lines, like light. That is why you can sit in front of an open fire with your face toasting gently and your back growing icicles.

The efficiency with which an object radiates or absorbs radiant heat depends on the color of its surface. A black object absorbs everything that comes to it, quickly and efficiently. The ability of a surface to radiate energy is closely related to its ability to absorb radiant energy, and the black object, when warm, is an excellent radiator. That is why the cooling fins of an engine are always painted black. A white object, on the other hand, neither absorbs nor radiates effectively. Worst of all is the silvery surface of many metals. The insulation in your attic includes a layer of aluminum foil, which prevents heat from radiating out in the winter or radiating in when it is hot outside.

Core Concept

Heat travels through metals by conduction; upward through fluids by convection; through empty space by radiation.

Try This Explain why a vacuum bottle is silvered, and why the space between its double walls is evacuated.

7-3. What's a Thermometer?

To make the concept of coldness-versus-hotness quantitative, we need some sort of device that changes observably when it is warmed up or cooled down. The most familiar example is the mercury–glass thermometer, illustrated in Figure 7-1. It consists of a glass tube with a very fine bore running throughout its length. At the bottom is a reservoir containing a quantity of liquid—often, but not always, mercury. Warming the bulb causes the liquid and the glass to expand, but the liquid expands much more than the glass does. The extra volume of liquid has no place to go but up into the bore of the tube. If the bore is thin enough, a small amount of extra volume will rise a good way into it.

Now all we have to do is mark a scale on the stem of our thermometer. Any sort of scale will do, provided only that it is clearly defined so that everyone knows what the numbers mean. We have to find a couple of fixed points that can be given arbitrary values and then simply divide up the stem into equal spaces.

The most convenient fixed points, points that can be carefully defined and used with any thermometer, are the transition temperatures between different phases of a substance—between liquid and solid or between liquid and gas. What happens, for example, if you put some ice into warm water in a completely insulated container? Since the ice is colder than the water, heat passes from the water into the ice. This makes some ice melt; it also makes the water cool down. This continues until the ice and the water

FIGURE 7-1

are in thermal equilibrium, and then no further change takes place. The crucial point is that there is only one temperature at which liquid water and ice can exist together in equilibrium, and we can use that point to define a temperature scale. That temperature is known as the *melting point* of ice.

Another fixed point can be found by heating water until it boils. At that temperature, both liquid water and water vapor can exist in equilibrium. This *boiling point*, however, is not always the same; it depends on the pressure of the vapor.

Now we can calibrate our thermometer—that is, fix the scale. Put the bulb into an ice–water mixture and wait for the mercury to stop falling. Then the thermometer will be in thermal equilibrium with the ice–water mixture, and therefore at the same temperature. Mark the mercury level 0°C. Then put the thermometer in boiling water at a pressure of 1 atmosphere, wait for equilibrium again, and mark it 100°C. Now divide the space between the two fixed points into 100 equal spaces, and the thermometer is finished. If you wish, you can extend the scale below zero and above 100°C, using the same spacing for the degree marks.

This procedure defines the Celsius temperature scale, which is universally used in laboratories and is in common use in most of the world. In the United States, we still cling to the old-fashioned Fahrenheit scale, in which the melting point of ice is 32°F and the boiling point of water is 212°F. A comparison of the two scales is shown in Figure 7-2.

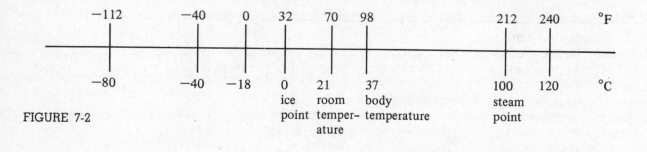

FIGURE 7-2

Core Concept

The Celsius scale defines 100 degrees between the melting point and the boiling point of water.

Try This On your mercury–glass thermometer, the distance from the melting point to the boiling point of water is 15.0 cm. What is the temperature of a liquid if the thermometer comes to equilibrium with it when the mercury stands 11.5 cm above the melting-point mark?

7-4. Temperature and Thermal Energy

If a rise in temperature indicates an increase in thermal energy, we might suspect that the change in temperature is proportional to the energy change:

$$\Delta H \propto \Delta T$$

However, there is another factor. It takes a lot more energy to heat a gallon of water to the boiling point than to heat a pint through the same change in temperature. When the temperature rises by a given amount, the energy added must also be proportional to the amount of material that was heated up. Therefore,

$$\Delta H \propto m \, \Delta T$$

And one more point. We have no reason to believe that all materials respond in the same way when heat is added to them. On the same flame, some materials will get hot a lot faster than others. We can make an equation out of this relationship by introducing a constant that is characteristic of the material whose thermal energy is being changed:

$$\Delta H = cm \, \Delta T \qquad \text{(Equation 7-4)}$$

where c is a constant, characteristic of a given material, called the *specific heat* of the substance. In the SI, it is expressed in joules per kilogram-Celsius degree, but this leads to extremely small numbers, and it is more usual to write it per gram: $J/g \cdot C°$.

This equation has been found by experience to work well over a large range of values. With extreme changes in temperature, adjustment is made by assigning different values to c. These values can be looked up in tables.

The specific heat of a substance has to be found by experiment. Add a measured amount of energy to a known mass of the material and see how much the temperature rises. This then makes it possible to use Equation 7-4 to calculate the specific heat. Examples are given in Sample Problems 7-1, 7-2, and 7-3.

Sample Problem

7-1 The specific heat of aluminum is 0.90 $J/g \cdot C°$. How much heat does it take to raise the temperature of a 350-g aluminum pot from 20°C to 85°C?

Solution   The temperature increase is 65 C°, and

$$\Delta H = cm \, \Delta T = \left(0.90 \frac{J}{g \cdot C°}\right)(350 \text{ g})(65 \text{ C}°) = 20,000 \text{ J}$$

Sample Problem

7-2 What is the specific heat of a metal if the addition of 15,000 J of heat raises the temperature of a 620-g sample of it from 15°C to 90°C?

Solution

$$c = \frac{\Delta H}{m \, \Delta T} = \frac{15,000 \text{ J}}{(620 \text{ g})(75 \text{ C}°)} = 0.32 \text{ J/g} \cdot C°$$

Sample Problem

7-3 To find the specific heat of a metal, you take 500 g of shot and drop it through a distance of 2.0 m, so that it is warmed up by the impact with the ground. After you have done this 1,000 times, the temperature of the shot has risen from 10°C to 38°C. What is the specific heat of the metal?

Solution   Looking at Equation 7-4, you have to know the mass of the sample (500 g), the increase in its temperature (28 C°), and the amount of added energy that produced this increase. The energy was added by lifting the shot 1,000 times. From Equation 5-12:

$$\Delta E = (1,000)(0.5 \text{ kg})\left(9.8 \frac{m}{s^2}\right)(2.0 \text{ m})$$

$$\Delta E = \Delta H = 9,800 \text{ J}$$

Now we can apply Equation 7-4:

$$\Delta H = cm \, \Delta T$$

$$9{,}800 \text{ J} = c(500 \text{ g})(28 \text{ C}°)$$

Which gives $c = 0.70$ J/g·C°.

Core Concept

Change in the thermal energy of an object, the heat added or removed, is proportional to the mass and temperature change and depends on the kind of material:

$$\Delta H = cm \, \Delta T$$

Try This How much heat must be added to a 250-g aluminum pan to raise its temperature from 20°C to 85°C. The specific heat of aluminum is 0.90 J/g·C°.

7-5. Counting Calories

Experiment has shown that the specific heat of water is very high, about 4.19 J/g·C°. The specific heat of water is used to define a unit of energy called the *calorie*. By definition, the specific heat of water is 1 cal/g·C°. Thus, it takes one calorie of heat to raise the temperature of each gram of water by 1 C°. Sample Problem 7-4 shows how this works.

Sample Problem

7-4 How much heat must be added to a cupful (250 g) of water to heat it from tap-water temperature (17°C) to the boiling point?

Solution Using Equation 7-4, we have

$$\Delta H = cm \, \Delta T$$

$$\Delta H = \left(1\frac{\text{cal}}{\text{g·C}°}\right)(250 \text{ g})(83 \text{ C}°)$$

$$\Delta H = 21{,}000 \text{ cal, or } 21 \text{ kcal}$$

There are a couple of difficulties with this definition of the calorie. In the first place, it is rather vague, since the specific heat of water varies somewhat at different temperatures. There is a newer and more precise definition of the calorie: 1 cal = 4.18605 J. For most purposes, however, it is accurate enough to take the specific heat of water as 1 cal/g·C°.

If you are counting calories in your diet, you are using a different calorie. It is more correctly called a kilocalorie, which is 1,000 cal or 4186.05 J. In this book we will say kilocalorie if that is what we mean.

Using the known value of the specific heat of water, we can find out lots of interesting things about other materials. We use the *method of mixtures*. The basic principle of this method is this: If you bring something hot into contact with water, the heat lost by the hot object must be equal to the heat gained by the water: $(cm \, \Delta T)_{object} = (cm \, \Delta T)_{water}$. There are six variables in this equation; if any five of them are known, the other one can be found.

In Sample Problem 7-5, we determine an unknown mass. Hot lead, at a known temperature, is added to cold water with a thermometer in it. When equilibrium is reached, both the lead and the water are at the same temperature, and we can easily find ΔT for both of them. Knowing the mass of the water and the two specific heats, we can solve for the unknown value of the mass of the lead.

In Sample Problem 7-6, the unknown is the initial temperature of the iron ingot; everything else is known. In Sample Problem 7-7, we are using the method of mixtures to find the specific heat of an unknown material. In each case, only one of the six variables is not known.

Sample Problem 7-8 introduces an additional complication. We do not know the final temperature, so we cannot put the ΔT's into the equation. The equation must be written with the final temperature T (*not* ΔT) as the unknown. Since the water cools down from 98°C to T, its $\Delta T = 98°C - T$. And the aluminum warms up from 20°C to T, so its $\Delta T = T - 20°C$.

Sample Problem

7-5 A handful of lead shot (specific heat 0.031 cal/g·C°) is heated to 95°C and dumped into 230 g of cold water at a temperature of 18°C. The mixture comes to thermal equilibrium at 32°C. What is the mass of the lead?

Solution While the lead cools down from 95°C to 32°C, the water warms up from 18°C to 32°C. The heat lost by the lead must be equal to the heat gained by the water:

$$(cm\ \Delta T)_{lead} = (cm\ \Delta T)_{water}$$

$$\left(0.031\,\frac{cal}{g\cdot C°}\right)(m)(63\ C°) = \left(1\,\frac{cal}{g\cdot C°}\right)(230\ g)(14\ C°)$$

which gives $m = 1,650$ g.

Sample Problem

7-6 To determine the temperature of a hot 1.8-kg iron ingot, it is dropped into 500 g of water at 20°C, and thermal equilibrium is reached at 89°C. If the specific heat of iron is known to be 0.115 cal/g·C°, what was the initial temperature of the iron?

Solution Heat lost by the iron equals heat gained by the water:

$$\left(0.115\,\frac{cal}{g\cdot C°}\right)(1,800\ g)(\Delta T) = \left(1\,\frac{cal}{g\cdot C°}\right)(500\ g)(69\ C°)$$

so $\Delta T = 167\ C°$. This is the amount that the temperature of the iron ingot dropped, so it started at 89°C + 167C° = 260°C.

Sample Problem

7-7 To determine the specific heat of a metal, 650 g of the metal are heated to 100°C and dropped into 240 g of water at 17°C. If the mixture reaches equilibrium at 38°C, what is the specific heat of the metal?

Solution The heat lost by the metal is equal to the heat gained by the water; both are equal to cm ΔT. Therefore:

$$\left(1\,\frac{cal}{g\cdot C°}\right)(240\ g)(21\ C°) = c_{metal}\ (650\ g)(62\ C°).$$

which gives $c_{metal} = 0.13$ cal/g·C°.

Sample Problem

7-8 What equilibrium temperature will be reached if you pour 450 g of hot water at 98°C into a 600-g aluminum pot at 20°C? The specific heat of aluminum is 0.22 cal/g·C°.

Solution We can write an equation stating that the heat lost by the water is equal to the heat gained by the aluminum:

$$(cm \, \Delta T)_{water} = (cm \, \Delta T)_{aluminum}$$

There is a difficulty. Since we do not know the final temperature, we cannot know ΔT. All we can do is put in the final temperature as our unknown; call it T. Then

$$\Delta T_{water} = 98°C - T, \text{ and } \Delta T_{aluminum} = T - 20°C$$

Our equation then becomes

$$\left(1\frac{cal}{g \cdot C°}\right)(450 \text{ g})(98°C - T) = \left(0.22\frac{cal}{g \cdot C°}\right)(600 \text{ g})(T - 20°C)$$

This simplifies to

$$44,100°C - 450T = 132T - 2,640°C$$

So $T = 80°C$.

Core Concept

The method of mixtures can be used to find many properties of a system; the calorie is the usual heat unit used in these problems.

Try This How much water at 20°C should you use to cool a 350-g iron pot from 195°C to 75°C? The specific heat of iron is 0.115 cal/g·C°.

7-6. Melting and Boiling

As we have seen, if you put a large amount of ice in warm water, heat will flow from the water to the ice until thermal equilibrium is reached at 0°C. The water cools down, but the ice does not warm up.

Our energy equation is in trouble. What happened to the heat that the water lost? As usual, we examine the situation to find out what change has taken place in the system. Easy: Some ice has melted. This implies that, at a temperature of 0°C, water contains more thermal energy than an equal mass of ice. In absorbing heat at 0°C, ice does not change temperature. Instead, it changes phase, from solid to liquid.

To investigate this phenomenon we might start with, say, 100 grams of very cold ice at −40°C. We will set it up in some sort of closed system and arrange to keep adding heat to it at a steady rate. Figure 7-3 is a graph of temperature as a function of time, showing what will happen. In the first minute, the ice warms up to 0°C. Then it starts to melt, and there is a mixture of ice and water at 0°C for the next 4 minutes. When all the ice has melted, the water begins to warm up, reaching 100°C in the next 5 minutes. Then it starts to boil, and as the water turns to steam, we will have a mixture of water and steam. The graph does not show it, but it would take 27 minutes for all the water to vaporize and the steam to begin to get hotter than 100°C.

Looking at the graph, we can calibrate it in terms of the rate at which heat is being added. It takes 5 minutes to raise the temperature of the water

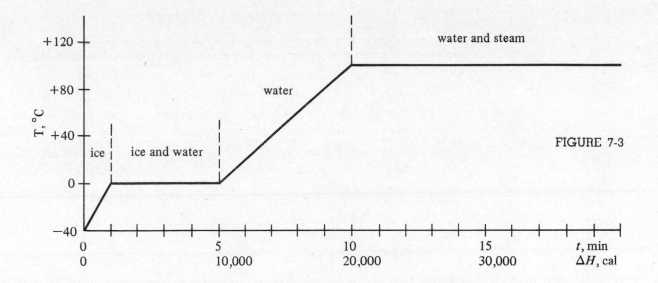

FIGURE 7-3

from 0°C to 100°C; Equation 7-4 tells us that this temperature increase for 100 grams of water calls for the addition of 10,000 cal. Therefore, we must be adding 2,000 cal of heat every minute. Now we can add another scale to the graph on the horizontal axis, showing the amount of heat that has been added to the system.

Between 1 and 5 minutes, we added 8,000 cal to our material. There was no temperature change; the only effect of this added heat was to melt the ice. Apparently, it takes 80 cal to melt each gram of ice, at 0°C. This quantity is called the *latent heat of fusion* of ice; it is 80 cal/g. If you want to find out how much energy it takes to melt any given amount of ice, use this equation:

$$\Delta H = mL \qquad \text{(Equation 7-6)}$$

See Sample Problems 7-9, 7-10, and 7-11.

Sample Problem

7-9 How much heat is needed to melt 300 g of ice at 0°C?

Solution From Equation 7-6,

$$\Delta H = mL = (300 \text{ g})\left(80\frac{\text{cal}}{\text{g}}\right) = 24,000 \text{ cal}$$

Sample Problem

7-10 You put ice cubes into 350 g of tap water at 18°C. How much ice melts?

Solution Heat lost by the water in coming to 0°C equals heat gained by the ice in melting:

$$(cm \ \Delta T)_{\text{water}} = (mL)_{\text{ice}}$$

$$\left(1\frac{\text{cal}}{\text{g·C}°}\right)(350 \text{ g})(18\text{C}°) = m\left(80\frac{\text{cal}}{\text{g}}\right)$$

So $m = 79$ g.

Sample Problem

7-11 You take 250 g of ice from the refrigerator at −8°C. How much heat does it absorb in warming up to room temperature at 22°C?

Solution First, the ice has to warm up to 0°C; the specific heat of ice is 0.55 cal/g·C°, so

$$\Delta H = cm\ \Delta T = \left(0.55 \frac{\text{cal}}{\text{g·C°}}\right)(250 \text{ g})(8\text{C°})$$

$$\Delta H = 1{,}100 \text{ cal}$$

Next, it has to melt, absorbing its heat of fusion:

$$\Delta H = mL = (250 \text{ g})\left(80\frac{\text{cal}}{\text{g}}\right)$$

$$\Delta H = 20{,}000 \text{ cal}$$

Finally, it must warm up to 22°C. Now it is water, so

$$\Delta H = cm\ \Delta T = \left(1\frac{\text{cal}}{\text{g·C°}}\right)(250 \text{ g})(22\text{C°})$$

$$\Delta H = 5{,}500 \text{ cal}$$

The total heat absorbed is the sum of these three quantities, or 26,600 cal = 26.6 kcal.

Ice is an excellent way to cool things down because it does not warm up as it absorbs heat. It will keep on taking heat from its surroundings until the things around it have reached the melting point of the ice, 0°C.

The process also works in reverse. To convert water to ice, you have to remove 80 cal for each gram that is to be frozen. That is what happens in the freezing compartment of your refrigerator. Your freezer, however, does not stop there; it continues removing heat from the ice until its temperature drops to about −5°C. To find out how much heat has to be removed, you would use Equation 7-4, with a value of 0.55 cal/g·C°.

Figure 7-3 shows that something similar happens at the second phase change, from water to steam at 100°C. Again, there is no increase in temperature as long as there is still some liquid water present. The *latent heat of vaporization* of water is very high: 540 cal/g. You can use Equation 7-6 to calculate heat changes in *any* phase change; for this one, the *L* stands for latent heat of vaporization. See Sample Problems 7-12 and 7-13.

Sample Problem

7-12 How much heat must be added to 180 g of water at 20°C to convert it into steam at 100°C?

Solution First, you have to heat the water to 100°C:

$$\Delta H = cm\ \Delta T = \left(1\frac{\text{cal}}{\text{g·C°}}\right)(180 \text{ g})(80\text{C°}) = 14{,}400 \text{ cal}$$

Then, you have to vaporize it:

$$\Delta H = mL = (180 \text{ g})\left(540\frac{\text{cal}}{\text{g}}\right) = 97{,}200 \text{ cal}$$

for a total of 112,000 cal, or 112 kcal.

Sample Problem

7-13 What final temperature is reached if 30 g of steam at 100°C is bubbled into 200 g of water at 12°C? All the steam condenses in the water.

Solution The steam must condense, giving up its latent heat of vaporization. Then it is water; now it cools down to the final temperature T, giving up some more heat. All this released heat goes into warming up the water:

$$(mL + cm\ \Delta T)_{steam} = (cm\ \Delta T)_{water}$$

$$(30\text{ g})\left(540\frac{cal}{g}\right) + \left(1\frac{cal}{g \cdot C°}\right)(30\text{ g})(100°C - T)\ = \left(1\frac{cal}{g \cdot C°}\right)(200\text{ g})(T - 12°C)$$

$$16,200\text{ cal} + 3,000\text{ cal} - \frac{30\ T\text{ cal}}{C°} = \frac{200T\text{ cal}}{C°} - 2,400\text{ cal}$$

$$T = 94°C$$

It is the extremely high value of the latent heat of vaporization that makes steam heat practical. A boiler adds 540 cal of heat every time it vaporizes a gram of water. In a radiator, the steam condenses back to liquid water, releasing all that latent heat to warm the room.

Core Concept

Change of phase is accompanied by a change in thermal energy with no change in temperature:

$$\Delta H = mL$$

Try This A 600-g iron ball (specific heat 0.115 cal/g·C°) at 240°C is dropped onto a very large cake of ice. how much ice melts?

7-7. A Better Temperature Scale

A bucket of water is at a temperature of 10°C. If its temperature doubles, what will the temperature be? Careful, now, the answer is *not* 20°C!

Let's see why not. Suppose I decide to do my calculation using the more familiar Fahrenheit scale. I find that 10°C = 50°F and double it to get 100°F, which is 38°C, not 20°C. What happened? The problem is that the two scales have different zeroes. On either scale, the zero point is not a real zero but a value selected arbitrarily, for convenience. That is why we have negative temperatures.

What we need is a temperature scale with a real zero—a scale that starts at the lowest possible temperature. Only then will it be possible to answer the question, how much is double a given temperature.

There is such a scale. It is defined on the basis of a relationship that has been known for a couple of centuries: When an enclosed sample of a gas is heated, its pressure rises. Further, the relationship between temperature and pressure is just about the same for all gases, as long as the pressure is fairly low and the gas is not too cold.

This is how the *Kelvin*, or *absolute*, temperature scale is defined: Take a small bottle containing hydrogen gas at low pressure; attach a pressure gauge to it. Immerse it in a mixture of ice and water and note the pressure. That pressure corresponds to a temperature of 273.15 kelvins (K). If you now immerse your constant-volume hydrogen thermometer in anything else (see Figure 7-4), the pressure will change. We define the Kelvin temperature this way:

$$T_K \propto p \qquad \text{(Equation 7-7a)}$$

Using this relationship, you can use your thermometer to find the Kelvin temperature of anything else. See Sample Problem 7-14.

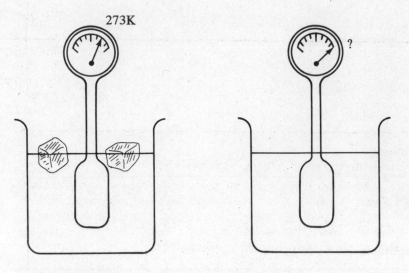

FIGURE 7-4

Sample Problem

7-14 The gauge of Figure 7.4 (calibrated to read absolute pressure) reads 1,200 mb when the thermometer is in ice water and 1,750 mb when it is in the other beaker. What is the temperature of the liquid in the other beaker?

Solution From Equation 7-7a, T/p is a constant, so

$$\frac{T_1}{p_1} = \frac{T_2}{p_2}$$

$$\frac{273 \text{ K}}{1,200 \text{ mb}} = \frac{T}{1,750 \text{ mb}}$$

$$T = 398 \text{ K}$$

That strange value of 273.15 K as the melting point of ice was chosen for a particular reason. It makes the kelvin equal to a Celsius degree. Thus, by definition, 0°C = 273.15 K. Since a kelvin equals a Celsius degree, the boiling point of water is just 100 K higher, or 373.15 K. In fact, you can always convert from degrees Celsius to kelvins this way:

$$T_K = T_C + 273.15 \text{ K} \qquad \textbf{(Equation 7-7b)}$$

Sample Problem

7-15 What is the Celsius temperature of the liquid in Sample Problem 7-14?

Solution

$$T_K = T_C + 273 \text{ K}$$

$$398 \text{ K} = 273 \text{ K} + T_C$$

$$T_C = 125°C$$

The Kelvin scale has a natural zero; there are no temperatures below 0 K. In fact, it is theoretically impossible ever to get anything that cold. The lowest temperatures yet achieved are under 0.001 K, but zero is unattainable. See Figure 7-5 for a comparison of the two scales.

Since the Kelvin scale has a natural zero, it can be used to calculate ratios between temperatures. Sample Problem 7-16 shows the correct way to find twice 10°C.

```
 0                                      273  294              373   K
 |_____|____|_____|_____
-273                                      0    21             100  °C
                                          ice  room           steam
                                          point                point
```

FIGURE 7-5

Sample Problem

7-16 How much, in degrees Celsius, is twice 10°C?

Solution   Temperature ratios must be calculated in kelvins, not degrees Celsius.

$$T_K = T_C + 273\ \text{K} = 10°\text{C} + 273\ \text{K} = 283\ \text{K}$$

Twice this is 566 K; to put it into degrees Celsius:

$$566\ \text{K} = T_C + 273\ \text{K}$$

$$T_C = 293°\text{C}$$

Core Concept

The pressure of an ideal gas at constant volume is proportional to its Kelvin temperature, which is equal to the Celsius temperature plus 273.15 K:

$$T_K \propto p; \qquad T_K = T_C + 273.15\ \text{K}$$

Try This   A gas in a bottle is at a temperature of 40°C. If it is cooled down until its pressure drops to half, what is its Celsius temperature?

7-8. Gases Expand

Consider the cylinder containing a sample of gas shown in Figure 7-6. The piston sits on top of the gas, adding its weight to the pressure of the atmosphere pushing down on the gas. Since the piston is free to move, the

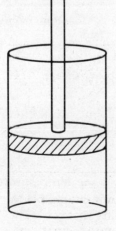 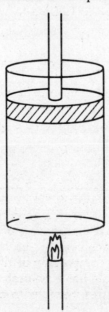

FIGURE 7-6

pressure on the gas will not change. Now we light a fire under the cylinder.

Like nearly anything else, a gas expands when heated. Gases are different from other materials, however, in a very important way. At constant pressure, the volume of a gas is approximately proportional to its Kelvin temperature:

$$V \propto T_K$$

Thus, if pressure is kept constant, the volume of a gas will shrink to half if its Kelvin temperature is halved.

If the pressure and volume of a given sample of gas are both changing, we can write an expression combining this equation with Equation 7-7a:

$$pV \propto T_K \qquad \text{(Equation 7-8)}$$

This equation is known as the *ideal gas law*, and Sample Problems 7-17 and 7-18 show how it can be applied. the law is very simple and easy to use. Unfortunately, it is only an approximation. An ideal gas is one that obeys this law, but no real gas fits it exactly. The fit is best when the gas is at high temperature and low pressure. Cooling a gas down toward zero kelvins will not make it shrink to nothing; it will liquefy long before that. And high pressure makes it liquefy at even higher temperatures. Further, the ideal gas law becomes an ever-poorer approximation as temperature drops and pressure rises. Still, it is a very good, useful approximation at ordinary temperatures and pressures.

Sample Problem

7-17 A cylinder contains 380 cm^3 of gas at 1.5 atm of pressure and 20°C. If it is compressed to a volume of 120 cm^3 and its pressure goes up to 4.6 atm, what will its temperature be?

Solution From Equation 7-8,

$$\frac{p_1 V_1}{T_1} = \frac{p_2 V_2}{T_2}$$

Warning! All temperatures must be in kelvins! So

$$\frac{(1.5 \text{ atm})(380 \text{ cm}^3)}{293 \text{ K}} = \frac{(4.6 \text{ atm})(120 \text{ cm}^3)}{T_2}$$

and $T_2 = 284$ K $= 11$°C.

Sample Problem

7-18 At sea level, 440 liters of helium are put into a balloon with the temperature at 28°C. if the balloon rises to a point where the pressure is 35 kPa and the temperature is −12°C, what volume does the helium occupy?

Solution The pressure at sea level is 100 kPa, and the temperature (from Equation 7-7b) is 301 K. At high altitude, the temperature is 261 K. Therefore, from Equation 7-8,

$$\frac{(100 \text{ kPa})(440 \text{ l})}{301 \text{ K}} = \frac{(35 \text{ kPa})(V)}{261 \text{ K}}$$

which gives $V = 1,100$ liters.

In applying the ideal gas law, there are three precautions you must be sure to observe: (1) Use only Kelvin temperatures. (2) The law works only as long as you do not add or remove any gas from the sample; you cannot use it to calculate temperatures and pressures in a bicycle tire when you blow it up. (3) Pressure must be *absolute* pressure. Many pressure gauges read gauge pressure—the difference between the pressure inside and the pressure outside. If your automobile tire is flat, your pressure gauge will read zero, even though the pressure in it is surely 1 atmosphere. Gauge pressure is 1 atmosphere less than absolute pressure. See Sample Problem 7-19.

Sample Problem

7-19 A 600-cm^3 sample of gas is in a cylinder under a gauge pressure of 1.8 atm. What will the pressure gauge read if the sample is compressed to 180 cm^3 at constant temperature?

Solution   With the temperature constant, pV is constant, so

$$p_1 V_1 = p_2 V_2$$

But the pressure in the sample starts out at 2.8 atm, 1 atm higher than the gauge reads. Therefore

$$(2.8 \text{ atm})(600 \text{ cm}^3) = p_2(180 \text{ cm}^3)$$

giving an absolute pressure $p_2 = 9.3$ atm. The gauge will read 1 atm less than this, or 8.3 atm.

For an ideal gas, Kelvin temperature is proportional to the product of pressure and volume

$$pV \propto T_K$$

Try This   A cylinder contains 300 cm^3 of gas at room temperature of 20°C and 1 atm of pressure—that is, it is open to the outside air. If it is sealed and plunged into boiling water, while a piston compresses the gas into a volume of 100 cm^3, what will a pressure gauge indicate for the pressure of the gas in the cylinder?

7-9. The Size of the Sample

We can write the ideal gas law, Equation 7-8, this way:

$$\frac{pV}{T_K} = \text{constant}$$

where the constant depends, somehow, on the size of the sample of gas. If we add some gas, or remove some, the value of pV/T_K changes. Let's take a look at that constant.

Experimental chemistry showed long ago that, if you take equal volumes of gas at the same temperature and pressure, the two samples will contain the same number of molecules, regardless of what kind of gas they are. (This is called *Avogadro's law*.) In other words, the value of the constant in the equation above depends *only* on the number of molecules in the sample. Then we can write the ideal gas law this way:

$$pV = nkT_K \qquad \text{(Equation 7-9)}$$

where n is the number of molecules in the sample and k is a universal constant of nature. It does not depend on the kind of gas in the sample; for a sample with a given number of molecules, pressure is determined only by the volume and temperature. The constant k is called the *Boltzmann constant*, and its value is

$$k = 1.38 \times 10^{-23} \frac{J}{molecule \cdot K}$$

It is often messy to apply this law, because it will come out right only if all quantities are expressed in SI units: newtons per square meter (pascals) for pressure, cubic meters for volume, and kelvins for temperature. See Sample Problems 7-20 and 7-21.

Sample Problem

7-20 How many molecules (mcl) are there in 2.0 liters of gas at 75°C if the absolute pressure is 3.0 atm?

Solution We need SI units throughout. The temperature is 75°C + 273 K = 348 K. Since there are 10^3 liters in 1 m^3, the volume is 2.0×10^{-3} m^3. And 1 atm is very nearly 1.0×10^5 Pa, so the pressure is 3.0×10^5 Pa. Now solve Equation 7-9 for n:

$$n = \frac{pV}{kT_K}$$

$$n = \frac{(3.0 \times 10^5 \text{ Pa})(2.0 \times 10^{-3} \text{ m}^3)}{(1.38 \times 10^{-23} \text{ J/mcl} \cdot \text{K})(348 \text{ K})}$$

$$n = 1.25 \times 10^{23} \text{ molecules}$$

Sample Problem

7-21 A 1.0-liter bottle at 20°C contains 1.5×10^{22} molecules of oxygen. What is the pressure of the gas?

Solution From Equation 7-9,

$$p = \frac{nkT}{V} = \frac{(1.5 \times 10^{22} \text{ mcl})(1.38 \times 10^{-23} \text{ J/mcl} \cdot \text{K})(293 \text{ K})}{1.0 \times 10^{-3} \text{ m}^3}$$

$$p = 6.1 \times 10^4 \text{ Pa}$$

Core Concept

For an ideal gas, the quantity pV/T depends only on the number of molecules in the sample:

$$pV = nkT_K$$

Try This What pressure is exerted by a sample of hydrogen containing 5.0×10^{22} molecules in a 0.50-liter bottle at 30°C?

7-10. The Why of the Gas Law

Theoretical physicists puzzled for years over the gas law, looking for a way to explain it. One model has shown itself to be extremely successful, both intuitively and mathematically. It is called the *kinetic molecular theory of gases*.

The model assumes the following: (1) A gas consists of a very large number of extremely small molecules, with great empty spaces between them; (2) the molecules are moving at random in all directions; (3) they interact with each other only through perfectly elastic collisions, which means that the total kinetic energy is conserved through the collision; (4) they exert pressure on the walls of their container by making elastic collisions with them.

This model accounts nicely for the relationship between the pressure and the number of molecules in the sample. Surely, if you double the number of molecules, there will be twice as many collisions with the walls, so the pressure will double. Then

$$p \propto n$$

How does volume affect the pressure? If you double the volume, every molecule will have twice as far to travel between collisions with the wall and so will collide half as often, so the pressure on the walls will drop to half. Therefore,

$$p \propto \frac{1}{V}$$

The speed of the molecules influences the pressure. The faster any molecule is moving, the more often it will hit any given wall; as it bounces from wall to wall, it returns to any given wall twice as often if it is going twice as fast. The molecules are traveling at various speeds, and the pressure is proportional to the average speed:

$$p \propto v_{av}$$

And one more. A faster-moving molecule hits the wall harder than a slower one; a molecule with a larger mass hits it harder than one with less mass. It can be proved from this that the impulse the molecules exert against the wall is proportional to their momenta (that's the plural of momentum). Again, we need the average, so

$$p \propto (mv)_{av}$$

Now we can combine all these proportionalities:

$$p \propto (n)\left(\frac{1}{V}\right)(v_{av})(mv)_{av}$$

Detailed mathematical analysis, allowing for the fact that the molecules move in three dimensions, tells us that the proportionality constant in this relationship is $\frac{1}{3}$. Then, finally, the equation becomes

$$p = \frac{nmv_{av}^2}{3V}$$

which can be written

$$pV = \frac{2}{3}n\left(\frac{mv_{av}^2}{2}\right)$$

Now, the average kinetic energy of the molecules is $(mv_{av}^2)/2$, so our equation becomes

$$pV = \frac{2}{3}n\bar{E}_{kin}$$

where the bar over the E means "average."

Comparing this equation with Equation 7-9 suddenly gives us a whole new insight into the meaning of temperature. We find that

$$\frac{2}{3}\bar{E}_{kin} = kT$$

or, in its more usual form,

$$\bar{E}_{kin} = \frac{3}{2}kT \qquad \textbf{(Equation 7-10)}$$

which says that temperature depends *only* on the average kinetic energy of the molecules. See Sample Problem 7-22.

Sample Problem

7-22 (a) At what temperature is the average kinetic energy of the molecules of oxygen equal to 6.8×10^{-22} J? (b) What is the temperature of hydrogen molecules with the same average kinetic energy?

Solution (a) From Equation 7-10,

$$T = \frac{2}{3}\frac{\bar{E}_{kin}}{k} = \frac{2(6.8 \times 10^{-22} \text{ J})}{3(1.38 \times 10^{-23} \text{ J/mcl} \cdot \text{K})} = 33 \text{ K}$$

(b) The same.

Does this relationship apply only to gases? By no means. Consider what happens when these gas molecules strike the walls of their container. They must surely set the molecules of the walls into motion. When equilibrium is reached, the molecules of the walls will be at the same temperature as the gas molecules—and they will also have the same average kinetic energy. Of course, they will not be bouncing around, like gas molecules; they will simply be oscillating around some central position. See Sample Problem 7-23.

Sample Problem

7-23 What is the average kinetic energy of the molecules of *anything* at a room temperature of 300 K?

$$\bar{E}_{kin} = \frac{3}{2}\left(1.38 \times 10^{-23}\frac{\text{J}}{\text{mcl} \cdot \text{K}}\right)(300 \text{ K})$$

$$\bar{E}_{kin} = 6.2 \times 10^{-21} \text{ J/mcl}$$

Core Concept

Temperature is the average random kinetic energy of molecules:

$$\bar{E}_{kin} = \frac{3}{2}kT$$

Try This What is the temperature of a gas if the average random kinetic energy of its molecules is 3.5×10^{-21} J?

7-11. Molecular Energy

Armed with this concept of temperature, we can do a lot of theoretical physics. If we know the mass of the molecules of a gas, for example, we can find out how fast they are going at any given temperature, as in Sample Problem 7-24. The only difficulty is in expressing the mass in an SI unit. The usual unit for molecular mass is the dalton (Dl); it takes 6.02×10^{26} Dl to equal one kilogram.

Sample Problem

7-24 What is the average speed of the oxygen molecules in a room at a temperature of 300 K? The mass of an oxygen molecule is 32 Dl.

Solution From Equations 7-10 and 6-2,

$$\bar{E}_{kin} = \frac{3}{2}kT = \frac{1}{2}mv^2$$

Solving for v gives

$$v = \sqrt{\frac{3kT}{m}}$$

We need the value of m in SI units:

$$32 \text{ Dl} \times \frac{\text{kg}}{6.02 \times 10^{26} \text{ Dl}} = 5.3 \times 10^{-26} \text{ kg}$$

Now solve for v:

$$v = \sqrt{\frac{3(1.38 \times 10^{-23} \text{ J/mcl} \cdot \text{K})(300 \text{ K})}{5.3 \times 10^{-26} \text{ kg}}}$$

$$v = 480 \text{ m/s}$$

Do the units check out right?

Now we can get some insight into this mysterious thing called latent heat. If water and ice can both exist at a temperature of 0°C, their molecules must have the same average kinetic energy at the melting point. The extra energy of the water, the latent heat of fusion, is potential energy—which has no effect on temperature. It takes work to separate a water molecule from an ice crystal, and the work done is the increase in its potential energy. See Sample Problem 7-25.

Sample Problem

7-25 How much work must be done to remove a single water molecule, whose mass is 18 Dl, from its place in an ice crystal?

Solution It takes 80 cal/g. You can convert this into calories per dalton just by using ordinary conversion factors. Then, multiply by 18 Dl per molecule to get the energy per molecule:

$$\left(80 \frac{\text{cal}}{\text{g}}\right)\left(\frac{\text{kg}}{6.02 \times 10^{26} \text{ Dl}}\right)\left(\frac{10^3 \text{ g}}{\text{kg}}\right)\left(\frac{18 \text{ Dl}}{\text{mcl}}\right)$$

$$= 2.4 \times 10^{-21} \text{ cal/mlc}$$

If you want the answer in joules, just multiply by 4.19 J/cal.

Just what happens when the kinetic energy of an object is converted to internal energy? Suppose a blob of metal falls a substantial distance. When it strikes the ground all of its kinetic energy is converted to thermal energy, and its temperature rises. But all the molecules had plenty of kinetic energy before the object hit; after all, they were all moving at the speed of the object. However, that organized kind of kinetic energy of molecules does not push thermometers up. When the object strikes the ground, all the kinetic energy becomes randomized. The molecules are moving just as fast, but now in random directions instead of all in the same direction. It is only the *random* motion that counts in calculating temperature.

The calculation of molecular energy gives us a way to find the specific heat of certain materials. Noble gases, like helium, neon, and argon, have molecules consisting of a single atom each. In that case, *all* the energy added becomes random kinetic energy of molecules; the gas has no internal potential energy. With Equation 7-10, we can relate the added energy to the temperature increase and calculate the specific heat. See Sample Problem 7-26.

Sample Problem

7-26 What is the specific heat of helium, a noble gas whose molecular mass is 4 Dl?

Solution   From Equation 7-4, the specific heat is

$$c = \frac{\Delta H}{m \, \Delta T}$$

Since all the added heat adds to the temperature, Equation 7-10 tells us the relationship between heat added and temperature change:

$$\frac{\Delta H}{\Delta T} = \frac{3}{2}k$$

Combining these two equations gives

$$c = \frac{3}{2} \times \frac{k}{m}$$

All we have to do is substitute in the mass (4 Dl) and convert units to get the answer in grams:

$$c = \frac{3}{2} \times \frac{1.38 \times 10^{-23} \text{ J/mcl} \cdot \text{K}}{4 \text{ Dl}} \times \frac{6.02 \times 10^{26} \text{ Dl}}{10^3 \text{ g}}$$

$$c = 3.1 \text{ J/g} \cdot \text{K} = 0.74 \text{ cal/g} \cdot \text{C}°$$

In other kinds of materials, only part of the energy added becomes the random kinetic energy of the molecules. Measurement of specific heat can then give clues as to how the added energy is partitioned. In metals, it turns out that only about half of the added energy makes the molecules move faster. The other half increases potential energy; the molecules move farther apart, and the metal expands.

Core Concept

Kinetic molecular theory provides insights into the molecular behavior of materials.

Try This   What is the average speed of the hydrogen atoms in a hydrogen–oxygen mixture if the oxygen molecules are averaging 2,400 m/s? Atomic masses: oxygen, 32 Dl; hydrogen, 2 Dl.

Summary Quiz

Supply the missing word or phrase:

1. If two objects are placed together and neither one undergoes any change, they have the same _____ .
2. Energy traveling spontaneously from a warm object to a cold one is called _____ .
3. No heat will pass from one object to another in contact with it if they are in a state of _____ .
4. Through _____ , heat travels best by radiation.
5. Through _____ , heat travels best by conduction.
6. Convection occurs in fluids that are _____ at the bottom.
7. A mercury–glass thermometer works on the principle that, when heated, mercury _____ more than glass.
8. The definition of 0°C is the equilibrium temperature between water and _____ .
9. For a given object, the change in its temperature is proportional to the _____ added or removed.
10. The heat needed to raise the temperature of a gram of a substance by 1 C° is called the _____ of the substance.
11. The specific heat of _____ is 1 cal/g · C°.
12. It takes _____ joules to equal 1 calorie.
13. Latent heat is the energy per unit mass absorbed or released in the process of changing _____ .
14. It takes _____ calories to convert a gram of water at 100°C into steam at 100°C.
15. At constant volume, the _____ of an ideal gas is proportional to its Kelvin temperature.
16. To find Kelvin temperature, add _____ to Celsius temperature.
17. The SI temperature unit, equivalent to the Celsius degree, is the _____ .
18. Gauge pressure is _____ less than absolute pressure.
19. For an ideal gas, the product of _____ and _____ is proportional to the Kelvin temperature.
20. The ideal gas law is a good approximation at _____ pressure and _____ temperature.
21. The value of pV/T_K, for any sample of gas, depends only on the _____ of the sample.
22. The average random kinetic energy of molecules is called _____ .
23. The kinetic molecular theory of gases assumes that all collisions between gas molecules are completely _____ .
24. Pressure of a gas is produced by the impact of _____ as they strike the walls of the container.
25. Latent heats represent a change in the _____ energy of molecules.
26. The _____ heat of a material tells us how heat added to its molecules is partitioned among the different kinds of molecular energy.

Problems

1. In a mercury–glass thermometer, the length of the mercury column is 3.5 cm when the bulb is in ice water, 19.0 cm when it is in boiling water, and 12.2 cm when it is in warm water. What is the temperature of the warm water?
2. A lead bullet (specific heat = 0.031 cal/g · K) going 460 m/s strikes a tree. If all the heat generated remains in the bullet, how much does its temperature rise?

3. If a waterfall is 110 m high, how much warmer is the water at the bottom than at the top?
4. How much heat does it take to heat a 580-g aluminum pot (specific heat = 0.22 cal/g · C°) from 20°C to 180°C?
5. To measure the specific heat of an unknown metal, 760 g of it is heated to 95°C and dumped into 300 g of cold water at 5°C. If the mixture reaches equilibrium at 28°C, what is the specific heat of the metal?
6. To heat 220 g of water from 20°C to 65°C, an iron ball (specific heat = 0.115 cal/g · C°) at 140°C is dropped into the water. What mass of iron is needed?
7. How much heat is needed to convert 200 g of ice at 0°C to steam at 100°C?
8. A batch of hot copper shot with a mass of 350 g (specific heat = 0.091 cal/g · C°) is dropped onto a cake of ice at 0°C, and 75 g of ice melts. What was the temperature of the copper?
9. What equilibrium temperature will be reached if 300 g of water at 15°C is poured into a hot 580-g copper vessel at 145°C. (Specific heat of copper = 0.091 cal/g · C°).
10. A constant-volume hydrogen thermometer registers an absolute pressure of 950 mb when in ice water and 620 mb in a freezing mixture. What is the temperature of the freezing mixture in (a) K; and (b) °C?
11. If a constant-volume hydrogen thermometer registers 1,400 mb in ice water, what will it read in boiling water?
12. Atmospheric air is admitted to a cylinder with a volume of 2.5 liters and sealed in. What will the gauge pressure be if a piston then compresses the air into a volume of 1.0 liter at constant temperature?
13. At the beginning of the compression stroke of a Diesel engine, the cylinder contains 600 cm^3 of air at atmospheric pressure and a temperature of 22°C. At the end of the stroke, the air has been compressed to 50 cm^3 and the absolute pressure is 40 atm. What is the temperature?
14. A liter of helium under a gauge pressure of 2 atm and a temperature of 22°C is heated until both the pressure and the volume have doubled. What is its temperature?
15. How many molecules of air are there in a 1-liter flask at 20°C if the pressure in it is 0.3 atm? (1 atm = 1 × 10^5 N/m^2).
16. A cylinder contains 500 cm^3 of air at 1 atm pressure and 20°C. Air is pumped in to double the mass of air in the cylinder, and the piston is pushed in until the volume is reduced to 200 cm^3. If the temperature is kept constant, what is the pressure in the cylinder?
17. What is the average kinetic energy of the molecules of boiling water?
18. In a mixture of hydrogen (molecular mass = 2 Dl) and oxygen (molecular mass 32 Dl), what is the ratio of the speed of the hydrogen molecules to the speed of the oxygen molecules?

CHAPTER 8

Making Waves

8-1. What Is a Wave?

As you sit on the shore and watch the waves roll in, you can see them pounding against the rocks and roiling the sand. They carry a lot of energy, transporting it from far out at sea. Yet a boat 100 yards from shore is not carried in with the waves. It just bobs up and down. Waves keep coming in to shore, but the water merely moves up and down, or back and forth. The waves deliver energy, not water, to the shore.

To create a simplified model of wave motion, grasp one end of a very long rope and shake it up and down, as in Figure 8-1. If there is a marker

FIGURE 8-1

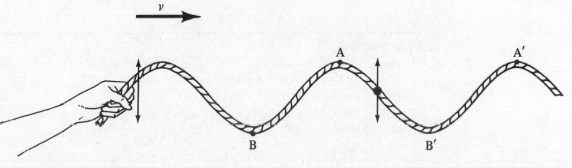

153

on the rope, some distance from your hand, you will see it move up and down exactly as your hand does. The marker follows the motion of your hand precisely, but with a time delay. Every point on the rope is pulled by the one behind it, and every point pulls on the next point in line. Thus energy is transmitted from one end of the rope to the other, but the points on the rope merely move up and down.

To talk about waves, we will need appropriate words to describe them. Here are some:

The *period* (T) of a wave is the length of time it takes any point to go through one full cycle, back and forth. It will ordinarily be expressed in seconds. Often, it is more convenient to express this time relationship as the *frequency* (f) of the wave, the number of cycles per second. This is simply the reciprocal of the period; if the period is $\frac{1}{10}$ of a second, the frequency is 10 cycles per second:

$$f = \frac{1}{T} \qquad \text{(Equation 8-1)}$$

The standard unit of frequency is the hertz (Hz), equal to one cycle per second. See Sample Problem 8-1. Since each point on the rope is pulled by the one before it, all points must have the same frequency.

Sample Problem

8-1 If a point on a wave makes a full cycle in 0.2 s, what is the frequency of the wave?

Solution The period (T) is 0.2 s; so

$$f = \frac{1}{T} = \frac{1}{0.2 \text{ s}} = \frac{5}{\text{s}}$$

which is read "5 cycles per second," or 5 Hz.

Transmission of energy down the rope takes time, so the oscillation of each point lags a little behind that of the one in front of it. We express this lag as a difference in *phase* of the points on the rope. If you go far enough, you will find a point that moves exactly the way your hand does; the lag is one full cycle. Points that move synchronously, reaching their peaks simultaneously, are said to be *in phase*. If two points are not in phase, they will be out of phase by some definite fraction of a cycle. This is expressed in degrees. If they are half a cycle apart, one point is at its peak when the other is at its low extreme. Then they will be said to be 180° out of phase; a quarter-cycle difference is a phase difference of 90°, and so on. See Sample Problems 8-2 and 8-3.

Sample Problem

8-2 If the period of a wave is 0.20 s, what is the phase difference in the vibration of two points if one reaches its maximum displacement 0.04 s before the other?

Solution The fraction of a cycle separating the two vibrations is 0.04/0.20; since a full cycle is 360°, the phase difference is

$$\frac{0.04}{0.20} \times 360° = 72°$$

Sample Problem

8-3 What is the frequency of a wave if two points 90° out of phase reach their peaks 0.10 s apart?

Solution The time difference of 0.10 s represents a phase difference of 90°/360°, or just one-fourth of a cycle. Therefore a full cycle takes 0.40 s, which is the period of the wave. Then, by Equation 8-1,

$$f = \frac{1}{T} = \frac{1}{0.40 \text{ s}} = 2.5 \text{ Hz}$$

When you shake that wave in the rope, you may vibrate your hand over a large distance or a small distance. In the wave, the distance between the equilibrium position of a point on the rope and its greatest displacement from that point, as it vibrates, is called the *amplitude* of the vibration. The greater the amplitude of a wave, the more energy it carries. The energy, in fact, is proportional to the square of the amplitude. In a perfectly elastic, ideal medium, the amplitude of the wave will not change as it travels. In any real medium, some of the energy will be converted to thermal energy, and the amplitude diminishes with distance from the source. This reduction in wave energy as the wave travels is called *damping*.

Core Concept

Waves are described in terms of period, frequency, amplitude, phase, energy, and damping:

$$f = \frac{1}{T}$$

Try This What is the phase difference between two points if they reach their peak displacements 0.05 s apart when the frequency is 7.5 Hz?

8-2. Traveling Waves

Note the two points marked A and A' in Figure 8-1. At the moment the picture was made, both were at the peak of a crest. Therefore, they must be in phase with each other. Similarly, the points B and B' are in phase, since both are at the bottom of a trough at the same moment. The distance between A and A', or between B and B', or between any other pair of points in phase, with no other such points in between them, is called the *wavelength* of the wave.

As the crest travels from A to A', the point A' goes through one full cycle, from crest to trough and back to crest again. This gives us an important relationship between period and wavelength: Since the wave travels one wavelength (λ, the Greek letter lambda) in the time of one period, the speed of the wave (v) is given as

$$v = \frac{\lambda}{T}$$

This can be put into its more usual form by applying Equation 8-1:

$$v = f\lambda \qquad \text{(Equation 8-2)}$$

See Sample Problems 8-4 and 8-5 for applications.

Sample Problem

8-4 What will be the wavelength of the wave in a rope if its speed in that rope is 12 m/s and the end of the rope is shaken with a frequency of 6.0 Hz?

Solution From Equation 8-2,

$$\lambda = \frac{v}{f} = \frac{12 \text{ m/s}}{6.0/\text{s}} = 2.0 \text{ m}$$

Note that Hz = 1/s.

Sample Problem

8-5 What is the speed of the wave in a rope if two points out of phase by 180° are 35 cm apart and the frequency is 5.0 Hz?

Solution Two points out of phase by 180° are half a wavelength apart, so the wavelength is 70 cm. Then, from Equation 8-2,

$$v = f\lambda = (5.0 \text{ Hz})(0.70 \text{ m}) = 3.5 \text{ m/s}$$

In the wave we have been looking at so far, the particles of the rope are moving at right angles to the rope, while the wave travels along the rope. Any wave in which the particle motion is perpendicular to the direction of travel of the wave is called a *transverse* wave. The wave in the rope is transverse whether the rope is shaken up and down, from side to side, or in any other direction perpendicular to the direction in which the wave is traveling.

Figure 8-2 illustrates another kind of wave. The medium in which the wave travels is a long, stretched spring. The hand that makes the waves is moving back and forth, parallel to the length of the spring. When the hand moves to the right, it creates a *compression;* moving to the left, it makes a *rarefaction.* The separate particles of the spring vibrate back and forth, parallel to the velocity of the wave, as the compressions and rarefactions follow each other down the spring. This is called a *longitudinal* wave.

FIGURE 8-2

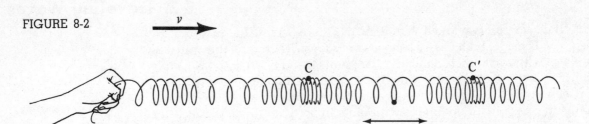

In a longitudinal wave, two points of greatest compression will be in phase with each other, and the distance between them is a wavelength. C and C' are a wavelength apart.

In either kind of wave, the velocity of the wave depends primarily on the nature of the medium—the rope or the spring. In some kinds of waves, velocity increases a little as the frequency rises, but it is usually a small effect. Also, the amplitude of the wave may affect the speed slightly, but usually not much. Velocity depends on the kind of medium, and frequency depends on the nature of the disturbance; together they determine the wavelength according to Equation 8-2.

In any wave, transverse or longitudinal, velocity depends primarily on the nature of the medium and is equal to the product of frequency and wave-length:

$$v = f\lambda$$

Try This What is the velocity of a wave if the frequency is 6.0 Hz and two points 90° out of phase with each other are 25 cm apart?

8-3. Sound Waves

Sound travels through space as a wave, transmitted by vibrations of air molecules. To see how this works, look at Figure 8-3, which shows successive stages in the vibration of a tuning fork.

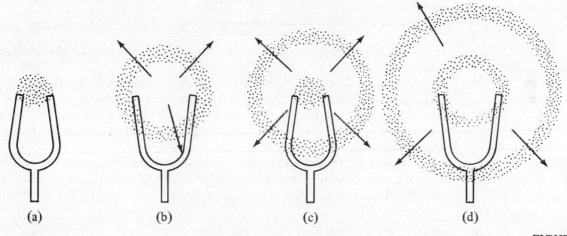

(a)　　　　(b)　　　　(c)　　　　(d)

FIGURE 8-3

At (a) the two prongs of the fork have moved closer to each other, creating a region of compressed air between them. Since the pressure here is higher than it is outside this region, the air in the compression zone expands outward. Meanwhile, the tuning fork prongs move farther apart, as at (b). This creates a low-pressure region, a rarefaction, between the prongs. The rarefaction zone follows the compression zone out from the tuning fork—and the process repeats itself as long as the fork continues to vibrate. Thus, the compression and rarefaction zones follow each other out, spreading radially.

When a wave consists of zones of compression and rarefaction, we know that the particles are vibrating parallel to the direction in which the wave travels. In this case, the direction of travel at any point is directly away from the source—radially outward. The wave is longitudinal.

The speed with which a sound wave travels in air depends only on the temperature. If the piccolo and the bassoon are playing together in the orchestra pit, the sounds will arrive at your ear together, no matter how far you are from the stage; frequency does not affect the wave velocity. But waves do travel faster in warmer air. The speed at any temperature can be found from this equation:

$$v_{air} = 331\frac{m}{s} + \left(0.6\frac{m}{s}\right)T_C \qquad \text{(Equation 8-3)}$$

where T_C is the Celsius temperature. See Sample Problem 8-6.

Sample Problem

8-6 How fast will a wave travel in air at a temperature of 15°C?

Solution   From Equation 8-3,

$$v_{air} = 331\frac{m}{s} + \left(0.6\frac{m}{s}\right)(15)$$

$$v_{air} = 340 \text{ m/s}$$

Unlike the wave in the rope or the spring, the amplitude of a sound wave diminishes drastically as the wave travels. This is because the wave is spreading out in three-dimensional space instead of traveling in a single line along a one-dimensional medium. Each crest is a sphere of compression, growing as it travels. Such a spherical crest or trough, or any other sphere of points in phase with each other, is called a *wave front*. The area of this wave front increases as the square of the radius, but the total energy in it (neglecting damping) does not change. Thus, the amplitude of the wave must drop off at greater distances from the source.

Core Concept

A sound wave is a longitudinal wave spreading through three-dimensional space at a speed that depends only on the air temperature:

$$v_{air} = 331\frac{m}{s} + \left(0.6\frac{m}{s}\right)T_C$$

Try This   With the temperature at 24°C, a tuning fork vibrates at 320 Hz. What is the wavelength of the sound wave it produces?

8-4. Music and Noise

What you hear depends on the properties of the sound wave that enters your ear. A loud sound represents a high-amplitude wave, one with lots of energy.

Musicians use a letter code to describe the *pitch* of a musical sound. The pitch of a sound represents the frequency of the wave. When an American orchestra tunes up, the oboe sounds the note that is called A-440—A on the musical scale, and a frequency of 440 Hz. All the rest of the instruments use this as a basis for tuning. There is nothing sacred about this relationship between pitch and frequency; European orchestras do it differently. And even within an orchestra, the note called A on an oboe is called something else on a clarinet.

Definition of a musical scale begins with an interval called an octave. This is the interval between middle C and high C, for example. An increase in pitch of one octave represents a doubling of the frequency of the wave. See Sample Problem 8-7.

Sample Problem

8-7 The lowest note on a piano is an A with a frequency of 27.5 Hz. What is the frequency of the A three octaves higher?

Solution   Each octave up is double the frequency, so 27.5 Hz must be doubled three times, to get 220 Hz.

Notes sound harmonious together when their frequencies are in simple whole-number ratios. The diatonic scale is built on such ratios. The group of three notes called a major triad, such as C–E–G or F–A–C, has its frequencies in the ratio 4:5:6. Figure 8-4 shows the frequencies of the notes of the diatonic scale in the key of C, with A taken as the standard orchestral value of 440 Hz. Other major triads can be formed by introducing sharps and flats into the scale.

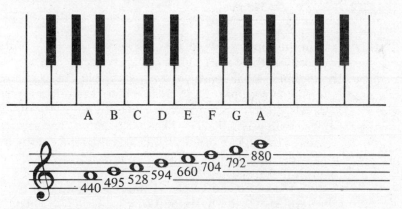

FIGURE 8-4

A flute playing C sounds very different from a violin playing the same note. The difference is in the *tone quality* of the sound. The reason for the difference is that, when a violin string is bowed, it produces a great many frequencies simultaneously. The lowest, called the *fundamental*, is the one that specifies the pitch. The other frequencies are called the *overtones* of the note. Every instrument produces its characteristic mix of overtones, and that is what gives each its special character. A flute produces nearly a pure sine wave with few overtones, especially in its upper register, while a violin's voice has many overtones.

The difference between a violin and a drum lies in the kinds of overtones. The overtones of a violin are simple whole-number multiples of its fundamental. When a cello plays a low A at 110 Hz, the string is also producing waves at 220 Hz, 330 Hz, 440 Hz, 550 Hz, and so on. We get a definite sense of pitch, represented by 110 Hz, the fundamental. But when a drum is struck, its overtones do not form such a simple series. If a drum produces its fundamental tone at 50 Hz, the overtone series looks like this: 79.7 Hz, 106.8 Hz, 114.8 Hz, 145.9 Hz, 179.9 Hz, and so on. In this sort of series, no definite sense of pitch is produced. This is noise.

Core Concept

Loudness of a sound corresponds to amplitude of the wave; pitch to frequency; tone quality to the mix of overtones.

Try This With the temperature at 22°C, a violin plays middle C at 264 Hz. What is the wavelength of the second overtone of the sound wave produced in air?

8-5. Diffraction

Waves are not hard-surface objects traveling in definite paths, like bullets or baseballs. They have a strong tendency to bend around corners. This phenomenon is called *diffraction*.

FIGURE 8-5

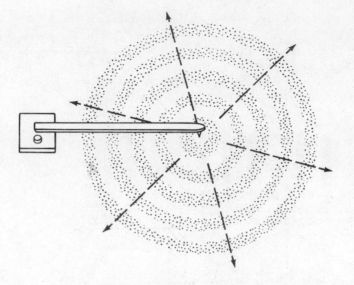

To study diffraction, we can make waves visible. Our model is the ripple produced on the surface of water by a point that is made to oscillate by means of a motor. This produces a series of circular waves, as in Figure 8-5. Alternate crests and troughs follow each other radially outward from the oscillator, forming traveling circular wave fronts.

If you try to block the passage of these waves, as in Figure 8-6, you will note that you cannot form a sharp shadow zone behind the barrier. The waves bend over into the shadow region. This wave diffracted into the shadow region drops off in amplitude gradually as you go farther into the shadow region.

FIGURE 8-6

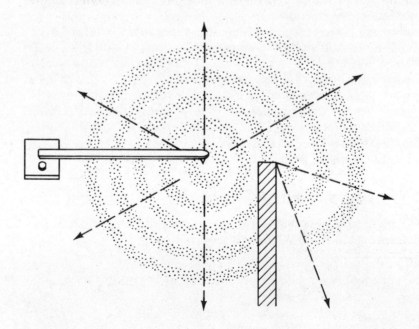

The amount of diffraction depends strongly on the wavelength. In Figure 8-7, the same source and barrier are used, but the frequency of the source has been speeded up so that the wavelength of the wave is smaller. In this case, the amplitude of the wave in the shadow zone drops off much more quickly. Longer waves diffract much more than short ones.

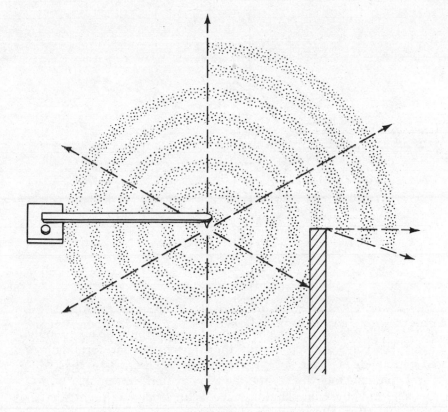

FIGURE 8-7

In Figure 8-8, a small barrier has been used—it is not much larger than a wavelength. Here the shadow region is very small; the waves diffract around the barrier and reunite on the other side. The net effect is just as though there were no barrier there at all. To block a wave, you need a barrier substantially larger than a wavelength. That is why a breakwater, designed to prevent the ocean's waves from washing sand down the beach, has to be so long.

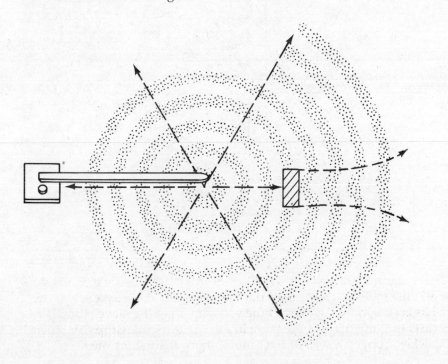

FIGURE 8-8

FIGURE 8-9

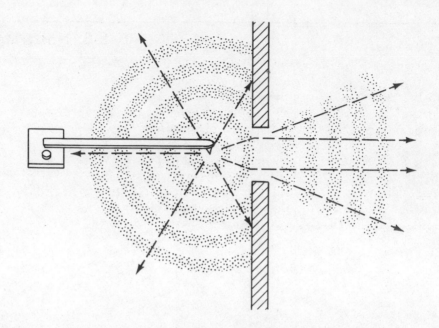

Figure 8-9 shows what happens when the wave comes to an opening in a barrier. The wave diffracts around the edges of the slit, forming arcs centered on the edges. If the slit is narrower, as in Figure 8-10, these diffracted arcs dominate the pattern, and the waves emerge on circular fronts. A narrow slit forms the same kind of wave as an oscillating-point source.

Since diffraction affects longer waves most, shorter waves tend to travel in straight lines, while longer ones bend around all corners. That is why a hi-fi speaker has only one large speaker ("woofer") for the low frequencies, but several small ones ("tweeters") pointing in different directions for the high frequencies.

FIGURE 8-10

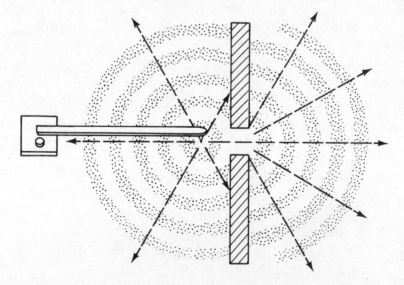

Core Concept

Waves diffract or bend around corners, and the effect is greater when the wavelength is longer.

Try This If you stand to one side of an open door listening to the music of an orchestra coming through the doorway, in what way will the sound you hear be distorted?

As you drive along in your car, you pass by a factory as its whistle blows. At the instant you pass the factory you hear a sudden drop in the pitch of the whistle. This is known as the *Doppler effect*.

Figure 8-11 shows why this happens. The concentric circles around the whistle represent zones of compression, spreading outward from the whistle. If you sit still, they will pass by you at a frequency equal to the frequency of the whistle. But if you are moving toward the whistle, your ear will intercept more of those compressions every second than are being produced by the whistle. You will hear a higher frequency than the whistle is producing. When you pass the factory, you will be moving away from the whistle, and the effect is just the reverse. You will hear a lower frequency.

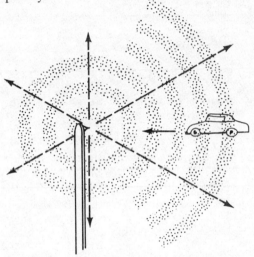

FIGURE 8-11

There is a similar Doppler frequency shift when it is the source of the sound that is moving. Again, you hear a frequency higher than that of the source as it approaches you, and lower as it recedes. But the reason is different, and the frequency shift is not exactly the same. Figure 8-12 shows the compression zones surrounding a source moving to the right. Each zone is a sphere, as usual, centered on the point at which it was produced. However, since the source is moving to the right, each compression originates farther to the right than the one before it. The result is that

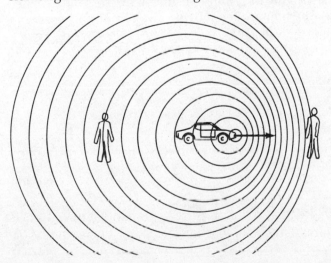

FIGURE 8-12

the wavelength of the sound is shorter in the direction in which the sound source is traveling, and longer behind it. Therefore the frequency shifts upward in front and downward behind.

Both kinds of Doppler shifts can be incorporated into a single formula:

$$f' = f\left(\frac{v - v_\text{o}}{v - v_\text{s}}\right)$$

(Equation 8-6)

where f is the frequency of the source and f' is the frequency heard by the observer. The velocities are v for the sound wave, v_o for the observer, and v_s for the source. This equation looks simple enough, but it is very tricky to apply because you have to be careful about using the proper signs for v_o and v_s (v is always positive). Do it this way: Make a diagram showing the source to the *left* of the observer. Then the v's are positive to the right and negative to the left. See Sample Problems 8-8 and 8-9.

Sample Problem

8-8 A whistle sounds a 530-Hz note when the temperature is 5°C. What frequency will you hear if you are approaching the whistle at 22 m/s?

Solution First we need the speed of sound, from Equation 8-3:

$$v = 331 \text{ m/s} + (0.6 \text{ m/s})(5) = 334 \text{ m/s}$$

The whistle is at rest, so $v_\text{s} = 0$. The speed of the observer is to the left, negative. Therefore,

$$f' = (530 \text{ Hz})\left(\frac{334 \text{ m/s} + 22 \text{ m/s}}{334 \text{ m/s}}\right) = 565 \text{ Hz}$$

Sample Problem

8-9 A train going 40 m/s sounds its 260-Hz horn. What frequency will be heard by someone (a) at rest behind the train; and (b) coming toward the train at 22 m/s? The air temperature is 26°C.

Solution First we need the speed of sound in air. From Equation 8-3,

$$v = 331 \frac{\text{m}}{\text{s}} + \left(0.6 \frac{\text{m}}{\text{s}}\right)(26) = 347 \text{ m/s}$$

For part (a), imagine viewing this from a position with the source to the left of the observer. The train is then moving away from the observer, going left. So its velocity is negative; $v_\text{s} = -40$ m/s. Now, using Equation 8-6, we get

$$f' = (260 \text{ Hz})\left[\frac{347 \text{ m/s} - 0}{347 \text{ m/s} - (-40 \text{ m/s})}\right]$$

$$f' = (260 \text{ Hz})\left(\frac{347}{387}\right) = 233 \text{ Hz}$$

For (b), you have to look at the train from the other side, so it is to the left of the other observer. Now $v_\text{s} = +40$ m/s, and $v_\text{o} = -22$ m/s. Therefore,

$$f' = (260 \text{ Hz})\left[\frac{347 \text{ m/s} - (-22 \text{ m/s})}{347 \text{ m/s} - (+40 \text{ m/s})}\right] = 311 \text{ Hz}$$

Equation 8-6 breaks down completely if the source is moving faster than sound. In that case, something quite different happens, as can be seen in Figure 8-13. All the compression zones overlap each other, and the sound energy is concentrated mainly in one big compression, the cone-shaped envelope of all the compression zones. If an airplane flies overhead at a speed faster than sound, that enormous cone of compression trails along behind it, intersecting the ground at some distance behind the plane. Where the cone touches the ground, there is a "sonic boom"; windows break, children start to cry, and alarmed citizens call police stations to find out what caused the explosion.

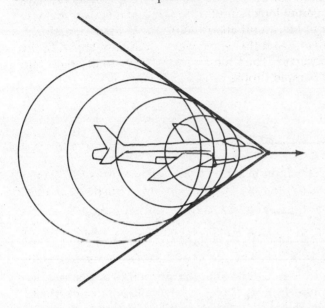

FIGURE 8-13

Core Concept

If an observer and a source of sound are in relative motion, the observed frequency is different from the emitted frequency:

$$f = f'\left(\frac{v - v_{\mathrm{o}}}{v - v_{\mathrm{s}}}\right)$$

Try This With the temperature at 10°C, a car sounding its horn at 240 Hz is going 25 m/s down the highway. If a policeman on a motorcycle is chasing it at 35 m/s, what frequency does the policeman hear?

8-7. Interference

One distinctive feature of waves is that it is possible for two waves, arriving simultaneously at the same place, to add up to nothing. This can occur with two sound waves if they arrive at some point exactly 180° out of phase with each other. Then one of them is producing a compression while the other is producing a rarefaction, and the net result is that nothing at all happens. This process is called *destructive interference*.

To show this experimentally, get a twin pair of speakers and set them a short distance apart. This has to be done in an open field so there will be no reflection from walls. Connect both speakers to the same signal generator, so that they are producing a steady, single-frequency sound in phase with each other, as shown in Figure 8-14. A microphone moved around

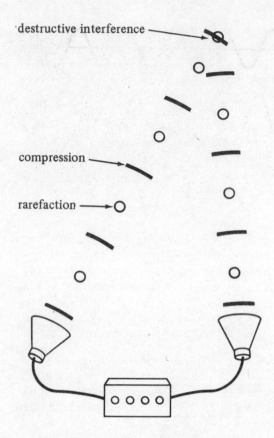

destructive interference

compression

rarefaction

FIGURE 8-14

some distance from the speakers will be able to find points of destructive interference where the waves from the two speakers arrive out of phase with each other. The microphone will pick up no sound there.

One way to locate such a point is to do a little geometry. If a point is just a half-wavelength farther from one speaker than it is from the other, the wave from the farther speaker will arrive a half-cycle later than the one from the near speaker. Then the two waves will be a half-cycle out of phase with each other and will interfere destructively. See Sample Problem 8-10.

Sample Problem

8-10 Two speakers, producing identical sound waves in phase in an open field, feed into a microphone. If the microphone can detect no sound when it is 4.7 m from one speaker and 5.2 m from the other, what is the wavelength of the sound?

Solution If interference is destructive, the simplest solution is that one speaker is a half-wavelength farther from the microphone than the other speaker. Therefore 5.2 m − 4.7 m = 0.5 m is a half-wavelength, and the wavelength of the sound is 1.0 m.

What happens to the energy of the sound waves? Surely it cannot just disappear! It turns up at a point of *constructive interference*. At any point equidistant from the two speakers, the two waves will arrive in phase and will reinforce each other. The sound is exceptionally loud there.

An easy way to hear the effect of interference is to get a couple of tuning forks, violin strings, or any other source of a musical sound tuned very slightly apart from each other. If both are sounded together, the two waves arriving at your ear will go alternately in and out of phase with each

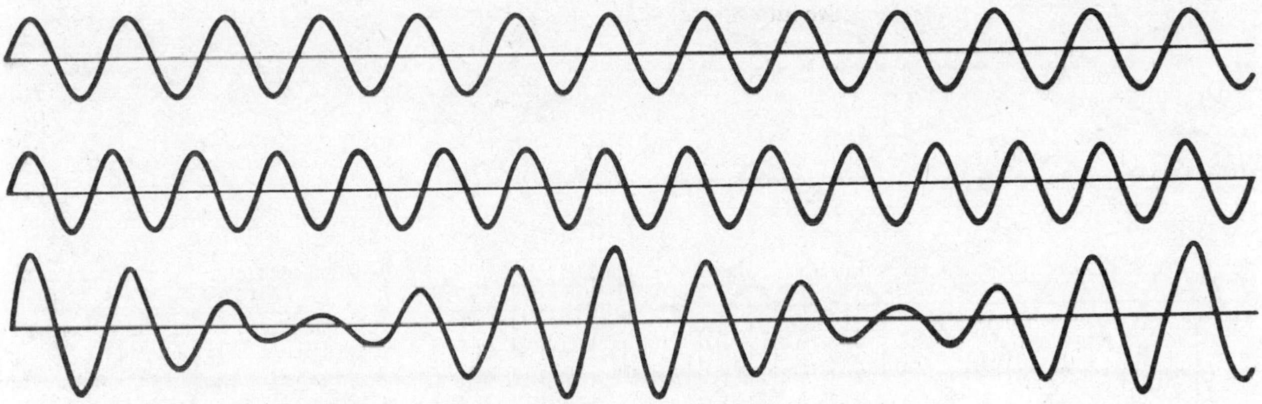

FIGURE 8-15

other. This can be seen in Figure 8-15, which shows the pressure variation at your ear due to the simultaneous arrival of two waves of slightly different frequencies. You will hear *beats*—waawaawaawaa—as the waves alternate constructive and destructive interference. The beat frequency will be the difference between the two sound-wave frequencies. See Sample Problem 8-11. Listening for beats is an extremely sensitive way to match frequencies, and musicians use it often in tuning instruments.

Sample Problem

8-11 A piano tuner adjusts a string to sound C-264 Hz. In trying to match a second string to it, he sounds both together and hears three beats per second. What is the frequency of the other string?

Solution The difference between the two frequencies must be 3 Hz, so the other one is either 261 Hz or 267 Hz.

Core Concept

Waves interfere constructively at points where they arrive in phase and destructively at points where they arrive out of phase.

Try This Two speakers produce sound waves 2.0 m long in phase. If a point of destructive interference is 12.5 m from one of the speakers, how far is it from the one that is farther away?

8-8. Standing Waves

Remember the rope we used to demonstrate waves? Let's see what happens when the wave gets to the end of the rope.

First, assume that the end of the rope is tied to some fixed support. The moving rope cannot make the support shake, and the energy of the wave has nowhere to go except back into the rope. The wave is reflected from the fixed end, passing back along the rope in the opposite direction. Now we have two identical waves in the rope, one going to the right and the other to the left.

To see what happens in this case, study Figure 8-16 carefully. The uppermost wave is moving to the right, the second one is traveling to the left, and the bottom wave is the sum of the other two, added point by point. Successive pictures, from left to right, show what the rope looks like at intervals of one-eighth of a cycle. There are five points on the rope, marked by black dots, that are not moving at all. These are places where

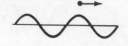

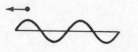

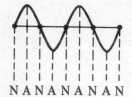

NANANANAN   NANANANAN   NANANANAN   NANANANAN

FIGURE 8-16

the two waves interfere with each other destructively. These points are called *nodes*, and they are spaced a half-wavelength apart down the whole length of the rope. The fixed end, unable to move, must be a node.

Halfway between the nodes are points of maximum vibration where the two waves interfere constructively. These points are *antinodes*. They are a half-wavelength apart. Alternate antinodes are in opposite phase, so that the distance between two antinodes in phase is one wavelength. See Sample Problem 8-12.

Sample Problem

8-12 If a wave with a wavelength of 1.2 m traveling in a rope is reflected from a fixed end, how far from the end are the first two antinodes?

Solution The end is fixed, so it must be a node. The first antinode is a quarter-wavelength in from the end, or 0.3 m, and the next one is a half-wavelength farther, at 0.9 m from the end.

This pattern of nodes and antinodes is called a *standing wave*. Within a given loop of a standing wave—from one node to the next—all points are in phase, but the amplitude varies from zero at the nodes to a maximum at the antinode.

Core Concept

A standing wave, consisting of nodes and antinodes alternating each quarter-wavelength, is formed when two equal traveling waves pass through each other in opposite directions.

Try This A rope is fastened at one end, and the other end is shaken with a frequency of 5.0 Hz. If the speed of the wave in the rope is 15 m/s, how far from the attached end are (a) the first node; and (b) the first antinode?

8-9. Making Music

When a guitar player plucks a string, he confidently expects a note of a certain frequency to emerge. If he is any good, it will. The frequency of the sound that the guitar produces must be the same as the vibration frequency of the string, since that is what produces the sound wave. And when a string is plucked, it will vibrate with a number of frequencies, which depend on the length of the string and the speed of the wave in it.

This is what happens when a string is plucked: Waves of many frequencies start traveling in the string in both directions. When these waves reach the ends of the string, they are reflected back. Most of them die out almost at once. The only ones that remain are those that form stable standing waves in the string. The wavelengths of the standing waves are limited by the *boundary conditions* of the string: A standing wave can form only if its wavelength is such that it forms a node at each end of the string. This limitation is imposed because the ends of the string are clamped, so that the string cannot vibrate there.

The possible wavelengths are illustrated in Figure 8-17. The longest possible standing wave is one with nodes at each end and an antinode in the middle of the string. Since the nodes of a standing wave are a half wavelength apart, the wavelength of this standing wave is twice the length of the string. The frequency of this wave—and therefore of the sound wave that it produces—can be found by applying Equation 8-2; it is the speed of the wave divided by the wavelength, as illustrated in Sample Problem 8-13.

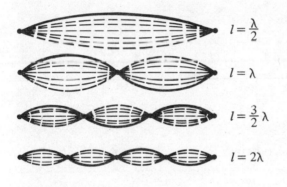

$l = \frac{\lambda}{2}$

$l = \lambda$

$l = \frac{3}{2}\lambda$

$l = 2\lambda$

FIGURE 8-17

Sample Problem

8-13 What is the fundamental frequency sounded by a violin string 45 cm long if the speed of the wave in the string is 280 m/s?

Solution   At the fundamental frequency, the only nodes in the string are at the ends. Since these are a half-wavelength apart, the wavelength of the wave is twice the length of the string, or 0.90 m. Then, from Equation 8-2,

$$f = \frac{v}{\lambda} = \frac{280 \text{ m/s}}{0.90 \text{ m}} = 311 \text{ Hz}$$

Other frequencies are possible because there are other standing waves that can fit into the string. The second sketch in Figure 8-17 shows another standing wave that meets the boundary conditions. It also has nodes at both ends, but in this case, it also has a node in the middle of the string. Its wavelength is thus the same as the length of the string. With half the wavelength of the first standing wave, it has twice the frequency. The other sketches show the basic rule that determines the wavelengths of the standing waves in a string: the length of the string must be some whole number of half-wavelengths as calculated in Sample Problem 8-14. To see what frequencies are produced by a string vibrating according to this rule, see Sample Problem 8-15.

Sample Problem

8-14 What are the wavelengths of the four longest waves that can stand in a string 60 cm long?

Solution For the longest wave, the string vibrates in a single loop, with a node at each end. In this oscillation, the string is $\frac{1}{2}$ wavelength long, so

$$60 \text{ cm} = \frac{1}{2}\lambda \qquad \lambda = 120 \text{ cm}$$

Next, there is a node at each end and one in the middle, making the string 1 wavelength long:

$$60 \text{ cm} = \lambda$$

Vibrating in three loops, it is $\frac{3}{2}$ wavelength long:

$$60 \text{ cm} = \frac{3}{2}\lambda \qquad \lambda = 40 \text{ cm}$$

And in four loops:

$$60 \text{ cm} = 2\lambda; \qquad \lambda = 30 \text{ cm}$$

Sample Problem

8-15 What are the four lowest frequencies of a 60-cm string if the speed of the wave in it is 240 m/s?

Solution For each of the possible wavelengths, from Equation 8-2,

$$f = \frac{v}{\lambda}$$

Using the wavelengths calculated in Sample Problem 8-14, we have

$$f_1 = \frac{240 \text{ m/s}}{1.20 \text{ m}} = 200 \text{ Hz}$$

$$f_2 = \frac{240 \text{ m/s}}{0.60 \text{ m}} = 400 \text{ Hz}$$

$$f_3 = \frac{240 \text{ m/s}}{0.40 \text{ m}} = 600 \text{ Hz}$$

$$f_4 = \frac{240 \text{ m/s}}{0.30 \text{ m}} = 800 \text{ Hz}$$

Thus, the frequencies that can be produced by a string depend on the length of the string and on the speed of a wave in it. This wave velocity increases as the tension in the string increases, as anyone who has ever tuned a violin or a guitar knows. Also, the speed of the wave depends on the thickness and density of the string. The low notes of a piano or a cello are produced by thicker, heavier strings. When a string is plucked or stroked, it produces, simultaneously, all those frequencies that are made by waves that can stand in the string. Sample Problem 8-15 shows that these frequencies form an overtone series in which all the frequencies are in simple, whole-number ratios. That is why the sound produced has a musical quality and a definite sense of pitch, which is determined by the lowest vibration frequency of the string.

Core Concept

The natural vibration frequencies of a string are specified by the boundary condition that there must be a node at each end.

Try This What are the first three frequencies produced by a plucked string 30 cm long if the speed of the wave in the string is 280 m/s?

8-10. Pipes

An air column has its own natural vibration frequencies, and these serve as the basis for all wind instruments. The air in the pipe is set into vibration at one end, by some means or other. In a clarinet or an oboe, it is done by making a reed vibrate. The musician who plays a trumpet or a French horn does it by vibrating his lips against the mouthpiece. In a flute or an organ pipe, the vibration is produced by turbulent air passing across a narrow slit.

The boundary conditions in an organ pipe are not like those of a string. At the ends of the pipe, the trapped air is open to the outside air. Since the air is not constrained at the ends, rapid vibration is possible there, and the ends are antinodes, not nodes.

It is difficult to illustrate the standing wave pattern inside an organ pipe. It is a sound wave, so it is longitudinal, and its speed can be determined from the temperature alone, using Equation 8-3. Wavelength is determined by the boundary condition that there must be an antinode at each end. As you can see by the symbolic representation in Figure 8-18, this produces the same rule as for a string: The length of the pipe must be a whole number of half-wavelengths. See Sample Problem 8-16.

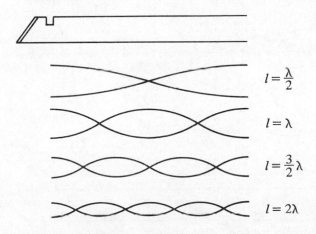

$$l = \frac{\lambda}{2}$$

$$l = \lambda$$

$$l = \frac{3}{2}\lambda$$

$$l = 2\lambda$$

FIGURE 8-18

Sample Problem

8-16 With the temperature at 15°C, what are the three lowest frequencies sounded by an open organ pipe 40 cm long?

Solution The speed of the wave in the pipe is

$$v = 331\frac{m}{s} + \left(0.6\frac{m}{s}\right)(15) = 340 \text{ m/s}$$

For the fundamental:

$$40 \text{ cm} = \frac{1}{2}\lambda; \qquad \lambda = 0.80 \text{ m}; \qquad f_1 = \frac{340 \text{ m/s}}{0.80 \text{ m}} = 425 \text{ Hz}$$

First overtone:

$$40 \text{ cm} = \lambda; \qquad f_2 = \frac{340 \text{ m/s}}{0.40 \text{ cm}} = 850 \text{ Hz}$$

Second overtone:

$$40 \text{ cm} = \frac{3}{2}\lambda; \qquad \lambda = 0.2667 \text{ m}; \qquad f_3 = \frac{340 \text{ m/s}}{0.2667 \text{ m}} = 1,275 \text{ Hz}$$

The frequency of a wind instrument is changed by altering the length of the pipe. In a trumpet or a French horn, the buttons open valves that bring additional pipe lengths into the air column. In a flute or a bassoon, the buttons open holes in the side of the pipe, thereby establishing new antinodal points.

Some organ pipes are closed at the end, and the picture in these is quite different. The closed end is a node; the end that is energized is an antinode. Thus, the fundamental frequency, produced by the longest wave that can stand in the pipe, is fixed by the need for the wave to go from a node to an antinode—a quarter-wavelength. Therefore, at the fundamental frequency, the pipe is a quarter-wavelength long. The first overtone occurs when the pipe is $\frac{3}{4}\lambda$; the next at $\frac{5}{4}\lambda$; etc. (See Figure 8-19 and Sample Problem 8-17.)

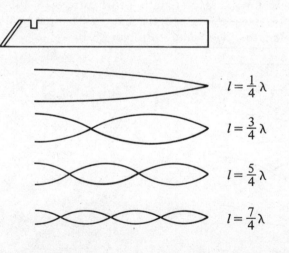

FIGURE 8-19

$l = \frac{1}{4}\lambda$

$l = \frac{3}{4}\lambda$

$l = \frac{5}{4}\lambda$

$l = \frac{7}{4}\lambda$

Sample Problem

8-17 What are the fundamental and first overtone frequencies of a closed organ pipe 35 cm long when the temperature is 18°C?

Solution   The speed of the sound wave in the pipe is 331 m/s + (0.6 m/s)(18) = 342 m/s. At the fundamental, the pipe is $\frac{1}{4}$ wavelength long, so

$$\frac{\lambda}{4} = 0.35 \text{ m}; \qquad \lambda = 1.40 \text{ m}$$

Then the frequency is

$$f = \frac{v}{\lambda} = \frac{342 \text{ m/s}}{1.40 \text{ cm}} = 244 \text{ Hz}$$

At the first overtone, the pipe is $\frac{3}{4}$ wavelength long, so the wavelength is $\frac{1}{3}$ as much as the fundamental. This makes the frequency three times as much, or 732 Hz.

Natural vibration frequencies of a pipe are fixed by the speed of sound in air and the boundary conditions of a node at a closed end and an antinode at an open end.

Try This At a temperature of 22°C, what are the first three frequencies of an open organ pipe 22 cm long?

8-11. Resonance

Every elastic object has its own special set of natural vibration frequencies. When you tap on a glass, knock on the door, or drop a screwdriver, the sound you hear is fixed in pitch by the natural frequencies of the object.

The natural vibration frequencies depend on two things: the speed of the wave in the object and the wavelengths that can stand stably in it. The speed is determined by the kind of material—its elasticity, its density, its thickness, and so forth. The wavelengths are determined by the boundary conditions. An edge that is open and unconstrained must be an antinode; any point that is clamped must be a node. The boundary conditions of a drumhead, for example, are that there must be nodes all along its edge. The tuning fork of Figure 8-20 has a node at the handle and antinodes at the ends.

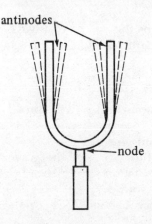

antinodes

node

FIGURE 8-20

Every object has a special sensitivity to stimulation at its natural vibration frequencies. If you depress the pedal on a piano and strike any note, the vibration of the string will set into vibration all the other strings that have that natural frequency. A fine wine glass can be made to vibrate so energetically that it will crack, just by exposing it to sound at one of its own natural frequencies. This phenomenon is called *resonance*.

In a violin, there are many resonances. The upper plate has some, and there is a whole set of short-wavelength resonances between the f-holes. There are other resonances in the lower plate, in the trapped air mass, and even in the bridge. It is these resonances that make the difference between a Stradivarius and a cheap fiddle.

Core Concept

In every object, resonant frequencies are determined by the ratio of wave velocity to the wavelengths of standing waves, which are fixed by boundary conditions.

Try This What are the two lowest resonant frequencies of a tuning fork if each tine is 18 cm long and the speed of the wave in it is 420 m/s? (These answers will be approximate; they take no account of the thickness of the metal.)

Summary Quiz

Fill in the missing word or phrase:

1. The length of time it takes a point on a medium carrying a wave to go through one full cycle is the _____ of the wave.
2. The reciprocal of a wave's period is its _____ .
3. The hertz is a measure of _____ .
4. Two points on a wave that are a half-wavelength apart are _____ degrees out of phase with each other.
5. The maximum displacement of any point on a wave is called the _____ of the wave at that point.
6. In a real wave, the process by which a wave loses energy as it travels is called _____ .
7. The distance between two successive points in phase is the _____ of the wave.
8. The product of frequency and wavelength gives the _____ of a wave.
9. In a longitudinal wave, the motion of the particles is _____ to the wave velocity.
10. The nature of the medium in which it travels largely determines the _____ of a wave.
11. A sound wave consists of alternate compressions and _____ spreading through space.
12. The speed of a sound wave in air depends only on _____ .
13. The pitch of a musical sound depends on the _____ of the wave.
14. The frequency of a note two octaves below A-440 is _____ Hz.
15. Overtones determine the _____ of a musical sound.
16. A drum produces no definite sense of pitch because its _____ do not form a series with integral ratios.
17. Diffraction around corners is greater when the _____ is larger.
18. Because of diffraction, a narrow slit acts like a _____ of waves.
19. If you are moving toward a source of sound, the _____ you perceive will be higher than that emitted by the source.

20. The wavelength of a sound wave varies from place to place if the _____ is moving.
21. Destructive interference occurs when two waves arrive at a point _____ with each other.
22. Beat frequency is the _____ of the frequencies of two sounds.
23. A point of maximum destructive interference is called a _____ .
24. In a standing wave in a rope, the distance from one antinode to the next is a _____ .
25. Stable standing waves can form in a string only at frequencies at which there is a _____ at each end of the string.
26. At the fundamental frequency, the length of a vibrating string is _____ wavelength.
27. The closed end of an organ pipe must be a _____ of the standing wave in the pipe.
28. To make an object resonate, it must be energized at one of its own _____ .

Problems

1. A wave in a rope travels at 12 m/s, and the wavelength of the wave is measured at 1.2 m. Find (a) the frequency of the wave; and (b) its period.
2. A wave in the ocean is 30 m long. A point on the wave front passes two boats that are 12 m apart. What is the phase difference between the oscillations of the boats?
3. In a longitudinal wave in a spring, the distance from a compression to the nearest rarefaction is 35 cm. If the frequency of the wave is 4.0 Hz, how fast is the wave traveling?
4. How fast does a sound wave travel in air at a temperature of −6°C?
5. What is the wavelength of a sound wave produced by a violin string vibrating at 640 Hz if the temperature is 26°C?
6. What is the frequency of the A two octaves below A-440?
7. If a major triad is formed by the notes A–C#–E, what is the frequency of C#?
8. If a police car approaches you with its siren going at 380 Hz, traveling 28 m/s when the air temperature is 18°C, what frequency will you hear?
9. With the temperature at −12°C, a car is speeding down the highway at 30 m/s. If a police officer on a motorcycle is chasing it at 37 m/s with the siren producing a signal at 260 Hz, what frequency will the driver of the car hear?
10. With the temperature at 10°C, two loudspeakers are producing a 680-Hz tone in phase with each other. If you locate a node 15.8 m from the nearer speaker, how far are you from the other one?
11. You are tuning a 12-string guitar and adjust one string to sound 220 Hz; when you strike this along with its matching string, you hear beats at a frequency of 4 Hz. You find that, if you tighten the matching string a little, the beats disappear. What was the frequency of the matching string?
12. A rope that transmits a wave at 8.0 m/s is shaken at one end at a frequency of 6.0 Hz. If the other end is fastened down, how far from that end are the first two antinodes?

13. Determine the fundamental and the first two overtones of a violin string 35 cm long if the speed of the wave in it is 180 m/s.
14. Determine the fundamental and the first two overtones of a closed organ pipe 35 cm long at a temperature of 20°C.
15. Determine the fundamental and the first overtone of a tuning fork whose tines are 18 cm long if the speed of the wave in the tuning fork is 60 m/s.
16. To tune a piano on the equal-tempered scale, a piano tuner wants to adjust the high E to 1,318.5 Hz and does so by listening for beats that this string makes with the second overtone of A-440. When the adjustment is right, what beat frequency will the piano tuner hear?

CHAPTER

9

It's Electric

9-1. Another Kind of Force

When you take off a nylon shirt, why does it tend to cling to your body? How is it that you can pick up tiny bits of paper with a comb after you have used it on your hair? What produces that shock you get when you touch a doorknob after walking across a wool rug?

These are manifestations of a kind of force that is in many ways similar to the force of gravity. For one, it seems to act at a distance, through empty space. You can show this with an apparatus like that of Figure 9-1, which consists of a Ping-Pong ball coated with graphite and suspended from a very light thread. If you comb your hair and then bring the comb near the ball, you will find that you can attract the ball to the comb, from a distance.

But the force is not gravity. First of all, this force is far stronger than gravity. The gravitational attraction between a comb and a Ping-Pong ball is far too small to be detected; you need something the size of the earth or the moon, anyway—before you can find a detectable gravitational force. And, second, there seems to be something rather temporary about this new kind of force, for it would not be found if you failed to comb your hair before you tried the experiment. And if you wait a while, it will probably disappear. The comb must be given an *electric charge* before it will attract the ball.

177

FIGURE 9-1

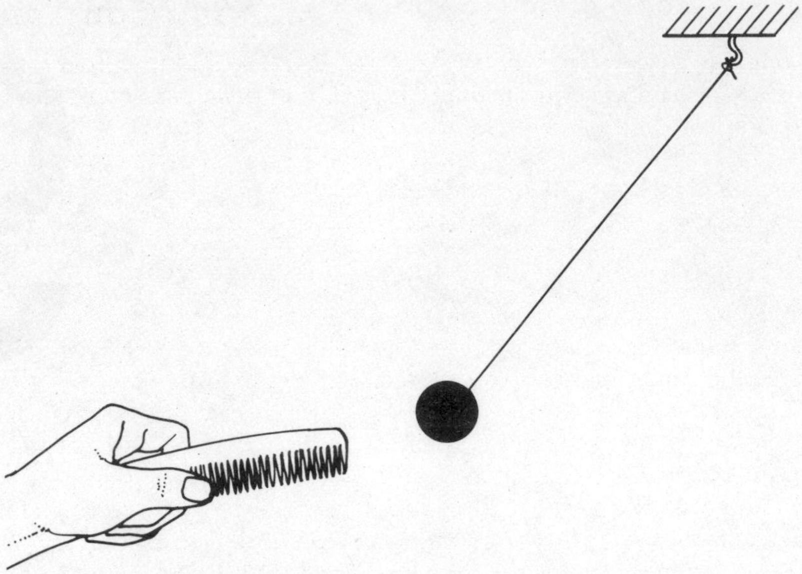

Now stroke the ball with the charged comb and see what happens. The ball jumps away from the comb; there is force of repulsion between the ball and the comb, something that never happens when the force is gravitational. If you hang two Ping-Pong balls side by side and stroke them both with the comb, they will repel each other, as shown in Figure 9-2.

Now suppose you charge the second ball by a different method: Take a dry glass rod and stroke it with the kind of thin sheet plastic that is used to wrap food. Then stroke the second Ping-Pong ball with the glass rod. The two balls, one charged from the rubber comb and the other from the glass rod, will now attract each other.

We seem to have two different kinds of charge. If two objects are both charged from a rubber rod, or both from a glass rod, they repel each other. If you charge one from glass and the other from rubber, they attract.

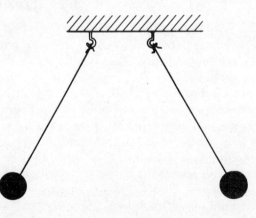

FIGURE 9-2

Core Concept

Similar charges repel; dissimilar charges attract.

Try This Suppose you stroke a vinyl tile with fur and find that it repels a Ping-Pong ball that had been charged from a rubber rod. What could you conclude about the charge on the vinyl?

Charge one ball from rubber and the other from glass, and then touch the balls to each other. Sometimes, both charges will disappear completely. The two charges eliminate each other.

We can invoke a well-known mathematical rule to express this situation. The rule is this: If two quantities add up to zero, one is positive and the other is negative. It does not matter which of our two kinds of charge we call positive and which negative, as long as we are consistent. The usual convention, adopted long ago and never changed, is that the charge on a rubber rod is negative and that on a glass rod positive.

When you stroke a rubber rod with fur, the rubber becomes negative. And the fur becomes positive! You started with two neutral objects—total charge zero. After the stroking, you have a positive object and a negative one. The total charge is still zero. Similarly, when the glass rod becomes positive, the plastic film you stroked it with becomes negative. You cannot create a charge of one kind without producing the opposite kind at the same time. In any such process, the total amount of charge does not change. This statement constitutes the *law of conservation of electric charge*. It is as fundamental to the theory of physics as the other conservation laws: mass–energy, momentum, and angular momentum.

When you stroke a rubber rod with fur, you may appear to be creating charge, but you are not. You are merely separating existing charges from each other. Everything is made of atoms, and every atom contains both positive and negative charges. Since the two kinds of charge are present in equal amount, every atom is neutral. The negative charges can be removed; when you stroke a rubber rod with fur, you simply transfer some of the negative particles from the fur to the rubber. These mobile negative charges, the *electrons*, give the rubber rod a negative charge. The dearth of electrons on the fur makes the fur positive.

Metals and certain other materials contain large numbers of electrons that move freely within the substance. If you prepare two oppositely charged Ping-Pong balls and then connect them through a metal rod, electrons from the negative ball will run into the rod; at the other end, electrons will run out of the rod into the positive ball. Both balls will be neutralized, since the charge can flow easily through the metal from one end to the other. A substance that allows charge to pass through it is called a *conductor*. (The graphite on the Ping-Pong balls is a conductor.)

An *insulator* is a material, such as rubber, plastic, or wood, that will not allow charge to flow through it. A charge placed at one end of a rubber rod stays there, even though all the accumulated electrons are repelling each other. Wires are wrapped in rubber, and tool handles in plastic, to keep the electric charge in the metal where it belongs.

In solids, it is only electrons that can move from place to place. But salt water, or the gas inside your fluorescent lamp, is an excellent conductor because it contains both negatively and positively charged atoms (ions) that can move freely under the influence of an electric force.

Core Concept

Charging consists of separating positive and negative charges, while the total charge remains unchanged.

Try This Explain why you cannot charge one end of a steel rod and leave the other end uncharged.

9-3. Charged Conductors

When a charge is placed on a conductor, the excess electrons can move freely through it. They repel each other, so they spread out until every electron is in equilibrium under the repulsive forces of all the others. Unless the repulsion is very strong, however, they cannot leave the surface because they are attracted by the stationary positive charges in the metal.

At equilibrium, *all* the excess charge is on the surface of the conductor. This is the arrangement when the electrons have repelled each other until they are as far apart as possible. There is never a charge inside a charged conductor. No matter how strongly charged the surface is, any charge added to the inside will immediately move to the outside.

This is how the Van de Graaff generator (Figure 9-3) is able to build up such a large charge. The moving belt picks up a small charge as it passes over a plastic pulley. The charge is carried up to metal fingers inside the spherical dome at the top. These fingers pick it up and deliver it to the inside of the dome. It immediately moves to the surface, so that more charge can be added to the inside. Large Van de Graaff generators may have domes up to 4 meters in diameter and can build up enormous charges on their surface.

FIGURE 9-3

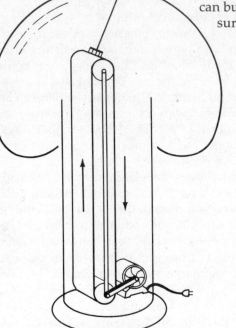

FIGURE 9-4

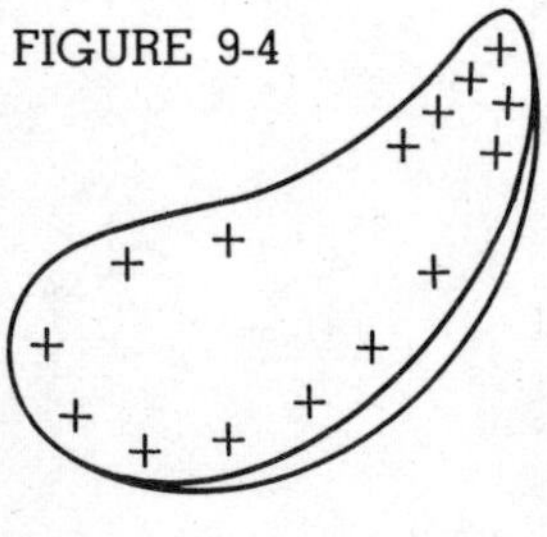

The dome of a Van de Graaff generator is always spherical and must be smoothly polished. If these conditions are not met, the charge will not be distributed uniformly over the surface of the dome. On a conductor, charge will concentrate wherever the curvature of the surface is greatest. A sharp point concentrates charge enormously. This is how a lightning rod operates. Since it is pointed, all the excess charge accumulates at the point, and that is where lightning strikes. Figure 9-4 shows how charge is distributed over the surface of an irregular conductor.

Core Concept

On a charged conductor, all charge is on the surface, and it concentrates at points of highest curvature.

Try This Explain why one safe place from which to operate the controls of a large Van de Graaff generator is inside the dome.

Can charge be measured? Standards of measurement must be established, and they can best be defined in terms of the forces that the charges exert on each other.

Sample Problem 9-1 shows how a simple vector analysis enables us to measure the electric force exerted by the comb on the Ping-Pong ball of Figure 9-5. The tension in the string is exactly balanced by the (vertical) weight of the ball and the (horizontal) electric repulsion of the comb.

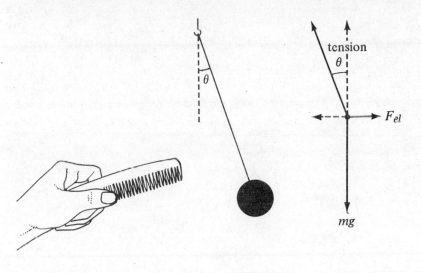

FIGURE 9-5

Sample Problem

9-1 The 2.5-g Ping-Pong ball of Figure 9-5 hangs at the end of a thread 40 cm long. When repelled by the comb, it is deflected 7.0 cm from its original position. How great is the electric force?

Solution   As the vector diagram shows, the ratio of the electric force to the weight is the tangent of the angle that the string makes with the vertical. The sine of this angle is 7.0/40 = 0.1750; its tangent is 0.1777. Therefore,

$$\frac{F_{el}}{mg} = 0.1777$$

$$F_{el} = (0.0025 \text{ kg})\left(9.8 \ \frac{\text{m}}{\text{s}^2}\right)(0.1777)$$

$$F_{el} = 4.4 \times 10^{-3} \text{ N}$$

What we need is a way to determine how the force between two charged objects depends on the amount of charge. Fortunately, we can find this experimentally because there is a way to divide a charge into known fractions: (1) Coat two Ping-Pong balls with graphite to make them conductive, so that they will share equally any charge they may have; (2) charge them and touch them together so that they have the same charge; (3) measure the forces on them by the angle of deflection when they are suspended; (4) touch one ball with the finger—this allows its entire charge to pass off into your body; (5) touch the two balls together again, so that each now has half of its original charge; (6) measure the force again. Repeat many times, with as much variation as possible.

These experiments show that the force between the charges is proportional to the product of the charges and inversely proportional to the square of the distance between them. This is called *Coulomb's law:*

$$F_{el} = \frac{kq_1q_2}{r^2}$$

(Equation 9-4)

in which the q's are the charges and r is the distance between them. The constant k is a universal constant whose value depends on how the units are defined. In the SI, the unit of charge is the *coulomb* (C) and $k = 9.00 \times 10^9$ N $\cdot$ m^2/C^2.

The coulomb is an extremely large unit of charge, and the nanocoulomb (nC = 10^{-9} C) is more convenient in experiments of the kind described above. Sample Problems 9-2 and 9-3 give rather typical values for such an experiment. When you create a spark by touching a doorknob after walking on a wool rug, you are discharging yourself of a charge that is probably no more than 50 nC. A good-sized bolt of lightning discharges about 5 C.

Sample Problem

9-2 What is the force of attraction between two Ping-Pong balls whose centers are 10.0 cm apart if the charges on them are +12 nC and −15 nC, respectively?

Solution Applying Equation 9-4,

$$F_{el} = \frac{kq_1q_2}{r^2}$$

$$F_{el} = \frac{\left(9.00 \times 10^9 \; \frac{\text{N} \cdot \text{m}^2}{\text{C}^2}\right)(12 \times 10^{-9} \text{ C})(-15 \times 10^{-9} \text{ C})}{(0.100 \text{ m})^2}$$

$$F_{el} = -1.6 \times 10^{-4} \text{ N}$$

The negative sign indicates a force of attraction.

Sample Problem

9-3 Two small, equal conducting spheres are charged, touched together, and then separated until their centers are 12 cm apart. If they now repel each other with a force of 3.0×10^{-5} N, how much charge do they have?

Solution Since they are equal, they will share whatever charge they have equally. Call the charge on each q. Then Equation 9-4 becomes

$$F = \frac{kq^2}{r^2}$$

From which

$$q = r\sqrt{\frac{F}{k}} = 0.12 \text{ m} \sqrt{\frac{3.0 \times 10^{-5} \text{ N}}{9.0 \times 10^9 \text{ N} \cdot \text{m}^2/\text{C}^2}}$$

$$q = 6.9 \times 10^{-9} \text{ C}, \qquad \text{or} \qquad 6.9 \text{ nC}$$

The form of Coulomb's law (Equation 9-4) is exactly the same as the form of Newton's law of gravitation (Equation 4-8). It just substitutes charge for mass and uses a different universal constant. But there are important differences. First, mass is always positive, so the gravitational force always has the same sign. Since charge can be either positive or negative, the force can be positive (repulsion) or negative (attraction). Also, note that the constant k is larger than G by a factor of 10^{20}! The electrical force is much too strong ever to be mistaken for gravity.

Core Concept

Electric force is proportional to the product of charges and inversely proportional to the square of the distance between them:

$$F = \frac{kq_1q_2}{r^2}$$

Try This Two graphite-coated Ping-Pong balls are charged, touched together, and then separated by 15 cm. If each exerts a repulsive force of 6.0×10^{-5} N on the other, how much charge is there on each ball?

9-5. Electric Fields

Both Newton's law of universal gravitation and Coulomb's law of electric force describe the interaction between points—point masses or point charges. This means that they will work well as long as the distance between the objects is much larger than the objects themselves. However, in many cases it is difficult or impossible to apply Coulomb's law. Two small, charged objects exert substantial forces on each other. Each excess (or deficient) electron on one object exerts forces on all the charges on the other object. To find the total force that one object exerts on the other, we would have to calculate each of these forces separately, using Coulomb's law, and then add them all up. Impossible!

There is a way around this difficulty, a technique for calculating electric forces that works very well in certain cases, even for small objects close together. We have to start by separating the electric interaction into two parts: (1) a charged object affects the space around itself, creating an *electric field* in that space; and (2) the electric field exerts a force on any other charge placed in it. In this approach, we do not deal with attraction or repulsion of charges. It is the field that exerts the force.

An electric field is a property of space. It can be detected by placing a positive test charge at any point in space. If a force is detected, acting on the charge, then there is an electric field at that location. By placing our test charge at many points, we can represent, using vectors, the properties of an electric field in any region, as in Figure 9-6. Each vector represents the

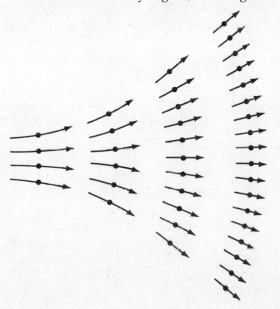

FIGURE 9-6

field at the point where it is drawn, that is, the force that would act on a unit positive charge placed there.

Figure 9-7 shows a simpler way to represent the same field. Like the vectors, the *field lines* represent the direction and magnitude of the field at any point. The direction is simply the direction in which the field line points at any position. The magnitude is the line density; the lines are closest together where the field is strongest. Thus, any positive charge placed in the field will be pushed in the direction of the field line that passes through that location. A negative charge will be pushed in the opposite direction. At any given point in space, the force the field exerts on a charge is proportional to the charge.

FIGURE 9-7

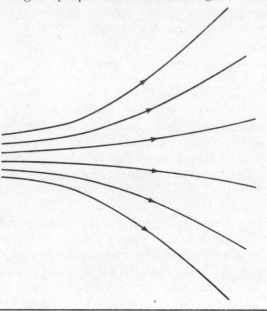

Core Concept

An electric field may be mapped by means of field lines, indicating the direction and magnitude of the force exerted by the field on a positive test charge at any point.

Try This A charge of $+5$ nC placed at a certain location in an electric field experiences a force of 2×10^{-5} N pushing it to the west. What are the direction and magnitude of the force that the field at the position will exert on a charge of -15 nC?

9-6. The Field Rules

Any charged object is surrounded by a region of space in which a small test charge will experience a force. The charged object creates a field in the space around itself.

We can visualize the properties of these fields by using field lines. They must obey certain rules:

1. Field lines point away from positive charges and toward negative charges, since positive objects will repel our positive test charge. In fact, every field line starts on a positive charge and ends somewhere in a negative charge.
2. The more charge on an object, the more field lines attached to it. This can be made quantitative by drawing a number of field lines proportional to the charge.

If the object is a conductor, there are further restrictions because the charges on a conductor will distribute themselves until all are in equilibrium. In this condition the following additional rules apply:

3. There is no field inside the object.
4. All field lines are perpendicular to the surface of the object. If any line were not, it would have a component parallel to the surface, and the charge in the surface would be set in motion.
5. The field lines are closest together at the surface where there are points and edges, since this is where the charge is most concentrated.

Using these rules, we can plot some fields. Figure 9-8 shows the field in the neighborhood of an irregular object with a positive charge, of two oppositely charged spheres, and of two negatively charged spheres. All are conductors, so they must obey all five rules.

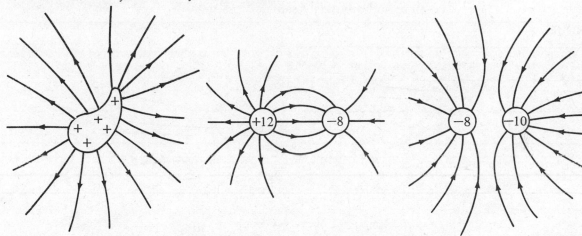

FIGURE 9-8

One field that is of special interest is the field between two oppositely charged flat plates set close together and parallel to each other. Since they are flat, the charge on them is evenly distributed (except near the edges), and so are the field lines. The lines are perpendicular to both plates—they run straight and parallel from one plate to the other, as shown in Figure 9-9. This is a *uniform field*. A charge placed anywhere between the plates will experience the same force, as long as it is not near the edge.

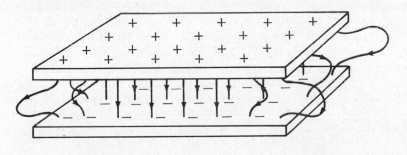

FIGURE 9-9

Core Concept

Charged objects are surrounded by electric fields, which obey certain definite rules.

Try This Using the known rules that control the formation of electric field lines, explain why the lines around a negatively charged sphere are radially inward and uniformly spaced.

9-7. Electrostatic Induction

A small metal needle, pivoted in the center like a compass needle, can be used to indicate the direction of an electric field. It works because an electric field will separate the charges in it.

How this works is illustrated in Figure 9-10. The field pushes positive charge in one direction and negative charge in the other. The positive charges cannot move, but the negative charges (electrons) move up the field, making one end negative and leaving the other end with an excess of positive charge. This effect, the separation of charge by the action of an electric field, is called *electrostatic induction.* Once the charges have been separated, the field exerts forces on the charges in opposite directions. As long as the forces are not lined up, they will result in a torque that tends to pull the needle into line with the field. The needle will oscillate around this position and eventually come to rest there.

FIGURE 9-10

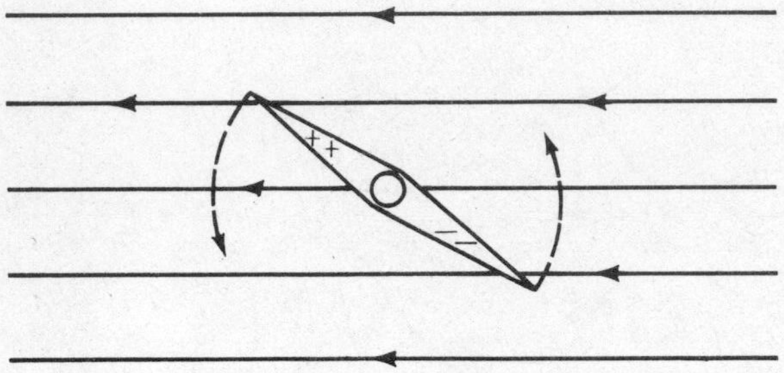

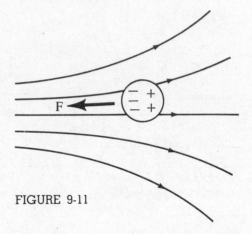

FIGURE 9-11

Electrostatic induction accounts for the commonly observed attraction of a neutral object to any charged object. The little ball in Figure 9-11 is in a *nonuniform* electric field. The field induces a charge separation in the ball. As shown, the positive end is in a weaker field than the negative end, so the force pushing the object up the field is stronger than the force on the positive end pushing it the other way. The object will move toward the stronger field. That is why you can pick up bits of paper with your comb.

If you place two conductors in a field, in contact with each other, they will acquire opposite charges by induction. You can separate them and move them out of the field, and you will have a ball with a positive charge and another with a negative charge, as in Figure 9-12.

If the field is strong enough, electrostatic induction will produce a spark. The field separates the positive nucleus of air atoms from their negative electrons and sends them shooting through the air in opposite directions. After this happens once, the particles accelerate in the field until

FIGURE 9-12

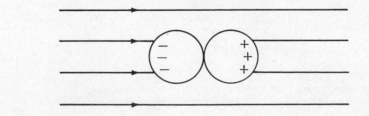

they hit another atom—causing it to split into positive and negative parts which repeat the process. In a very short time, air atoms in the trillions of trillions are giving up their electrons, and their tremendous speed agitates the surrounding air atoms to incandescence. Spark.

Core Concept

An electric field will separate the positive and negative charges of a neutral object.

Try This Explain why a neutral object placed between a pair of parallel charged plates may rotate but will not experience any net force.

9-8. Electricity has Energy

To separate negative and positive charges from each other, work must be done against the force of attraction. Therefore separated charges are in a high-energy state. When the charges are brought together again, energy must be released. It may be in the form of a spark. When you plug in a lamp and turn it on, you are using this electric potential energy, converting it into heat and light.

A charge in an electric field has energy, just as a mass in a gravitational field has gravitational potential energy. Consider a small test charge in the electric field around an irregular charged object, as in Figure 9-13. The field is pushing it in the direction of the field lines. If some external force pushes it the other way, against the direction of the field, work is being done on it. Therefore, the positive charge has its greatest electric potential energy at the upper end of the field. Since the field pushes a negative charge the other way, its potential energy is greatest at the low end of the field. If released, positive charges fall down the field, and negative charges fall up it, losing potential energy as they go.

It takes no work to move a charge perpendicular to the field lines. This is like carrying a mass horizontally in a gravitational field; the force is perpendicular to the displacement. Equation 5-3 tells us that, when this is done, there is no change in the energy of the object, since the cosine of the angle between force and displacement is zero.

This gives us another method of plotting the geometry of an electric field. Figure 9-14 represents the same field as Figure 9-13, but a series of *equipotential* lines have been drawn on it. Each of the lines is drawn so that it is perpendicular to the field lines at all points. When a charge moves along an equipotential, its displacement is always perpendicular to the field lines, and thus perpendicular to the force on it. Therefore, it takes no work to move a charge along an equipotential. An equipotential is a line along which a charge always has the same electric potential energy.

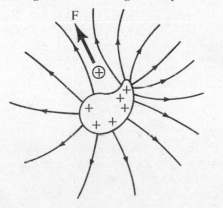

FIGURE 9-13

FIGURE 9-14

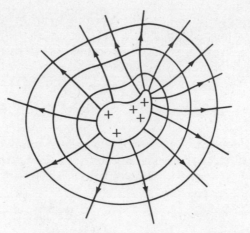

Note that the surface of any charged conductor is always equipotential, since the field lines must be perpendicular to the surface. Between the parallel plates of Figure 9-9, where the field is uniform, the equipotentials form a series of planes parallel to the plates and equally spaced between them. In a uniform field, this is the only shape that equipotentials can have so as to be everywhere perpendicular to the field lines. See Sample Problem 9-4.

Sample Problem

9-4 A pair of charged plates like those of Figure 9-9 are 3.0 cm apart, and the upper plate is negative. The field in between the plates exerts a force of 2.0×10^{-5} N on a charged object between the plates, pushing it downward. (a) How much work would it take to move the object from the lower plate to the upper plate? (b) How much work would it take to move the object 6 cm horizontally? (c) Is the charge positive or negative?

Solution (a) From Equation 5-3, $W = \mathbf{F}\,\Delta\mathbf{s}\cos\theta$; here the force is in the direction of motion, so $\cos\theta = 1$. Therefore,

$$W = (2.0 \times 10^{-5}\text{ N})(0.030\text{ m}) = 6.0 \times 10^{-7}\text{ J}$$

(b) $\theta = 90°$ and $\cos 90° = 0$, so no work is done.
(c) The upper (negative) plate is repelling the charge, so the charge must also be negative.

Core Concept

Equipotentials are surfaces of equal potential energy for any given charge and are always perpendicular to the field lines.

Try This If the field of Figure 9-9 exerts a force of 6.5×10^{-5} N on a charge, and the plates are 5.0 cm apart, what is the change in the electric potential energy of the charge as it moves from the left end of the lower plate to the right end of the upper plate?

9-9. Potential

The gravitational potential energy of an object depends on its mass and its location in a gravitational field. Similarly, its electric potential energy depends on its charge and its location in an electric field.

The crucial question about the location of an object in an electric field is this: On what equipotential is it located? It takes a definite amount of energy to move each unit of charge from one place to another. We define the *potential difference* between two points in the field as the work that would have to be done to move a unit charge from one of the points to the other:

$$\Delta V = \frac{\Delta E_{el}}{q} \qquad \text{(Equation 9-9)}$$

where ΔV is the potential difference, ΔE_{el} is the difference in potential energy, and q is the charge. The SI unit of potential difference is the joule per coulomb, which is called a *volt* (V). See Sample Problems 9-5, 9-6, and 9-7 for examples of the use of this relationship.

Sample Problem

9-5 How much work does it take to move a charge of $+30$ nC through a potential difference of 6.0 V?

Solution   From Equation 9-9:

$$\Delta E = q\,\Delta V = (30 \times 10^{-9}\ \text{C})\left(6.0\ \frac{\text{J}}{\text{C}}\right) = 1.8 \times 10^{-7}\ \text{J}$$

Sample Problem

9-6 A charge of -4.5 nC is moved in an electric field from a position where the potential is $+2.0$ V to another where the potential is -15 V. (a) Is the charge gaining or losing energy? (b) Must work be done on it, or does it release energy? (c) What is its change in energy?

Solution   (a) A negative charge moving to a lower potential is gaining energy.
(b) Since it is gaining energy, work must be done on it.
(c) $\Delta E = q\,\Delta V = (-4.5 \times 10^{-9}\ \text{C})(-17\ \text{V}) = +7.7 \times 10^{-8}\ \text{J}$.

Sample Problem

9-7 How much is the energy change of a charge of -20 nC that moves in an electric field from an equipotential of $+3$ V to an equipotential of $+12$ V?

Solution   From Equation 9-9,

$$\Delta E_{el} = q\,\Delta V$$

In going from a potential of $+3$ V to one of $+12$ V, the charge is going through a potential difference $\Delta V = +9$ V. Therefore its energy change is

$$\Delta E_{el} = (-20 \times 10^{-9}\ \text{C})\left(+9\ \frac{\text{J}}{\text{C}}\right)$$

$$\Delta E_{el} = -1.8 \times 10^{-7}\ \text{J}$$

In moving to a higher potential, a *negative* charge *loses* energy.

The volt is a convenient unit for a number of reasons. For one thing, it can be measured directly by means of an instrument called a *voltmeter*. If you connect the terminals of a voltmeter to the terminals of an ordinary zinc–carbon dry cell battery, the voltmeter will read 1.5 volts. This means that the chemical activity inside the battery separates charge in such a way that it adds 1.5 joules of energy to each coulomb of charge it separates. The size of the battery does not matter; a big one will be able to provide more energy and do it more quickly, but the amount of energy per unit charge, the potential difference, is the same for a C cell as for an AA cell. Batteries that give a larger potential difference are made of combinations of cells, unless they use a different combination of chemicals to produce the electric energy. See Sample Problems 9-8 and 9-9.

Sample Problem

9-8 How much energy must a battery add to a charge of +35 C in a circuit to move it from a point at a potential of 60 V to another point where the potential is 83 V?

Solution   The potential difference is 83 V − 60 V = 23 V. Then the increase in energy of the charge is $\Delta E = q \, \Delta V = (35 \text{ C})(23 \text{ V}) = 805 \text{ J}$.

Sample Problem

9-9 To toast a couple of slices of bread, a toaster has to use 30,000 J of energy, drawn from a 110 V wall outlet. How much charge flows through the toaster?

Solution   From Equation 9-9:

$$q = \frac{\Delta E_{el}}{\Delta V}$$

$$q = \frac{30{,}000 \text{ J}}{110 \text{ J/C}} = 270 \text{ C}$$

Core Concept

Potential difference, in volts, is the difference in electric energy per unit charge:

$$\Delta V = \frac{\Delta E_{el}}{q}$$

Try This   If it takes 850 J of energy in the starter motor, supplied by a 12-V battery, to start your engine, how much charge flows through the starter motor?

9-10. Electrons and Such

With a voltmeter and a battery supplying an adjustable potential difference, there is a method of measuring extremely small charges. This experiment, when first performed, had a profound effect on the whole theory of physics.

The apparatus is illustrated in Figure 9-15. Two flat plates, a few centimeters apart, are set in a horizontal position and attached to opposite terminals of an adjustable electric energy source. A voltmeter measures the potential difference between them, and a switch allows this meter to be turned on and off. An atomizer produces tiny droplets of oil, which fall through a tiny hole in the upper plate into the uniform electric field between the plates. As the droplets fall, they are watched through a calibrated microscope. With a stopwatch, the experimenter can determine how long it takes a droplet to fall through a distance measured by the microscope.

To do the experiment, turn off the electric field and squeeze the atomizer bulb. When you see a falling droplet, time its passage between the marks in the microscope to determine the speed with which it is falling. Then turn on the electric field and adjust the potential difference until the droplet stands still. Read the voltmeter, and you will have enough information to calculate the charge on the droplet.

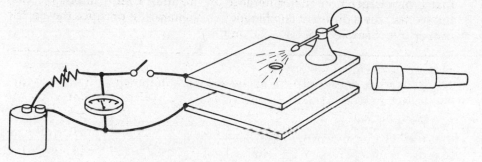

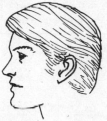

FIGURE 9-15

Here's how: First of all, when the electric field was turned off, the droplet was falling at its terminal velocity. Knowing the velocity, it is possible to calculate the viscous drag produced by its movement through the air. At terminal velocity, this is equal to the weight of the droplet. So, with the field turned off, its velocity tells us how much the droplet weighs.

Now turn the field on and adjust the potential difference between the plates until the droplet stands still. Now the electric force exerted on the droplet by the field is equal to the weight of the droplet, so we know how much the electric force is. We read the voltmeter and determine how much potential difference between the plates it takes to produce this much electric force.

If the droplet were allowed to travel the entire distance Δs between the plates, the change in its energy would be the work done on it, $F \Delta s$. Since F is its weight, we can say that $\Delta E = w \Delta s$. Also, from Equation 9-9, the change in its energy would be $q \Delta V$. Therefore,

$$w \Delta s = q \Delta V$$

and we know all the quantities in this equation except q. Solve for q, and you will have the charge on the droplet. See Sample Problem 9-10.

Sample Problem

9-10 What is the electric charge on an oil droplet whose weight is 2.9×10^{-15} N if it is held stationary by the electric field between two horizontal plates separated by 5.0 cm when the potential difference between the plates is 90 V?

Solution The electric force must be equal to the weight. The force times the distance between the plates must be equal to the work that would have to be done to move the charge from one plate to another, which is (from Equation 9-9) equal to $q \Delta V$. Therefore

$$(2.9 \times 10^{-15} \text{ N})(0.050 \text{ m}) = q \left(90 \, \frac{\text{J}}{\text{C}}\right)$$

from which $q = 1.6 \times 10^{-18}$ C.

The most remarkable outcome of this experiment was the discovery that every value of charge measured by this method turned out to be a definite integral multiple of the same small charge: 1.60×10^{-19} C. This implied that charges can be added or subtracted from the droplet only in quanta, or steps that are all the same size. This electric charge quantum is the charge on an electron.

Since this experiment was first performed, over 100 subatomic particles have been found, and every one of them is either neutral or has exactly one quantum of charge, positive or negative. Electric charge is not infinitely subdivisible, and all charges everywhere consist of some whole number of quanta.

Core Concept

Electric charges exist only in whole-number multiples of the electron charge, 1.60×10^{-19} C.

Try This In an oil drop experiment, a droplet is found to have a charge of $+8.0 \times 10^{-19}$ C. How many electrons has it lost?

9-11. Conduction

Why are metals such good conductors? Chemists tell us that what makes a metal a metal is the ease with which an electron can be removed from each atom. In fact, there is reason to believe that in the best conductors—silver, copper, gold, aluminum—every atom is short one electron, and these electrons are moving quite freely through the spaces between atoms. One fairly successful model treats the electrons mathematically like the atoms of a gas, bouncing around freely in all directions and rebounding from each other and from the atoms that make up the mass of the metal.

These free electrons have a surprisingly large charge. Take a cubic centimeter of copper, for example, which has a mass of 8.9 grams. Any chemist could tell you that this much copper contains 8.4×10^{22} atoms. If each atom provides one free electron, the total free charge in the copper is

$$(8.4 \times 10^{22} \text{ electrons})(1.60 \times 10^{-19} \text{ C/electron})$$
$$= 1300 \text{ C}$$

When the cubic centimeter of copper is uncharged, there is also 1,300 C of positive charge in it—in the atomic nuclei. If you charge a 1-cm^3 sphere of copper as much as possible, you might put an additional 20 nC of charge onto it—less than a hundred billionth of the free charge in it. This is like adding one drop of water to a rather large swimming pool.

Those electrons are bouncing around like mad. We can find out how fast they are going by applying the ordinary gas laws (Equation 7-10), which tell us that

$$\bar{E}_{\text{kin}} = \frac{3}{2}kT \qquad \text{(Equation 7-10)}$$

from which

$$\frac{1}{2}mv^2 = \frac{3}{2}kT \qquad \text{or} \qquad v = \sqrt{\frac{3kT}{m}}$$

The mass of the electron has been measured by a method that you will meet in Chapter 11; it is 9.1×10^{-31} kg. Using $T = 300$ K and

$$k = 1.38 \times 10^{-23} \text{ J/molecule} \cdot \text{K},$$

we get the average speed of the electrons,

$$v = 1.2 \times 10^5 \text{ m/s}.$$

This is fast enough to get from Chicago to New York in 8 seconds. This is all random motion, so no net movement of the electrons can be detected. At any given moment, just as many are going one way as the other.

Now suppose we stretch our cubic centimeter of copper out into a long wire, say 10 m long and 0.1 mm^2 in cross section. We attach the ends to the terminals of a 1.5-volt dry cell. The excess electrons on the negative terminal of the dry cell will now have room to spread out, repelling each other, and will move into the wire. At the same time, some of the free electrons in the wire will move into the positive terminal of the dry cell. This starts up the chemical action inside the cell, which continuously replaces the electrons at the negative terminal and removes them from the positive terminal. There will be a steady flow of charge through the wire, electrons traveling from the negative terminal of the battery to the positive.

The rate of flow of this charge can be measured. Under the conditions given, it comes to about 0.8 coulombs per second. This implies that the steady progression of the electrons down the wire is very slow. If there are 1,300 coulombs of free charge in the wire and they come out the end at only 0.8 C/s, the length of time it takes the last electron to travel the whole length of the wire is

$$\frac{1,300 \text{ C}}{0.8 \text{ C/s}} = 1,600 \text{ s}$$

If it takes that last electron this long to travel 10 meters then its *drift velocity* in the wire is 6 mm/s—a snail's pace on a good day.

Core Concept

Charge travels through a wire in the form of free electrons, in extremely rapid random motion and very slow drift velocity from the negative to the positive terminal of a battery.

Try This As the temperature increases, what happens to the (a) random velocity; and (b) drift velocity of the electrons in a wire?

Summary Quiz

Supply the missing word or phrase:

1. Two similar electric charges _____ each other.
2. Comparing electric and gravitational forces, it will be found that the _____ is stronger.
3. The charge on a rubber rod is designated as _____ .
4. When you comb your hair, the comb becomes negative because _____ have been transferred to it from your hair.
5. Metals are said to be good _____ because electric charge moves freely through them.
6. When a conductor is charged, all the charge is on its _____ .
7. To confine electric charge and prevent it from escaping, it can be surrounded by a(n) _____ .
8. On a charged conductor, the charge is most concentrated where the _____ is greatest.
9. The SI unit of electric charge is the _____ .
10. The electric force between two charges is proportional to their _____ .
11. Doubling the distance between two charges changes the force between them by a factor of _____ .
12. Electric field lines are _____ where the field is strongest.
13. An electric field will exert a force on a charge in the direction opposite the field if the charge is _____ .
14. In a graphical representation of the electric field around a charged object, the number of field lines is proportional to the _____ on the object.
15. Electric field lines start on _____ charges.
16. On a charged conductor, the field lines are _____ to the surface.
17. The field between a pair of oppositely charged plates, located close together, is _____ .

18. Electrostatic induction is the process by which a(n) _____ separates positive and negative charges from each other.
19. A charge moving through an electric field will have no change in its electric potential energy if it moves along a(n) _____ .
20. Equipotentials are always perpendicular to _____ .
21. Electric potential difference is measured in _____ .
22. A 6-V battery adds 6 J of energy to each _____ .
23. All electric charges are integral multiples of the charge on a(n) _____ .
24. Charge passes through a wire in the form of _____ .
25. The best conductors of electricity provide one free electron per _____ of metal.

Problems

1. Two identical conductive spheres are charged at +32 nC and −12 nC, respectively. If they are touched together and then separated, how much charge will each have?

2. Two conductive spheres are separated by 15 cm, measured center to center. Sphere A is charged to +25 nC and sphere B to +15 nC. (a) How much force does A exert on B? (b) How much force does B exert on A?

3. Two 3.0-g Ping-Pong balls are suspended side by side, with 10 cm between their centers, from strings 60 cm long. They are charged and then touched together. If they are then separated by 18 cm, how much charge is there on each ball?

4. At a certain location in space, the electric field exerts a force of 3.0×10^{-3} N on a Ping-Pong ball bearing a charge of 12 nC. What is the charge on an object in that position if the electric force on it is 2.5×10^{-5} N?

5. A 1-cm plastic sphere and a 5-cm plastic sphere, both neutral, are placed in an electric field and then separated from each other. If the small sphere acquires a charge of −12 nC, what is the charge on the large one?

6. How much work must be done to move a charge of +220 nC from a place where the electric potential is +30 V to another position where the potential is +5 V?

7. How much energy is delivered to a motor if 6.0 C passes through it from a 24 V battery?

8. How much energy is there in the spark produced when a Van de Graaff generator terminal charged to 240 nC discharges to a terminal at a potential 110,000 V lower?

9. Two large, flat metal plates are set parallel to each other, separated by 12 cm, and connected to opposite terminals of a 240 V power supply. A plastic sphere carrying a charge of +2.0 nC is placed at the positive plate and released. Find (a) the energy lost by the sphere in falling to the other plate; and (b) the force exerted on the sphere by the electric field between the plates.

10. How many excess electrons are there on a Ping-Pong ball that has a charge on it of −8.0 nC?

11. How much energy is lost by an electron when it is part of the discharge spark of a Van de Graaff generator that had been charged to a potential of 120,000 V above the ground potential?

12. A cubic centimeter of copper, containing 1,300 C of free charge, is drawn into a very thin wire 150 m long and connected to the terminals of a battery. An ammeter indicates that charge enters one end of the wire at the rate of 0.50 C/s. What is the drift velocity of the electrons in the wire?

CHAPTER 10

The Power in the Wire

10-1. Making Electricity Useful

During the past century, electricity has changed our lives. Learning how to use it has endowed us with a crucial faculty that we have never had before: the ability to transport large amounts of energy over long distances quickly and efficiently.

Electric charge has always existed. We make it useful by doing work on it to separate the negative charges from the positive charges, thus using the charge to store energy. When the charges come back together and neutralize each other, the energy must be released. The trick we have learned is how to use this released energy in a variety of ways, often at a great distance from the point at which the original separation of charge took place.

Although we can separate charges by stroking a rubber rod with fur, this is not a practical way to store large amounts of energy in the separated charges. What we need is a means of separating large amounts of charge and of doing it *continuously*. We need a gadget that will keep adding energy to electric charge while that energy is being used somewhere else.

There are two kinds of devices in common use that can do this: the battery and the generator. In a battery, a chemical process removes electrons from one terminal and deposits them on another, thus adding energy to the charge. A generator does the same thing by taking advantage of magnetic phenomena; you will see how this works in Chapter 11.

195

Any source of electric energy must get the energy from somewhere. The battery gets energy from the chemical reaction. The generator gets energy from the work done in turning it. Any such source can be rated in terms of the electric energy it produces for each unit of charge separated. This quantity—the energy per unit charge converted from some other form to electrical—is called the emf (pronounced ee-em-eff) of the source. The letters emf stand for "electromotive force," but this is a very poor name, left over from the days when electricity was not well understood. The emf of a source is not a force; it is a potential, a value of energy per unit of charge. It is measured in joules per coulomb (volts) and expressed as

$$\mathscr{E} = \frac{\Delta E_{el}}{q}$$
(Equation 10-1)

The common zinc–carbon dry cell, or the newer alkaline cell, produces an emf of 1.5 volts regardless of its size (Figure 10-1). A large cell produces more energy by separating more charges, but the energy added to each coulomb separated depends only on the kind of chemical reaction that does the job. If your calculator uses a 9-volt battery, it is made of six dry cells, each adding 1.5 joules of electric energy to each coulomb that passes through it.

FIGURE 10-1

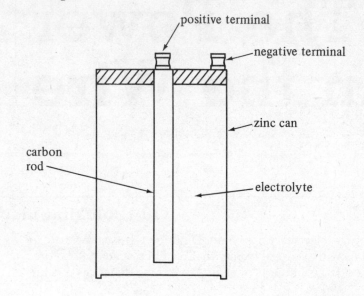

Core Concept

The emf of a source, measured in volts, is the energy per unit charge converted from some other form to electrical: $\mathscr{E} = \Delta E_{el}/q$.

Try This How many dry cells are there inside the 22.5-volt battery used to run a hearing aid?

10-2. Producing Electric Current

A battery, with a concentration of electrons at one terminal and an excess positive charge at the other, is a storehouse of electric energy. If a metal wire is connected from one terminal to the other, electrons will run off the negative terminal into the wire, and out of the wire into the positive terminal. All the electrons in the wire will shift position, drifting toward the higher potential. As they travel, they lose electric energy, which is converted into random motion of molecules and electrons inside the wire. The wire gets hot.

With an instrument called an *ammeter* we can measure the rate at which the charge is flowing out of the battery, through the wire, and back into the battery. Figure 10-2 shows how this meter can be connected to measure the rate of flow of charge, and the diagram, using standard symbols, is the way the circuit is represented schematically. The ammeter indicates the rate at which the charge is flowing through it in coulombs per second, or *amperes* (A). This quantity is called *current* and is represented by the symbol *I*:

$$I = \frac{\Delta q}{\Delta t}$$ (Equation 10-2)

See Sample Problem 10-1 for the use of this equation.

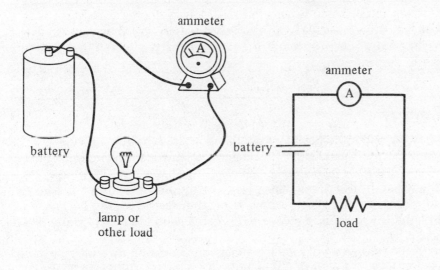

FIGURE 10-2

Sample Problem

10-1 If a light bulb uses 0.50 A of current, how much charge flows through it in 5 min?

Solution   Five minutes is 300 s, so from Equation 10-2,

$$\Delta q = I \, \Delta t = (0.50 \text{ A})(300 \text{ s}) = 150 \text{ C}$$

Compared with static charges, the current, even that in a flashlight bulb, is enormous. Charge on a Van de Graaf generator is measured in billionths of a coulomb, and we see a spectacular display of fireworks when it discharges. Yet the ammeter in Figure 10-2, registering the current when a small dry cell is used to operate a flashlight bulb, might well be a couple of amperes. But all this charge does not accumulate anywhere. The current flows out of the bulb as fast as it flows in—and the same can be said of the wires, the ammeter, and the battery. The ammeter tells us the rate at which the charge is flowing through every part of the circuit.

A current may consist of either kind of charge in motion. In a wire, it is the negative electrons that are moving; in a cyclotron it might be positive protons; in a plating solution or a battery, both kinds of charge move, in opposite directions. We define the direction of a current as the direction of flow of *positive* charge. Electrons flowing northward produce many of the same effects as positive charges flowing southward; in both cases the ammeter tells us that the direction of the current is southward. In the circuit of Figure 10-2, the current flows out of the positive terminal of the battery, through the ammeter, through the bulb, and then back into the negative terminal of the battery and through the battery to the positive terminal.

10-3. Delivering the Energy

Charge flowing out of the positive terminal of a battery is at a high energy level; when it enters the negative terminal, its energy is low. The function of the battery is to boost the energy level of the charge.

Somewhere in the circuit, the energy of the flowing charge is converted into other forms. When a bulb is lit, the electric energy becomes heat and light. If you want to measure this conversion of energy, you will need a *voltmeter*.

Figure 10-3 shows a voltmeter in place for measuring the energy lost by the current as it passes through the bulb. A voltmeter, unlike an ammeter, is designed in such a way that very little current goes through it. Generally, we will be able to treat the voltmeter as though it is not there at all, ignoring any current it uses. The terminals of a voltmeter are connected to two different points in the circuit, and it registers the difference of potential between these two points.

FIGURE 10-3

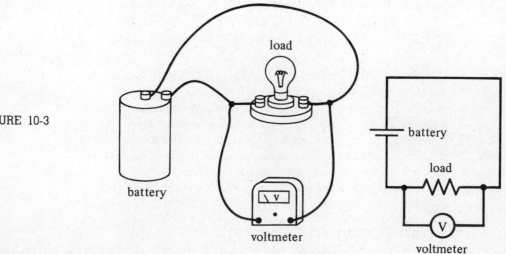

Potential difference between two points is the energy difference per unit charge; if the voltmeter in the circuit reads 1.5 volts, it means that every coulomb that passes through the bulb loses 1.5 joules of energy. Therefore, the total energy converted in the bulb, from Equation 9-9, is

$$\Delta E_{el} = q\, \Delta V$$

For an example see Sample Problem 10-2.

Sample Problem

10-2 If a light bulb operates on 110 V, how much energy does it convert from electrical to other forms when 150 C of charge passes through it?

Solution From Equation 9-9,

$$\Delta E = q\, \Delta V = (150 \text{ C})(110 \text{ V}) = 16,500 \text{ J}$$

The ammeter and a clock can tell us the amount of charge that passes through the bulb, since (Equation 10-2)

$$q = I\, \Delta t$$

Consequently, the total amount of energy converted to heat and light in the bulb is

$$\Delta E_{el} = I\, \Delta t\, \Delta V$$

See Sample Problem 10-3 for this calculation.

Sample Problem

10-3 A light bulb operating on 110 V has 0.50 A of current passing through it. How much energy does the bulb convert in 5 min?

Solution From Equation 10-2, the charge passing through the bulb is $I\, \Delta t = (0.50 \text{ A})(300 \text{ s}) = 150$ C. From Equation 9-9, the energy it converts is

$$q\, \Delta V = (150 \text{ C})(110 \text{ V}) = 16,500 \text{ J}$$

If you examine the bulb carefully, you might find that it bears a label reading, perhaps, "1.5 V, 4 W," which is read as "1.5 volts, 4 watts." We previously met the watt as a unit of power, equal to one joule per second, and we used power as a measure of the rate of doing work. Now we must use the word in its more general sense—to mean the rate of any kind of energy conversion. The label on the bulb indicates that, if the bulb is connected to a 1.5-volt battery, it will convert 4 joules of electric energy into other forms every second.

Since power is the rate of energy conversion, it is equal to $\Delta E_{el}/\Delta t$. Therefore the equation above becomes

$$P = I\, \Delta V \qquad\qquad \textbf{(Equation 10-3)}$$

See Sample Problems 10-4 to 10-7.

Sample Problem

10-4 What is the power rating of the light bulb of Sample Problem 10-3?

Solution From Equation 5-10, as modified, power is

$$\frac{\Delta E_{el}}{\Delta t} = \frac{16,500 \text{ J}}{300 \text{ s}} = 55 \text{ W}$$

In electrical circuits, power can also be found from Equation 10-3:

$$P = I \, \Delta V = (0.50 \text{ A})(110 \text{ V}) = 55 \text{ W}$$

Sample Problem

10-5 An electric iron is rated at 900 W on a 120-V circuit. How much current does the iron draw from the outlet?

Solution   From Equation 10-3,

$$I = \frac{P}{\Delta V} = \frac{900 \text{ W}}{120 \text{ V}} = 7.5 \text{ A}$$

Sample Problem

10-6 A wall outlet, operating on 120 V, delivers 4 A to an immersion heater for 10 min. If the heater is used to heat 2,000 g of water, how much does the temperature of the water rise?

Solution   The energy delivered is

$$\Delta E = I \, \Delta t \, \Delta V = (4 \text{ A})(600 \text{ s})(120 \text{ V}) = 288,000 \text{ J}$$

According to Equation 7-4, this must be equal to $cm \, \Delta T$:

$$(288,000 \text{ J})\left(\frac{\text{cal}}{4.19 \text{ J}}\right) = \left(1 \, \frac{\text{cal}}{\text{g} \cdot \text{C}°}\right)(2,000 \text{ g})(\Delta T)$$

Therefore $\Delta T = 34 \text{ C}°$.

Sample Problem

10-7 At a certain moment, all the lights and appliances in your home are using 65 A on the 120-V circuit. What is the rate of energy consumption?

Solution   The rate of energy consumption is power, and from Equation 10-3,

$$P = I \, \Delta V = (65 \text{ A})(120 \text{ V}) = 7,800 \text{ W}$$

or 7.8 kW (kilowatts).

Core Concept

Electric power, the rate of conversion of electric energy into other forms, is equal to the product of current and potential difference:

$$P = I \, \Delta V$$

Try This   How much current flows through a 600-W toaster in use on a 120-V circuit?

10-4. How Much Current?

When you connect a wire to a battery, the amount of current in the wire depends on two things: the nature of the wire and the potential difference provided by the battery. A copper wire has a lot more current in it than an iron wire of the same dimensions; a linen thread has no current at all. There are good conductors and bad conductors.

Also, for a given wire, the current will be greater if the potential difference between its ends is greater. A wire, or any other electrical device, has a property called *resistance*, which tells how much current it will have for any given value of potential difference between its ends:

$$R = \frac{\Delta V}{I}$$

(Equation 10-4)

where I is current in a given conductor, R is resistance, and ΔV is potential difference between the ends. For examples, see Sample Problems 10-8 through 10-11. Note that the unit of resistance, the volt per ampere, is called an *ohm* and is represented by Ω, the Greek letter omega.

Sample Problem

10-8 When a wire is connected to a 9-V battery, the current is 0.020 A. What is the resistance of the wire?

Solution From Equation 10-4,

$$R = \frac{\Delta V}{I} = \frac{9\ V}{0.020\ A} = 450\ \Omega$$

Sample Problem

10-9 How much is the current in a light bulb whose resistance is 350 Ω when the bulb is connected to a 100-V outlet?

Solution From Equation 10-4,

$$I = \frac{\Delta V}{R} = \frac{110\ V}{350\ \Omega} = 0.31\ A$$

Sample Problem

10-10 In an electric circuit, a current of 0.025 A is flowing through a 2,200-Ω resistance. What is the potential difference between the ends of the object?

Solution From Equation 10-4,

$$\Delta V = IR = (0.025\ A)(2{,}200\ \Omega) = 55\ A$$

Sample Problem

10-11 A relay with a resistance of 12 Ω is in an electric circuit with one side at a potential of 85 V and the other at 71 V. How much current is in the relay?

Solution The potential difference across the relay is 85 V − 71 V = 14 V. From Equation 10-4,

$$I = \frac{\Delta V}{R} = \frac{14\ V}{12\ \Omega} = 1.2\ A$$

For a metal conductor, such as a wire, the resistance is constant at any given temperature. For other kinds of devices, such as transistors, solutions, plasmas, and vacuum tubes, the relationship between current and potential is much more complex, and resistance is not a constant.

A *resistor* is a conductor that is made especially to have a particular resistance. These devices are used in electronic circuits to control currents and potentials. Each resistor is calibrated according to its resistance; many are marked in kilohms ($= 10^3\ \Omega$) or megohms ($= 10^6\ \Omega$). Some are made of a clay–graphite mixture much like the "lead" in a pencil; others consist of a wire wrapped around a ceramic core. A variable resistor has a resistance that can be altered by turning a shaft. Many of the controls you turn when you use a radio or a television set are variable resistors. See Sample Problem 10-12.

10-12 In a circuit, a variable resistor is to be connected across a potential difference of 60 V to allow the current to be adjusted in the range from 40 mA to 100 mA (mA = milliamperes, or 10^{-3} A). Over what range must the resistance vary?

Solution   1 mA = 10^{-3} A; since $R = \Delta V/I$, the required range is

$$\frac{60 \text{ V}}{40 \times 10^{-3} \text{ A}} \quad \text{to} \quad \frac{60 \text{ V}}{100 \times 10^{-3} \text{ A}}$$

$$1,500 \ \Omega \quad \text{to} \quad 600 \ \Omega$$

Core Concept

Resistance, measured in ohms, is the ratio of potential difference to current, and it is constant for metals and some other materials at constant temperature:

$$R = \frac{\Delta V}{I}$$

Try This   A current of 0.15 A flows through a 20-Ω resistor away from a point in a circuit where the potential is 45 V. What is the potential at the other end of the resistor?

10-5. How Much Resistance?

Let's take a closer look at what goes on inside a wire in which there is a current. We ought to be able to figure out what determines the resistance of a wire and why it is constant.

Electrons drift down the wire, propelled by the electric field in the wire. Since fat wires have more electrons than skinny wires, we should expect that there will be more current in a fat wire than in a skinny one. The current is proportional to the cross-section area of the wire:

$$I \propto A$$

Current also depends on how fast the electrons are going, on their drift velocity. The faster they are moving, the greater the number of electrons passing a given point every second. The drift velocity depends directly on the force that the electric field exerts on the electrons.

We can get a useful relationship by considering the energy change of the electrons as they pass through the wire. From Equation 9-9, this must be equal to $q \, \Delta V$. This is also the work done on the charge by the field, so, from Equation 5-3, the energy change is also $F \, \Delta s$. Since Δs is the length of the wire (l), it follows that

$$q \, \Delta V = Fl$$

or $F/q = \Delta V/l$. But since the current is proportional to F/q, the force acting on the charges,

$$I \propto \frac{\Delta V}{l}$$

combining the two proportionalities gives

$$I \propto \frac{A \, \Delta V}{l}$$

One more factor enters. The number of electrons available to carry the current depends on the chemical composition of the wire and on its temperature. To allow for this factor, we can introduce a constant called the *resistivity* of the metal, represented by the Greek letter ρ (rho). Then we can write the equation

$$\rho I = \frac{A \, \Delta V}{l}$$

Now we can solve this equation for $\Delta V/I$, which is R, the resistance of the wire:

$$R = \frac{\rho l}{A} \qquad \text{(Equation 10-5)}$$

Now we see why the resistance of a wire is constant: The resistance depends only on the dimensions of the wire and on the material of which it is made. See Sample Problems 10-13 and 10-14.

Sample Problem

10-13 If the resistivity of copper is 1.72×10^{-6} $\Omega \cdot$cm, what is the resistance of a copper wire 35 m long and 0.025 cm^2 in cross section?

Solution From Equation 10-5,

$$R = \frac{\rho l}{A} = \frac{(1.72 \times 10^{-6} \; \Omega \cdot cm)(3{,}500 \; cm)}{0.025 \; cm^2}$$

$$R = 0.24 \; \Omega$$

Sample Problem

10-14 We would like to make a 600-Ω resistor out of Nichrome wire with a cross-section area of 3.0×10^{-3} cm^2. If the resistivity of Nichrome is 110×10^{-6} $\Omega \cdot$cm, how long should the wire be?

Solution From Equation 10-5,

$$l = \frac{RA}{\rho} = \frac{(600 \; \Omega)(3.0 \times 10^{-3} \; cm^2)}{110 \times 10^{-6} \; \Omega \cdot cm} = 1.64 \times 10^4 \; cm, \text{ or } 16.4 \; m$$

Temperature introduces a complication, since the resistivity of most materials varies with temperature. In metals, resistivity rises with temperature. When you turn on a light bulb, the current is very large at first but drops down to the rated value as the resistance of the filament increases. There are some metals whose resistance vanishes completely at temperatures near 0 K. If a current is started in a ring made of such a superconductor, it will flow around the ring forever. In semiconductors like carbon and silicon, the effect is exactly the opposite: Resistance decreases as the metals heat up.

Core Concept

Resistance depends on the resistivity of the metal and the dimensions of the wire:

$$R = \frac{\rho l}{A}$$

Try This What is the resistance of an aluminum wire 120 m long and 0.15 cm^2 in cross section at room temperature when its resistivity is 2.6×10^{-6} $\Omega \cdot$cm?

10-6. Big Batteries, Little Batteries

The shutter of my camera operates on a 1.5-volt dry cell about the size of a dime. In the laboratory, I often have occasion to use a 1.5-volt dry cell that stands 15 cm high and weighs over a pound. Why can't I use my camera battery in the laboratory?

The emf of a battery is the energy per unit charge converted from chemical to electrical. It is 1.5 V in both batteries. However, not all of this energy gets out of the battery. Some of it is converted into heat inside the battery itself.

If you connect a good voltmeter across the terminals of a dry cell, the voltmeter will indicate 1.50 volts. There is negligible current, since a good voltmeter has extremely high resistance. Even if the cell is dead, the voltmeter will indicate very nearly 1.50 volts. The voltmeter tells you how much energy would be lost by each coulomb of charge traveling from one terminal to the other. It says nothing about whether any charge actually moves.

If you connect an ammeter between the terminals, the story will be altogether different. Ammeters have extremely low resistance, and an ammeter is a short circuit, just as a short, thick copper wire would be. If the cell is one of those big #6 cells used in the laboratory, the ammeter will indicate about 30 A if the cell is fresh. If the cell is a tiny dime-sized one, the current might be only a thousandth of that value. If either battery is old and used up, you might not even be able to detect this short-circuit current.

With only the voltmeter attached, there is no current, and the meter registers the emf of the cell. With only the ammeter attached, the only resistance in the circuit is the *internal resistance* of the battery. The current is found, then, from Equation 10-4, using the emf of the cell instead of any measurable potential difference. A small cell has more internal resistance than a large one, and a dead cell has a great deal more internal resistance than a live one. When the cell is short-circuited, by an ammeter or a short, thick wire, all the energy converted by the chemical process goes into heating up the interior of the cell. See Sample Problem 10-15.

Sample Problem

10-15 A voltmeter connected across a small battery reads 9 V, and an ammeter reads 16 A. What is the internal resistance of the battery?

Solution The emf of the battery is 9 V, and the internal resistance is the only resistance in the circuit when the ammeter is used. Therefore

$$R = \frac{\mathscr{E}}{I} = \frac{9\ \text{V}}{16\ \text{A}} = 0.56\ \Omega$$

The current supplied by a battery depends on both the internal and external resistance of the circuit. This current can be found from the equation

$$I = \frac{\mathscr{E}}{R_{\text{int}} + R_{\text{ext}}}$$ (Equation 10-6a)

where $\mathscr{E}$ is the emf of the battery.

If you measure the potential difference across a battery supplying a lot of current, the voltmeter will read *less than* the emf of the battery. This terminal potential difference is the energy per unit charge supplied to the external resistance; it does not include the energy used up inside the battery itself, which is IR_{int}. The equation above can be written

$$IR_{int} + IR_{ext} = \mathcal{E}$$

and, since $IR_{ext} = \Delta V_{term}$, this becomes

$$\Delta V_{term} = \mathcal{E} - IR_{int} \qquad \text{(Equation 10-6b)}$$

See Sample Problem 10-16.

Sample Problem

10-16 The battery of Sample Problem 10-15 is connected to a resistance of 1.80 Ω. What will a voltmeter read if it is connected across the battery or the resistor?

Solution The current must be equal to the emf divided by the total resistance, according to Equation 10-6a:

$$I = \frac{\mathcal{E}}{R_{int} + R_{ext}} = \frac{9 \text{ V}}{0.56 + 1.80}$$

$$I = 3.8 \text{ A}$$

Then, according to Equation 10-6b,

$$\Delta V = \mathcal{E} - IR_{int} = 9 \text{ V} - (3.8 \text{ A})(0.56 \ \Omega)$$

$$\Delta V = 6.9 \text{ V}$$

Core Concept

The terminal potential difference of a battery is its emf minus IR_{int}, the energy lost per unit charge by the current inside the battery:

$$\Delta V_{term} = \mathcal{E} - IR_{int}$$

Try This How much is the current in a 3.0-V battery with an internal resistance of 0.20 Ω if the battery terminals are connected to an external resistance of 1.0 Ω?

10-7. Circuits

Current will flow only in a complete circuit consisting of a source of emf and a complete conducting path from its high-potential terminal to its low-potential terminal. The complete circuit shown in the circuit diagram of Figure 10-4 uses a battery as its source of emf. Starting at the high-potential end (the longer line of the battery symbol), the current passes through a resistance (such as a lamp), a switch, and an ammeter and then back to the battery. The voltmeter is connected between the ends of the resistance wire and registers the potential difference (the "IR drop") between the ends.

When the switch is open, electrons flow out of the battery, through the ammeter to the open switch. They can go no further. In a tiny fraction of a second, the potential on that side of the switch reaches the same level as the potential at the negative terminal of the battery, and there is no further flow of electrons. Similarly, the potential on the other side of the switch reaches the level of the potential at the positive terminal of the battery, and there is no current.

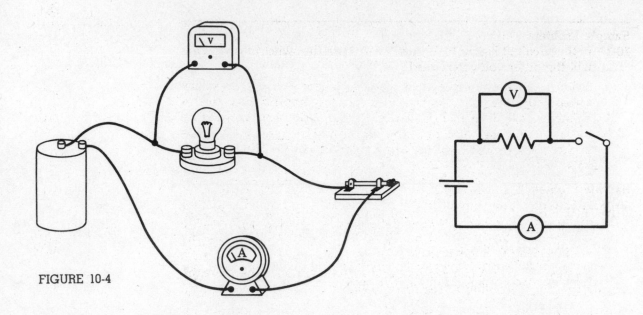

FIGURE 10-4

Now suppose we have four voltmeters in the circuit, as shown in Figure 10-5. V_1 reads the potential difference across the battery; with no current, this value is the same as the battery emf. V_2 reads the potential difference across the resistance, which is IR; with $I = 0$, there is no potential drop across the resistor. V_3 reads the potential drop across the switch; this value will be the same as V_1, since each side of the switch is at the same potential as the battery terminal to which it is connected. V_4 registers the potential drop across the ammeter; this value will be zero, since there is no current and no resistance in the ammeter.

Now what happens if the switch is closed? The ammeter will indicate the current in all parts of the circuit, which must be equal to the emf of the battery divided by the total resistance (internal plus external) of the whole circuit. The potential difference across the battery is its emf minus IR_{int}; across any other circuit element, the potential difference is IR. Since neither the switch nor the ammeter has any substantial resistance, V_3 and V_4 both read zero. V_1 and V_2 read alike. See Sample Problems 10-17 and 10-18 for detailed calculations.

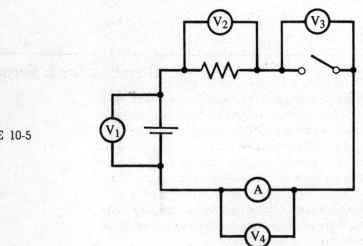

FIGURE 10-5

Sample Problem

10-17 In the circuit of Figure 10-5, $V_1 = 9.6$ V. With the switch closed, what will the other voltmeters read?

Solution Each voltmeter must read IR, where R is the resistance across which the voltmeter is connected. Since both a switch and an ammeter have negligible resistance, V_3 and V_4 are both zero. Since there is no resistance between the battery and the resistor, the ends of the resistor must be at the same potentials as the battery terminals, so V_2 reads 9.6 V.

Sample Problem

10-18 A battery with an emf of 12.0 V is connected in a circuit like that of Figure 10-5. If the internal resistance of the battery is 0.50 Ω and the external resistor is rated at 8.0 Ω, find (a) the current; and (b) the readings on the voltmeters.

Solution (a) The total resistance of the circuit is 8.5 Ω. Therefore the current is, according to Equation 10-6a,

$$I = \frac{12.0 \text{ V}}{8.0 \text{ }\Omega + 0.5 \text{ }\Omega} = 1.41 \text{ A}$$

(b) in the resistor, the potential difference is $\Delta V = IR = (1.41 \text{ A})(8.0 \text{ }\Omega) = 11.3$ V. The switch and the ammeter have zero resistance, so V_3 and V_4 are both 0. The drop across the internal resistance is $(1.41 \text{ A})(0.5 \text{ }\Omega) = 0.7$ V; V_1 reads the *difference* between the emf and the drop in the battery, or 11.3 V. This is the same as V_2.

Each coulomb of charge makes a complete circuit. Each is given, in the battery, an amount of electric energy equal to the emf of the battery. This energy is converted to heat in all the resistances of the circuit: some in the internal resistance of the battery, the rest in the external resistances of the circuit.

Core Concept

Current flows only in a complete circuit; the energy it gains in the emf of the battery is lost in the resistances of the circuit.

Try This In a circuit like that of Figure 10-5, V_1 reads 4.50 V when the switch is open. Closing the switch causes it to drop to 4.35 V, and the ammeter to read 0.12 A. Find (a) the internal resistance of the battery; and (b) the external resistance.

10-8. Circuits with Branches

In Figure 10-5, the current goes from the battery through one light bulb and then returns to the battery. Often, the current has more than one possible route on its way out of the battery and back into it again.

Consider, for example, Figure 10-6, consisting of a network of seven resistors, each with an ammeter for measuring the current in it. If you know the readings on three of the resistors, is it possible to figure out what the other currents are?

It is, if you remember a simple rule: Charge cannot accumulate anywhere, and its total quantity is conserved. This means that any current that flows into a point in the circuit must also flow out of it.

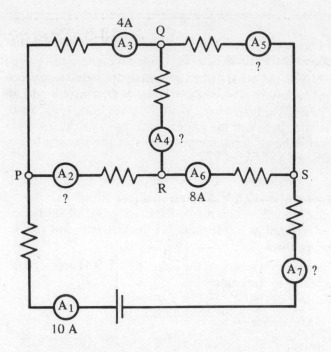

FIGURE 10-6

Ammeter A_1 shows that the current flowing into point P is 10 A, so 10 A must be leaving point P. And A_3 tells us that 4 A flows in the upper branch, to point Q. Therefore, the other 6 A must be going through A_2 to point R.

The current A_6, flowing out of point R, is 8 A, but only 6 A is coming to that point from P. Therefore, there must be 2 A coming down through A_4 from point Q to point P. That leaves 2 A going through A_5 and joining the 8 A from A_6 at point S. This adds up to 10 A that flows through A_7 and back into the battery—the same, of course, as the current leaving the other terminal of the battery.

Core Concept

The sum of the currents entering any circuit point is equal to the sum of the currents leaving it.

Try This In Figure 10-7, the ammeter A_1 reads 20 A, A_2 is 12 A, and A_3 is 9 A. Find the current in each of the seven resistors.

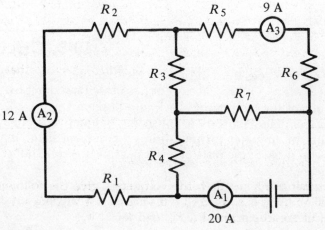

FIGURE 10-7

Figure 10-8 is another circuit that branches. This time voltmeters have been placed so as to measure the potential difference between the ends of each of the resistors. If we are given the readings on three of the voltmeters, can we figure out the other three?

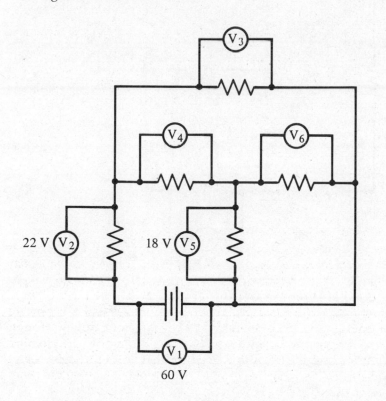

FIGURE 10-8

Yes, by applying one simple rule. Every coulomb of charge that leaves the high-potential end of the battery (the left end as shown) has to lose 60 joules of energy before it goes back into the other end of the battery. We know this because the voltmeter V_1 shows that the potential is 60 volts higher at the left side of the battery than at the right side; 60 V = 60 J/C. Then whatever path the charge takes through the circuit, all the potential differences in that path have to add up to 60 V.

Consider, for example, the complete path through the three resistors connected across voltmeters V_2, V_4, and V_5. They form a complete path, so their potential differences have to add up to 60 V. Since V_2 reads 22 V and V_5 says 18 V, the potential difference reading on V_4 must be 20 V.

Now look at another path: the readings on V_2 and V_3 must add to 60 V, so V_3 reads 28 V. And going through V_2, V_4, and V_6 tells us that V_6 reads 18 V.

Core Concept

Potential differences along any path through a complete circuit must add up to the terminal potential difference of the battery.

Try This In the circuit of Figure 10-9, the voltmeters give the following readings: $\Delta V_B = 50$ V; $\Delta V_1 = 20$ V; $\Delta V_2 = 5$ V; $\Delta V_4 = 4$ V; $\Delta V_5 = 18$ V. Find the potential differences across R_3, R_6, and R_7.

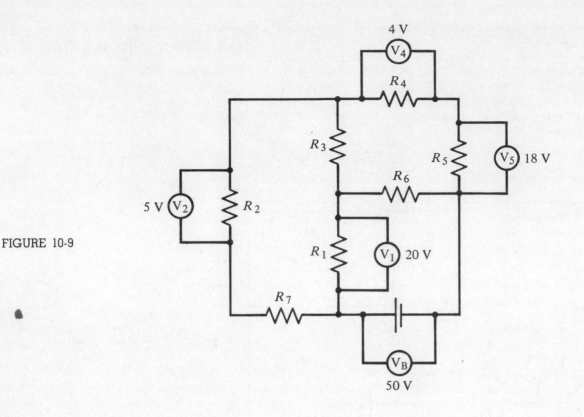

FIGURE 10-9

10-10. The Series Circuit

To determine the current in a circuit, you have to divide the emf of the source by the total resistance of the circuit. If the circuit contains a number of devices, the first task is to determine the total resistance of the combination of devices.

In a series circuit, like that of Figure 10-10, this is easy. There is no branching point in the circuit, so the same current must pass through every part of the circuit. An ammeter placed anywhere in the circuit will register this current. As the current passes through each conductor in turn, some of its energy is converted to heat.

Since there is only one path through the circuit, the total potential drop in the entire circuit, the terminal potential difference of the battery or

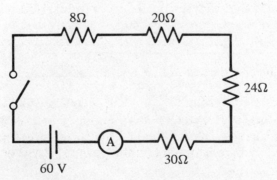

FIGURE 10-10

generator, must equal the sum of all the potential drops of the separate conductors:

$$\Delta V_{total} = \Delta V_1 + \Delta V_2 + \Delta V_3 + \cdots$$

So that

$$IR_S = IR_1 + IR_2 + IR_3 + \cdots$$

where R_S is the total resistance of a number of conductors connected in series. Since all the I's are the same,

$$R_S = R_1 + R_2 + R_3 + \cdots \qquad \text{(Equation 10-10a)}$$

See Sample Problems 10-19 and 10-20 for applications of this rule.

Sample Problem

10-19 In the circuit of Figure 10-10, find (a) the combined resistance; (b) the current; and (c) the potential difference across the 20-Ω resistor.

Solution (a) The combined resistance in a series circuit is the sum of the separate resistors: $8\ \Omega + 20\ \Omega + 24\ \Omega + 30\ \Omega = 82\ \Omega$.
(b) The current in the whole circuit is $\Delta V/R = (60\ \text{V})/(82\ \Omega) = 0.73\ \text{A}$.
(c) The potential difference across the 20-Ω resistor is $IR = (0.73\ \text{A})(20\ \Omega) = 15\ \text{V}$.

Sample Problem

10-20 Three resistors are connected in series to a 24-V battery, and an ammeter in the circuit reads 0.50 A. The first resistor is rated at 22 Ω, and the second at 8 Ω. Find (a) the total resistance; (b) the resistance of the third resistor; and (c) the potential difference across the third resistor.

Solution (a) The total resistance of the circuit is

$$R = \frac{\Delta V}{I} = \frac{24\ \text{V}}{0.50\ \text{A}} = 48\ \Omega$$

(b) In a series circuit, resistances add, so $22\ \Omega + 8\ \Omega + R_3 = 48\ \Omega$, and $R_3 = 18\ \Omega$.
(c) For any circuit element, $\Delta V = IR$, so $V_3 = (0.50\ \text{A})(18\ \Omega) = 9\ \text{V}$.

Which conductor dissipates the largest amount of energy? Since the rate at which energy is converted in a conductor is equal to $I\,\Delta V$ (Equation 10-3) and $\Delta V = IR$ (Equation 10-4), it follows that

$$P = I^2 R \qquad \text{(Equation 10-10b)}$$

While this equation is universally true, it is most conveniently applied in a series circuit in which all the I's are the same. The equation tells us that, in a series circuit, power is delivered in the greatest quantity to the largest resistance. This is why poor connections in a circuit, where the resistance is high, tend to overheat.

Core Concept

In series circuits, total resistance is the sum of all resistances:

$$R_S = R_1 + R_2 + R_3 + \cdots$$

Try This Three resistors, of 20 Ω, 40 Ω, and 60 Ω are connected in series to a 24-V battery of negligible internal resistance. Find (a) the current; (b) the potential difference across the 20-Ω resistor; and (c) the power dissipation in the 60-Ω resistance.

10-11. The Parallel Circuit

What is the resistance of a parallel circuit such as that in Figure 10-11? The chief characteristic of this circuit is that it branches; the current flows out of the battery, and some of it goes into each of the conductors.

Unlike a series circuit, in a parallel circuit it is possible to turn on some of the circuit devices and not the others. With all four switches open, there is no current at all, and the emf of the battery will register on a voltmeter connected across any of the switches. If the switch S_1 is closed, there will be a complete circuit through R_1. The current in R_1, and in the ammeter, will be the battery potential difference ΔV divided by R_1.

FIGURE 10-11

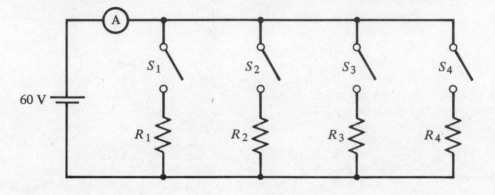

Now what happens when S_2 is closed? There is now an additional conducting path in the circuit. Since there is a complete path from the battery through R_2 and back to the battery, the entire battery potential difference will be across R_2. The current in R_2 must then be $\Delta V/R_2$. *Both* currents flow through the ammeter, which must now read the sum of the two currents. Closing S_3 and S_4 provides additional conducting paths, and their currents $\Delta V/R_3$ and $\Delta V/R_4$ add to the total current. Adding additional current paths can increase the current indefinitely, limited only by the internal resistance of the battery.

In a parallel circuit, every circuit element has the same potential difference as the source. Every time a new path is added, the total current goes up; each new current adds to that already coming out of the source. This means that every added circuit element *reduces* the total resistance of the circuit.

There is a rule for combining resistances in parallel. Since the current in each branch is $\Delta V/R$, and the total current is the sum of the branch currents,

$$\frac{\Delta V}{R_P} = \frac{\Delta V}{R_1} + \frac{\Delta V}{R_2} + \frac{\Delta V}{R_3} + \cdots$$

where R_P is the combined resistance of conductors in parallel. Since all the ΔV's are the same, this becomes

$$\frac{1}{R_P} = \frac{1}{R_1} + \frac{1}{R_2} + \frac{1}{R_3} + \cdots \qquad \textbf{(Equation 10-11)}$$

Sample Problems 10-21 and 10-22 apply this rule.

Sample Problem

10-21 In Figure 10-11, the same four resistors as in Figure 10-10 are connected in parallel. Find (a) the combined resistance; (b) the current in the battery; (c) the current in the 8-Ω resistor; and (d) the power dissipated in the 24-Ω resistor.

Solution   (a) From Equation 10-11,

$$\frac{1}{R_P} = \frac{1}{8\ \Omega} + \frac{1}{20\ \Omega} + \frac{1}{24\ \Omega} + \frac{1}{30\ \Omega}$$

$$\frac{1}{R_P} = \frac{15 + 6 + 5 + 4}{120\ \Omega}$$

Therefore $R_P = 4.0\ \Omega$.

(b) $I = \dfrac{\Delta V}{R} = \dfrac{60\ V}{4.0\ \Omega} = 15\ A$

(c) $I = \dfrac{\Delta V}{R} = \dfrac{60\ V}{8\ \Omega} = 7.5\ A$

(d) $I = \dfrac{\Delta V}{R} = \dfrac{60\ V}{24\ \Omega} = 2.5\ A$

$P = I\ \Delta V = (2.5\ A)(60\ V) = 150\ W$

Sample Problem

10-22 Three resistors are connected in parallel to a 24-V battery, and the battery current is 3.0 A. The first resistor is rated at 20 Ω and the second at 40 Ω. Find (a) the total resistance; (b) the resistance of the third resistor; and (c) the current in the third resistor.

Solution   (a) The total resistance is $R = \Delta V/I = 24\ V/3.0\ A = 8.0\ \Omega$.
(b) From Equation 10-11,

$$\frac{1}{8.0\ \Omega} - \frac{1}{20\ \Omega} + \frac{1}{40\ \Omega} + \frac{1}{R_3}$$

which becomes

$$\frac{1}{R_3} = \frac{1}{8.0\ \Omega} - \frac{1}{20\ \Omega} - \frac{1}{40\ \Omega}$$

$$\frac{1}{R_3} = \frac{5 - 2 - 1}{40\ \Omega}$$

$$R_3 = \frac{40\ \Omega}{2} = 20\ \Omega$$

(c) $I = \dfrac{\Delta V}{R} = \dfrac{24\ V}{20\ \Omega} = 1.2\ A$

In a household circuit, of course, all the lamps and appliances are connected in parallel. This way, each receives its rated potential difference of 115 volts and you can turn them on and off separately.

In a parallel circuit, the lion's share of the energy is delivered to the *smaller* resistances because they have the larger currents. Remember that the power is $I\ \Delta V$, and that all the ΔV's are the same.

Core Concept

In a parallel circuit, the inverse of the combined resistance equals the sum of the inverses of the separate resistances:

$$\frac{1}{R_p} = \frac{1}{R_1} + \frac{1}{R_2} + \frac{1}{R_3} + \cdots$$

Try this   Three resistors, of 20 Ω, 40 Ω, and 60 Ω, are connected in parallel to a 24-V battery of negligible internal resistance. Find (a) the combined resistance; (b) the current; (c) the power dissipated in the 40-Ω resistance.

10-12. Series-Parallel Combinations

The combined resistance of a complicated circuit, like that of Figure 10-12a, can often be computed by treating it as a set of series and parallel circuits. You can repeatedly make combinations according to Equation 10-10a and Equation 10-11 until you arrive at a single value for the resistance of the whole circuit.

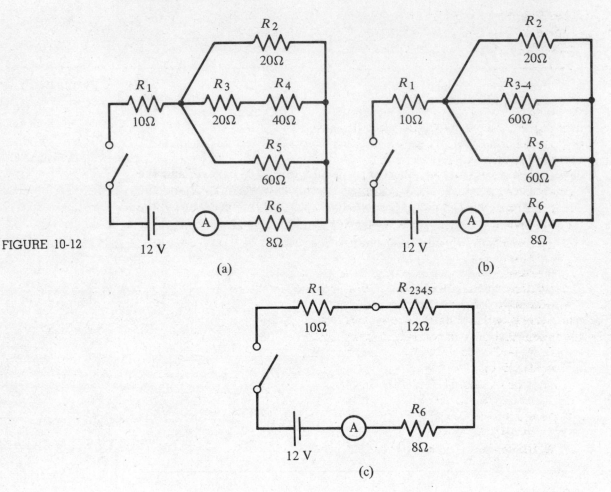

FIGURE 10-12

(a)

(b)

(c)

In this case, start by combining R_3 and R_4. Since they are in series with each other, their combined resistance is simply the sum of the separate resistances: $R_{34} = 60\ \Omega$. Replacing R_3 and R_4 by their 60-Ω equivalent makes the circuit look like that of Figure 10-12b.

Now we can calculate the combined resistance of the parallel combination of R_2, R_{34}, and R_5:

$$\frac{1}{R_{2345}} = \frac{1}{20\ \Omega} + \frac{1}{60\ \Omega} + \frac{1}{60\ \Omega}$$

which gives $R_{2345} = 12\ \Omega$. Now the equivalent circuit is Figure 10-12c.

R_{2345} is in series with R_1 and R_6, so these three can simply be added together to give $R_{total} = 30\ \Omega$. The current from the battery is therefore 12 V/30 Ω = 0.40 A.

To find the currents in each of the conductors, we first have to find the potentials. The entire current of 0.40 A flows through both R_1 and R_6.

The *IR* drop across R_1 is 0.4 A × 10 Ω = 4 V; through R_6 it is 3.2 V. This leaves 4.8 V across the parallel combination, which will be found across R_2, R_{34}, and R_5. It is now easy to find the current in each of these resistances.

Net resistance can be found by making appropriate combinations of series and parallel circuits.

Try this Find the net resistance of the circuit of Figure 10-13.

Core Concept

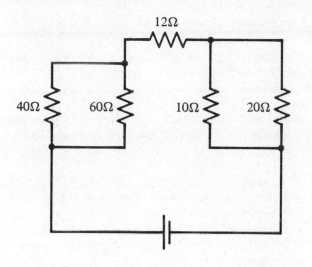

FIGURE 10-13

Summary Quiz

For each of the following, supply the missing word or phrase:

1. Electric energy is produced by the _____ of charges.
2. The electric energy produced by a source per unit of charge separated is called the _____ of the source.
3. The unit of emf is the _____ .
4. The rate of flow of electric charge is called _____ .
5. A coulomb per second is called a(n) _____ .
6. Current is measured with an instrument called a(n) _____ .
7. Energy converted per unit charge is measured with an instrument called a(n) _____ .
8. The rate at which energy is converted from one form to another is measured in _____ .
9. The product of the current in a circuit element and the potential difference across it expresses the _____ in that element.
10. The current in any circuit element depends on the potential difference across the element and on its _____ .
11. A volt per ampere is called a(n) _____ .
12. In a wire, the drift velocity of the electrons is proportional to the _____ between the ends of the wire.
13. Drift velocity is inversely proportional to the _____ of the wire.
14. The property of a metal that helps to determine the resistance of a wire made of that metal is called the _____ of the metal.
15. To measure the emf of a battery, you should connect to it a _____ and nothing else.

16. A small battery provides less current than a large one because the small battery has more _____ .
17. The more the _____ , the greater the difference between the emf of a battery and its terminal potential difference.
18. The purpose of a switch is to control the _____ in a circuit.
19. If there is no current, a voltmeter connected across a resistor will register _____ .
20. At any point in a circuit, the _____ coming in must equal that which leaves.
21. Between any two points in a circuit, the sum of all _____ is the same through any pathway.
22. Combined resistance is the sum of separate resistances provided that the various conductors are connected in _____ .
23. In a series circuit, the largest amount of power is delivered to the _____ resistance.
24. In a parallel circuit, each circuit element has the same _____ .
25. Adding new circuit branches in parallel increases the total _____ and _____ .
26. In a parallel circuit, the largest amount of power is delivered to the _____ resistance.

Problems

1. What is the emf of twelve 2.2-V storage cells connected in series?
2. The battery in Problem 1 delivers 4.0 A for 5.0 min. Find (a) the amount of energy the battery adds to each coulomb of charge; (b) the total amount of charge that passes out of the battery; (c) the total amount of energy supplied: (d) the power output.
3. In the circuit of Figure 10-14, which meter is correctly placed to measure (a) the current in the battery; (b) the potential difference across R_4; (c) the potential difference across R_3; (d) the current in R_2?

FIGURE 10-14

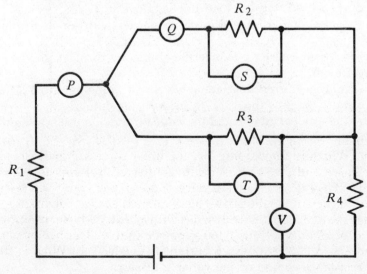

4. If an immersion heater delivers 11 A on a 115-V circuit, how long will it take to heat 1 cup of water (250 g) from 20°C to the boiling point?
5. What is the resistance, when hot, of a light bulb labeled "60 W, 115 V"?
6. A kilowatt-hour (kwh) is the energy of 1,000 W delivered for 1 hr. If your television set draws 4.0 A on a 120-V line, and electric energy costs you 8¢/kwh, how much does it cost to run the set for 8 hr?
7. What is the current in a 1,200-Ω resistor if its ends are at potentials of 180 V and 30 V, respectively?
8. What is the potential drop in a 40-Ω resistor if the current in it is 90 mA?
9. If your electric iron uses 900 W on a 115-V line, how much current does it draw?
10. How long must an aluminum wire be made to have a resistance of 0.50 Ω if the wire has a cross-section area of 0.10 cm^2? The resistivity of aluminum is 2.8×10^{-6} $\Omega \cdot$cm.
11. At what rate is heat produced in a copper wire 150 m long and 0.05 cm^2 in cross-section area connected to the terminals of a 12-V battery? The resistivity of copper is 1.72×10^{-6} $\Omega \cdot$cm.
12. What is the internal resistance of a battery if a voltmeter connected to its terminals reads 12 V and an ammeter in the same position reads 140 A?
13. What is the current in a 4.5-Ω resistor connected to a 22.5-V battery with an internal resistance of 0.8 Ω.
14. What is the terminal potential difference of the battery in Problem 13?
15. The emf of a battery is 24 V, and a voltmeter connected to its terminals reads 23.6 V when the battery is running a 300-W motor. Find (a) the resistance of the motor; and (b) the internal resistance of the battery.
16. In the circuit of Figure 10-15, $A_5 = 8$ A, $A_2 = 3$ A, and $A_4 = 2$ A. What are the readings on the other three ammeters?

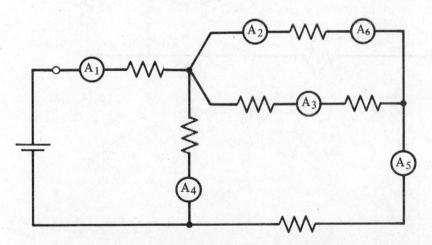

FIGURE 10-15

17. In the circuit of Figure 10-16, $V_2 = 6$ V, $V_3 = 8$ V, $V_5 = 3$ V, and $V_7 = 20$ V. What are the readings on the other three voltmeters?
18. A 12-Ω resistor and a 18-Ω resistor are connected in series to a 24-V battery of negligible internal resistance. Find (a) the combined resistance; (b) the current; (c) the power dissipation in each of the resistors.
19. A variable resistor is connected in series with a heater whose resistance is 20 Ω, to be operated from a 160-V power source. The heater is to operate at 850 W. (a) How much current is needed? (b) What is the required resistance setting of the variable resistor?

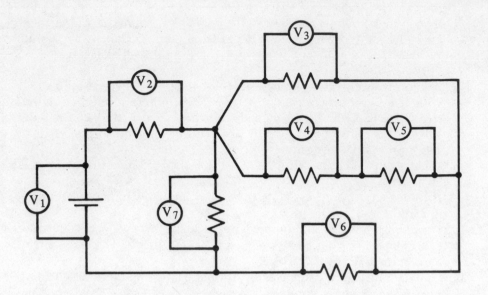

FIGURE 10-16

20. A 600-W toaster, a 150-W lamp, and a 40-W radio are all operating in parallel on a 120-V line. Find (a) the current in each device; and (b) the total current.
21. Determine the net resistance of 100 Ω, 250 Ω, and 400 Ω connected in parallel.
22. Find the combined resistance of the circuit shown in Figure 10-17.

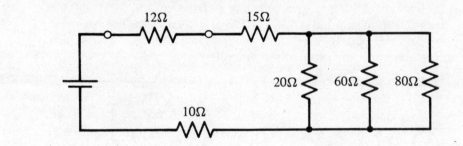

FIGURE 10-17

CHAPTER 11

Magnetism: Interacting Currents

11.1 The Force Between Wires

Like anything else, a wire is full of positive charges (nuclei) and negative charges (electrons). Since there are equal numbers of each, an electric field will have no net effect on the wire. Two wires placed near each other will not influence each other at all.

This changes if you send an electric current through both wires. Suddenly, a force acts on both wires. If the wires are parallel and the currents are in the same direction, the wires attract each other. If the currents are in opposite directions, the wires repel each other. No force is detected if the wires are perpendicular to each other. See Figure 11-1.

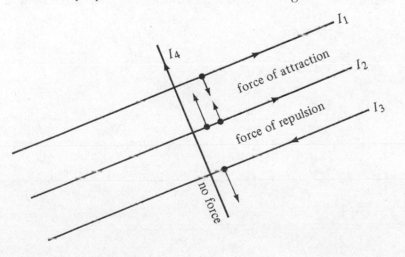

FIGURE 11-1

This is not the familiar electric force, since there is no accumulation of charge in the wires. The negatives and the positives still balance each other out and add to zero. The only difference is that the negative charges are moving! What we are observing is a different kind of force, one that acts between *moving* charges. It is called the *magnetic force*.

You have surely encountered this force before, in the form of permanent horseshoe magnets, or electromagnets, or the earth's effect on a compass needle. Just how these effects are related to those two wires pushing each other apart is not immediately obvious. It is true, nevertheless, that all magnetic interactions are forces acting between moving charges. We will soon work out the details.

Moving charge is electric current. In most circumstances, it is most convenient to describe magnetism as an interaction between currents, although there are at least two other approaches as well.

Core Concept

By the magnetic interaction, parallel currents attract and antiparallel currents repel.

Try This   A wire carrying a current is placed parallel to a beam of electrons flowing in the same direction as the current. Will the electron beam bend toward the wire or away from it?

11-2. Space Magnetism

Every navigator and every Boy Scout knows that there is something special about the space around the earth. A compass placed anywhere will somehow line itself up in a particular direction. One end (the N, or north-pointing end) winds up, after some oscillation, settling down and pointing north, more or less. The compass has detected a special property of the space around the earth: a *magnetic field*. We can define the direction of the magnetic field at any location as the direction pointed out by the N-end of a compass needle.

The N-end of a compass does not always point to geographic north. Place the compass near a current, and you can make it point in any direction at all. Figure 11-2 shows how compasses point in the neighborhood of a horseshoe magnet. The earth is by no means the only source of magnetic fields.

FIGURE 11-2

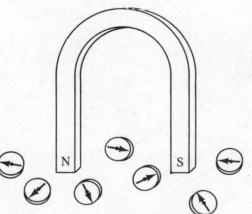

The precise, mathematical definition of a magnetic field is based on the force the field exerts on a current. The situation is rather complicated, because the force depends not only on the strength of the field and the size of the current, but also on the length of the wire and its orientation in the field. If the current is parallel or antiparallel to the field, there will be no force at all. The field exerts its maximum force when the current is perpendicular to the field. These relationships can be summarized this way:

$$\mathbf{F}_{mag} = (I\mathbf{l}) \times \mathbf{B} \qquad\qquad \text{(Equation 11-2)}$$

where I is the current, $\mathbf{l}$ is the length of the wire in the field, and $\mathbf{B}$ is the strength of the field. The $\times$ indicates that the formula yields the correct value of force only when the wire is perpendicular to the field.

The SI unit of magnetic field, which you can discover from Equation 11-2, is the newton per ampere-meter. It is called a *tesla*, abbreviated Ts. This is quite a large unit, and the millitesla (mTs = 10^{-3} Ts) is commonly used. The earth's magnetic field averages about 0.06 mTs. See Sample Problems 11-1 and 11-2.

Sample Problem

11-1 What is the magnetic field that exerts a force of 2.4×10^{-4} N on a current of 12 A in a wire 30 cm long set perpendicular to the field?

Solution Since the current is perpendicular to the field, Equation 11-2 can be written as

$$B - \frac{F}{Il}$$

Therefore

$$B = \frac{2.4 \times 10^{-4}\,\text{N}}{(12\,\text{A})(0.30\,\text{m})} = 6.7 \times 10^{-5}\,\text{N/A}\cdot\text{m}$$

Since 1 mTs is 10^{-3} N/A·m, the answer can be written as 0.067 mTs.

Sample Problem

11-2 A wire carrying 1.5 A has a length of 20 cm in a magnetic field of 40 mTs. If the wire is perpendicular to the field, how much force does the field exert on the wire?

Solution With the wire perpendicular to the field,

$$F = IlB = (1.5\,\text{A})(0.20\,\text{m})(0.040\,\text{Ts}) = 0.012\,\text{N}$$

The direction of the force on a current follows certain definite rules. It is always perpendicular to the field and perpendicular to the current. For

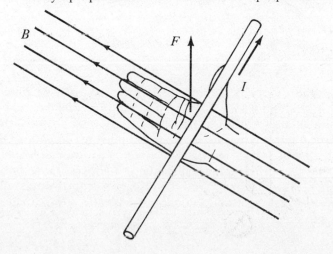

FIGURE 11-3

example, if a current runs eastward at the equator, where the earth's magnetic field is directly northward, the force on the current will be upward.

To find the direction of the force on a current, follow this rule, illustrated in Figure 11-3: Open your right hand flat and point the fingers in the direction of the field, with the thumb pointing the way the current flows. Now push with your palm. You are helping the field to push the current.

Core Concept

The magnetic force on a current, maximum when the current is perpendicular to the field, is perpendicular to both the field and the current and equal to the product of the field, current, and length of the wire:

$$\mathbf{F}_{mag} = (I\mathbf{l}) \times \mathbf{B}$$

Try This What are the direction and magnitude of the force that an upward-directed magnetic field of 2.0 mTs exerts on a north-flowing current of 10 A in a wire 15 cm long?

11-3. Loop Currents

Consider a rectangular loop of wire carrying a current through a magnetic field, as in Figure 11-4. If you apply the right-hand rule to each of the four sides of the loop, you will find that the force is always directly away from the center. The loop may stretch, but it will neither translate nor rotate.

FIGURE 11-4

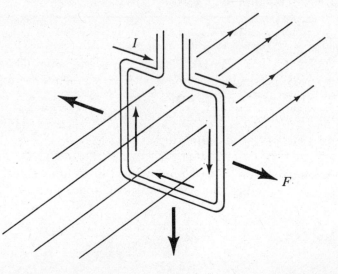

Now let's see what happens if the loop is oriented with its plane parallel to the field, as in Figure 11-5. The top and bottom wires are now, respectively, antiparallel and parallel to the field, so there is no force on them at all. The vertical wires have equal forces in opposite directions, as before, but the forces are no longer in line. The result is a torque, as calculated in Sample Problem 11-3. The loop will rotate. It will pass the equilibrium position of Figure 11-4, at which point the direction of the torque will reverse. The loop will oscillate around its equilibrium position until friction brings it to rest there.

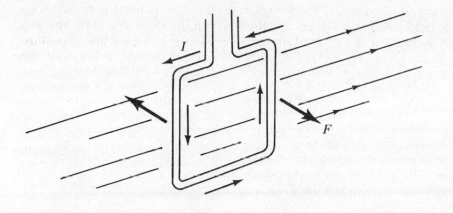

FIGURE 11-5

Sample Problem

11-3 A rectangular wire loop 6.0 cm high and 8.0 cm wide is suspended, with two sides vertical, in a horizontal magnetic field of 120 mTs, and it carries a current of 0.50 A. (a) How much is the force on each of the vertical sides? (b) When the plane of the loop is parallel to the field, how much is the force on the horizontal sides? (c) How much is the maximum torque exerted on the loop?

Solution (a) $F = IlB = (0.50$ A$)(0.060$ m$)(0.120$ Ts$) = 3.6 \times 10^{-3}$ N.

(b) In this position, the horizontal wires are parallel to the field and there is no force on them.

(c) When the plane of the loop is parallel to the field, the torque on each vertical wire, from Equation 2-11, is

$$\mathbf{F}r = (3.6 \times 10^{-3} \text{ N})(0.040 \text{ m}) = 1.44 \times 10^{-4} \text{ N·m};$$

the total torque is twice this, or 2.9×10^{-4} N·m.

A *solenoid* is a coil consisting of many loops in series. Figure 11-6 shows a solenoid at rest in its equilibrium position in a magnetic field. As in Figure 11-4, the plane of each loop is perpendicular to the field; the axis of the solenoid points in the direction of the field. If a solenoid, free to rotate, is placed in a field, it will oscillate and come to rest aligned with the field, just like the loop of Figure 11-4. The solenoid behaves like a compass. Can it be that a compass works as it does because it contains loop currents? In any event, we can, by analogy, designate the end of the solenoid that points in the direction of the field as its N-end.

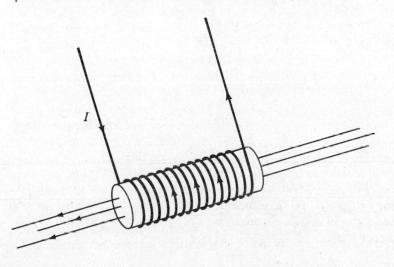

FIGURE 11-6

The torque on a loop current is used in a *Weston galvanometer* to make an extremely sensitive device for measuring currents, as shown in Figure 11-7. The solenoid, wrapped around an iron cylinder, is suspended between the poles of a permanent horseshoe magnet. Current in the solenoid produces a torque, and the solenoid rotates, carrying with it a needle that marks off positions on a scale. The greater the current, the more torque there will be on the coil. The rotation of the solenoid is counteracted by a spiral spring, so that the solenoid comes to rest when the current in it is just enough to produce a torque equal to that produced by the spring, but in the opposite direction. The needle connected to the solenoid and rotating with it marks off microamperes of current on a scale. This galvanometer movement, with the appropriate associated hardware, is the basis of many other instruments, including the familiar ammeters and voltmeters.

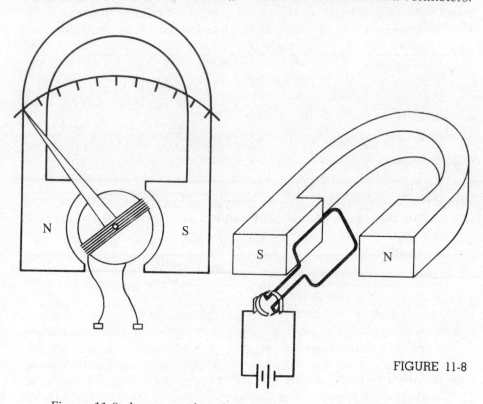

FIGURE 11-7

FIGURE 11-8

Figure 11-8 shows another device that makes use of the torque on a loop current: the electric motor. In this simple motor, the current is fed into the armature (the rotating solenoid) through two *brushes* that rest on a *split-ring commutator*. This is necessary so that the current in the armature will be reversed as the armature passes its equilibrium point. This keeps the torque constantly in the same direction, so that the armature will continue to spin instead of oscillating around its equilibrium point.

Core Concept

The torque on a loop current in a magnetic field has many practical uses.

Try This The loop of Figure 11-5 is 10 cm high and 20 cm wide, and it carries a current of 5 A in a field of 30 mTs. Find (a) the force acting on each of the vertical sides of the loop; and (b) the torque on the loop.

The picture on the face of your television set is painted, 30 times a second, by a thin beam of electrons. They come out of an "electron gun" in the back of the picture tube and are fired toward the face of the tube, where the beam produces a tiny spot. On its way to the tube face, the beam passes through magnetic fields that exert forces on the electrons, making the spot sweep across the screen from right to left and from top to bottom.

The force that a magnetic field exerts on a charged particle moving through it is simply expressed:

$$\mathbf{F}_{mag} = (q\mathbf{v}) \times \mathbf{B} \qquad \text{(Equation 11-4)}$$

where the "$\times$" indicates, as before, that the equation gives the correct result only if the velocity of the particle is perpendicular to the field. Sample Problems 11-4 and 11-5 are examples of the use of this equation.

Sample Problem

11-4 A pith ball with a charge of $+30$ nC is traveling at 12 m/s through a magnetic field of 50 mTs. If the ball is moving perpendicularly to the field, how much is the force on it?

Solution When the velocity is perpendicular to the field,

$$F = qvB = (30 \times 10^{-9} \text{ C})(12 \text{ m/s})(0.050 \text{ Ts}) = 1.8 \times 10^{-8} \text{ N}$$

Sample Problem

11-5 How strong a field is needed to exert a force of 5×10^{-11} N on a proton traveling perpendicularly through the field at 2.0×10^8 m/s? (See Appendix 3 for constants.)

Solution

$$B = \frac{F}{qv} = \frac{5 \times 10^{-11} \text{ N}}{(1.60 \times 10^{-19} \text{ C})(2.0 \times 10^8 \text{ m/s})} = 1.6 \text{ Ts}$$

A moving charge is an electric current, and the direction of the force on the charge follows the same rule that applies to the force on a current: the force is always perpendicular to the field and also perpendicular to the velocity of the particle. To find the direction of the force, use the usual right-hand rule for currents, with the thumb pointing the way the particles are traveling. One important proviso: If the particles are negative, use your left hand. See Sample Problems 11-6 and 11-7.

Sample Problem

11-6 What is the direction of the force that a vertical magnetic field, directed upward, will exert on an electron traveling eastward in it?

Solution Point your *left* fingers upward, since you are dealing with a negative charge. Rotate your hand until the thumb points east. Your palm will point northward, and that is the direction of the force.

Sample Problem

11-7 In the picture tube in your television set, a beam of electrons comes toward you from the rear of the tube and strikes the face of the tube, where it makes a bright spot. What must be the direction of the magnetic field in the tube to make the spot move from right to left across the screen?

Solution   Use your left hand because the particles are negative. With your thumb pointing toward you and your palm facing to the right, the fingers will point upward, in the direction of the field.

The rule that the force exerted by a magnetic field is always perpendicular to the velocity of the particle has an important consequence. The field can change the direction of the velocity but cannot make the particle speed up or slow down. Furthermore, as we learned in Chapter 4, a force exerted perpendicular to the direction of motion does no work and cannot increase the energy of the particle.

The most spectacular machines of modern physical research are giant particle accelerators. They are used to produce charged particles traveling at nearly the speed of light. These electrons, protons, and ions are used to investigate the nature of matter; they are fired into other particles to see what reactions occur. Some of these accelerators push protons through a pipe over a mile long, forming a perfect circle underground. In all these machines, the charged particles are accelerated and have their energy increased as they pass through electric fields. Magnetic fields are used to bend the beams of particles into the desired trajectories.

Core Concept

Charged particles moving through a magnetic field experience a force perpendicular to the field and perpendicular to their velocity:

$$\mathbf{F}_{mag} = (q\mathbf{v}) \times \mathbf{B}$$

Try This   What magnetic field would be needed to exert a force of 2.0×10^{-12} N on an electron moving perpendicular to the field at 2.5×10^8 m/s? (Remember the charge on an electron? See Appendix 3.)

11-5. Measuring Mass

If you can get some charged particles into a magnetic field, you can measure their mass. This works because the magnetic force on the particles is always perpendicular to their velocity. This means that the force is centripetal, as we saw in Chapter 4. Now we can make use of three known physical relationships to arrive at one that is useful.

Equation 4-2 gives us a general relationship between force and acceleration:

$$\mathbf{F} = \mathbf{ma}$$

Equation 1-11b tells us how the acceleration is related to the radius of the circle of travel whenever the acceleration is kept constantly perpendicular to the velocity:

$$a_c = \frac{v^2}{r}$$

And Equation 11-4 gives us the magnetic force that a field exerts on a charged particle moving through it perpendicularly:

$$F = Bqv$$

The conditions under which these equations apply exist in our stream of electrons. From the first and third,

$$ma = Bqv$$

Substituting the second:

$$\frac{mv^2}{r} = Bqv$$

which simplifies to

$$mv = Bqr \qquad \textbf{(Equation 11-5)}$$

Therefore, we can find the mass of a charged particle moving through a magnetic field if we know the strength of the field, the charge on the particle, the radius of the circle the particles travel in, and their speed. This is the way the mass of the electron was measured; it is also the theoretical basis of the mass spectrometer, an instrument that measures the masses of all kinds of charged particles. See Sample Problem 11-8.

Sample Problem

11-8 What is the mass of particle bearing a single elementary charge if it travels in a circle whose radius is 35 cm when it is going at 0.90 times the speed of light in a magnetic field of 8.0 Ts? (See Appendix 3 for constants.)

Solution From Equation 11-5,

$$m = \frac{Bqr}{v} = \frac{(8.0 \text{ Ts})(1.60 \times 10^{-19} \text{ C})(0.35 \text{ m})}{0.90(3.0 \times 10^8 \text{ m/s})} = 1.66 \times 10^{-27} \text{ kg}$$

The experimental determination of the mass of an electron uses the apparatus illustrated in Figure 11-9. A tiny filament is heated by a small electric current; when the filament is red hot, it emits electrons. These electrons are attracted to the brass cone that surrounds the filament, kept at a high positive potential. There is a little hole at the apex of this cone, and a beam of electrons emerges from this hole. These electrons have been

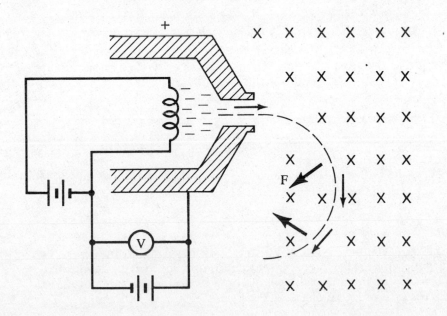

FIGURE 11-9

accelerated through a known potential difference, which can easily be read on the voltmeter. Equation 9-9 tells us that the amount of electric energy the electrons lose is $q\,\Delta V$. Since they are being accelerated in a vacuum, all this energy becomes kinetic energy, so we know the kinetic energy of the electrons as they leave the hole and from Equation 6-2:

$$\frac{1}{2}mv^2 = q\,\Delta V$$

These electrons then enter a magnetic field, directed into the page in the figure. The field exerts a force on them, pushing them into a circular path. A tiny amount of gas in the container produces a glow along the path of the electrons so the radius of their path can be measured. Then Equation 11-5 gives us mv, the momentum of the electrons.

If you know the kinetic energy and the momentum of anything, you can calculate both its mass and its velocity. For a typical example of this process, see Sample Problem 11-9.

Sample Problem

11-9 An electron is accelerated through a potential difference of 150 V and projected into a magnetic field of 25 mTs, where it goes into a circular path of radius 7.1 cm. What is its (a) kinetic energy; (b) momentum; (c) mass?

Solution   (a) The kinetic energy of the electron is the electric energy it loses in falling through the field:

$$q\,\Delta V = (1.60 \times 10^{-19}\ \text{C})(150\ \text{V}) = 2.4 \times 10^{-17}\ \text{J}$$

(b) Its momentum, from Equation 11-5, is

$$Bqr = (0.025\ \text{Ts})(1.60 \times 10^{-19}\ \text{C})(0.071\ \text{m}) = 2.84 \times 10^{-22}\ \text{kg·m/s}$$

(c) Since its momentum $p = mv$ and its kinetic energy $E_{kin} = \frac{1}{2}mv^2$, it follows that

$$p^2 = m^2v^2 \quad \text{(square the momentum equation)}$$

$$\frac{p^2}{2m} = \frac{mv^2}{2} = E_{kin} \quad \text{(divide both sides by } 2m\text{)}$$

$$m = \frac{p^2}{2E_{kin}} \quad \text{(solve for } m\text{)}$$

$$m = \frac{(2.84 \times 10^{-22}\ \text{kg·m/s})^2}{2(2.4 \times 10^{-17}\ \text{J})} = 1.7 \times 10^{-27}\ \text{kg}$$

Core Concept

The circular path in which charged particles travel in a magnetic field can be used to determine their masses:

$$mv = Bqr$$

Try this   A beam of protons, accelerated through a potential difference of 75 V, enters a magnetic field of 3.7 mTs and travels in a circle of radius 0.30 m. Find (a) the kinetic energy of the protons; (b) their momentum; (c) their mass. (Remember that the charge on a proton is the same in magnitude as the charge on an electron.)

11-6. How Do You Make a Field?

The earliest hint that there is some sort of relationship between electricity and magnetism was the observation that a current in a wire could deflect a nearby compass needle. It is electric currents that create magnetic fields.

Consider the large wire of Figure 11-10, carrying current I in an upward direction. To test the field around this current, we insert a small test current I' into the field as shown and rotate it until the force on it is a maximum. It will then be parallel to I, and I' will experience a force to the left, attracting it to I. This is according to the rules of interaction between currents that we have already discussed.

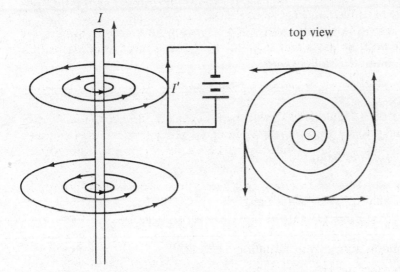

top view

FIGURE 11-10

What is the direction of the field produced by I? Our test current I' can tell us the field at its own location since we know that the current is upward and the force on it is to the left. Using the usual right-hand rule at I', we find that the field at that point is directed away from us, into the page. The field at all points near I is tangential, perpendicular to a line drawn from I. The magnetic field lines form rings around I, as shown. Compasses placed near I will point tangentially, their N-poles lining up along the circles concentric on I, as shown in Figure 11-11.

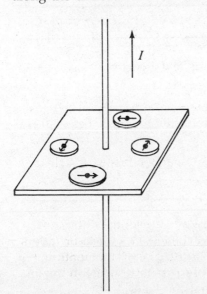

FIGURE 11-11

These magnetic field lines have an important and obvious difference from electric field lines. Magnetic field lines do not terminate—anywhere. Every line forms a loop, closing on itself. The distance between lines is greater as you go farther from the current that produces the field; this indicates, as usual, that the strength of the field diminishes with distance from I. The strength of the field can be calculated in this way:

$$B = \frac{2k'I}{r}$$

(Equation 11-6)

where I is the current that produces the field, r is the distance from that current, and k' is the universal constant known as the magnetic constant of free space. Its value is 10^{-7} N/A^2, exactly. To find the direction in which the field lines circulate around the wire, another right-hand rule is handy. Point your thumb in the direction of the current and wrap your fingers around the wire. The fingers point in the direction of circulation of the field lines. See Sample Problems 11-10 and 11-11.

Sample Problem

11-10 Find the direction and magnitude of the magnetic field 12 cm above a wire carrying a current of 6.0 A northward.

Solution   The magnitude of the field is

$$B = \frac{2k'I}{r} = \frac{2(10^{-7} \text{ N/A}^2)(6.0 \text{ A})}{(0.12 \text{ m})}$$

$$B = 1.0 \times 10^{-5} \text{ N/A·m} = 0.010 \text{ mTs}$$

To get the direction, point your right thumb northward along the wire and curl your fingers. Above the wire, your fingers point to the east.

Sample Problem

11-11 How much current is needed to produce a field of 0.25 mTs at a distance of 20 cm from the heavy wire that is carrying the current?

Solution   A lot. From Equation 11-6,

$$I = \frac{Br}{2k'} = \frac{(0.25 \times 10^{-3} \text{ Ts})(0.20 \text{ m})}{2 \times 10^{-7} \text{ N/A}^2} = 250 \text{ A}$$

Core Concept

Magnetic fields are produced by currents; the field lines form closed loops around the currents: $B = 2k'I/r$.

Try This   Find the direction and magnitude of the magnetic field 5.0 cm above a wire carrying 20 A to the west.

11-7. Stronger Fields

Look at the loop of wire carrying a current as shown in Figure 11-12. What is the direction of the field inside the loop?

Take out your right hand and apply the rule to the current on the left-hand side of the loop. You will find that the field inside the loop is

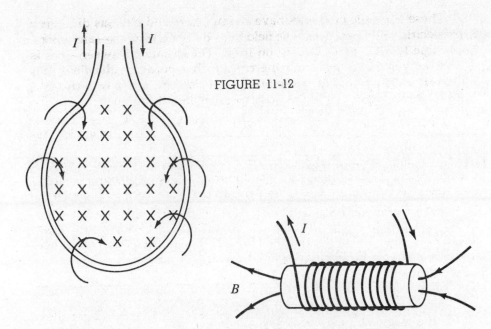

FIGURE 11-12

FIGURE 11-13

directed into the page, away from you. If you use this rule, in fact, for the current in any part of the loop, you will get the same answer. Within the loop, all fields point away from you. The fields due to the current in all parts of the loop add up to produce a strong field inside the loop. If you want a still stronger field, you can use a longer wire and coil it into a lot of loops instead of just one, as in Figure 11-13. In such a solenoid, the field is quite uniform, except near the ends.

To find the direction of the field inside a solenoid, wrap your hand around the solenoid with the fingers pointing in the direction of the current. Then your thumb points in the direction of the field lines inside the solenoid (see Figure 11-14). The field lines run through the center, parallel to the axis, as shown. If such a solenoid is suspended in an external field and is free to turn, it will come to equilibrium with its axis parallel to the external field. The end at which the field lines emerge will point in the direction of the external field, so that end is the N-pole of the solenoid. Your thumb points to the N-pole.

The strength of the uniform field near the center of a solenoid depends on the current and on how closely the wires are crowded together. This crowding can be expressed as the number of turns of wire per unit

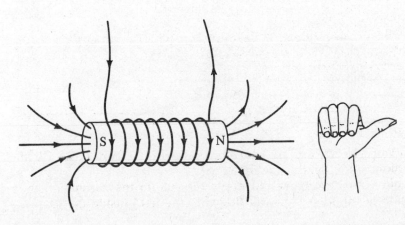

FIGURE 11-14

length of solenoid. The wire may be wrapped in more than one layer to increase this quantity. The field is given by the expression

$$B = 4\pi k'I\frac{n}{l}$$

(Equation 11-7)

where n/l is the number of turns of wire per unit length of solenoid. For applications, see Sample Problems 11-12 and 11-13.

Sample Problem

11-12 A solenoid is wound with wire in a single layer and connected to a battery to produce a field inside it that has a magnitude of 4.0 mTs. If the solenoid is rewound in two layers, making it half as long, how strong is the field inside it?

Solution The solenoid now has twice as many turns per unit length with the same current, so the field is twice as great: 8.0 mTs.

Sample Problem

11-13 Two layers of wire, with 40 turns each, are wound onto a coil form 15 cm long, and a current of 0.30 A is passed through the wire. What is the magnetic field inside the solenoid?

Solution

$$B = 4\pi k'I\frac{n}{l}$$

$$B = 4\pi(10^{-7}\ \text{N/A}^2)(0.30\ \text{A})\left(\frac{80}{0.15\ \text{m}}\right)$$

$$B = 2.0 \times 10^{-4}\ \text{N/A·m} = 0.20\ \text{mTs}$$

Another way of getting a uniform field—larger, but weaker—is with the Helmholtz coils illustrated in Figure 11-15. These two coils are parallel to each other, and just one coil radius apart.

In your television set, the neck of the picture tube has wrapped around it a *yoke*, consisting of five coils of wire. One is wound with the neck of the tube along the axis of the solenoid; the current in this coil focuses the electron beam to a fine spot on the screen. The other four coils are two pairs of Helmholtz coils. One pair, at the right and left of the neck of the picture tube, are the vertical deflection coils. The plane of the coils is vertical, so they produce a horizontal field that moves the spot up and down. The horizontal deflection coils, above and below the neck, move the spot from left to right and back.

FIGURE 11-15

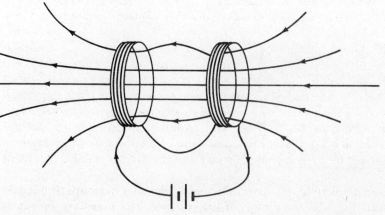

Strong magnetic fields are produced by circulating currents:

$$B = 4\pi k'I\frac{n}{l}$$

Try This A solenoid is tightly wound in four layers of wire with a diameter of 0.20 mm. How much current is needed to produce a field of 1.5 mTs?

11-8. Magnets—At Last!

A rod placed inside a solenoid might produce a drastic change in the magnetic field produced by the solenoid. Some materials weaken the field slightly; others make it a little stronger. And there are certain substances that can strengthen the field of a solenoid by a factor of hundreds, or even thousands. These are the *ferromagnetic* materials.

At room temperature, only four elements are ferromagnetic: iron, nickel, cobalt, and gadolinium. These elements are special because of a peculiar property of their atoms, due to the arrangement of the electrons in them.

Every electron is like a little solenoid, or a compass. It possesses a negative charge and it is spinning. A crude model of such an electron is shown in Figure 11-16; without too much oversimplification of the strange and complex properties of this object, we can think of it as a tiny ball of negative charge, spinning on an axis. Using the left hand (because the charge is negative), we can apply the solenoid rule of the last section to determine that the electron has an N-pole and an S-pole. Placed in an external magnetic field, each electron will tend to come to equilibrium with its axis lined up with the field.

Electrons in an atom are under another constraint: they generally occur in pairs. The members of the pair are spinning in opposite directions, so that their magnetic fields nullify each other. Ferromagnetic elements are unique in having three or four electrons, in every atom, spinning in the same direction. This turns the whole atom into a tiny magnet, with its N-pole and S-pole.

In an ordinary piece of iron, the tiny atomic magnets are oriented at random, so that there is no net magnetic effect. When the iron is placed in a magnetic field, many of the atoms rotate and line up with the field. In a strong field, it is possible to line up all the atoms in the iron. Whatever atoms are aligned with the external field add their own magnetic fields to the external field and thus make it much stronger. A really strong electromagnet of this sort can be used to move junked automobiles from one place to another.

Certain alloys, such as hard steel (iron + carbon) and alnico (aluminum + nickel + cobalt) are ferromagnetic and have the additional property that their atoms do not go back to their random magnetic orientation when the external field is turned off. These are the materials of permanent magnets. Permanent ferromagnets can also be made from certain ceramics that contain oxides of iron. The earliest known magnet, natural lodestone, is a material of this kind. The field of a permanent bar magnet, since it is formed by loop currents, is just like the field of a solenoid, as in Figure 11-17.

Just as in a solenoid, the magnetic field lines of a permanent magnet are continuous through the length of the magnet. The poles are points at

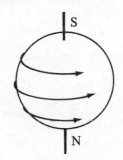

FIGURE 11-16

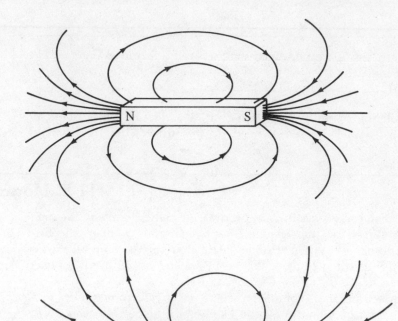

FIGURE 11-17

which concentrations of field lines enter or leave. If a bar magnet is cut, as in Figure 11-18, the field lines emerge at the cut ends, and two new poles are created.

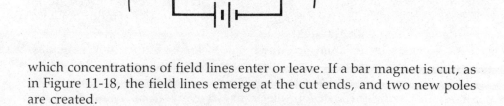

FIGURE 11-18

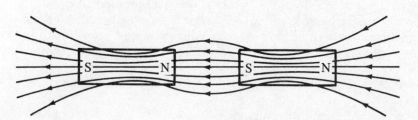

Ferromagnetic materials greatly increase the strength of a magnetic field because their individual atoms act as solenoids that can be rotated into alignment with the applied field.

Try This Draw the field lines in and near the horseshoe magnet of Figure 11-19.

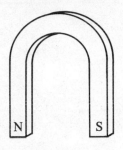

FIGURE 11-19

11-9. The Earth Is a Magnet

It is easy enough to plot the magnetic field of the earth. To explain it is quite another matter.

It is a most useful fact that, if a permanent magnet is suspended free to rotate, one end of it points north. This is only approximately true. The compass points to "magnetic north," which is a point now located at 70° north latitude, on the east coast of Greenland. A compass at that position would point straight down; magnetic field lines enter the earth there. Note that that spot in Greenland is the S-pole of the earth; it is the point where field lines enter, not leave!

As Figure 11-20 shows, the field lines are not greatly concentrated at the magnetic north pole. The pattern looks like the pattern of field lines produced by a magnet 1,000 miles long, embedded in the earth and tilted at an angle of 15° from its axis. The field is horizontal near the equator only.

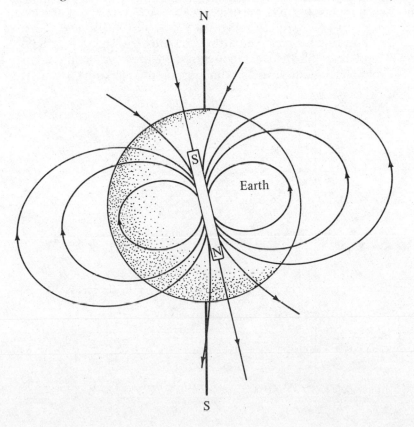

FIGURE 11-20

Elsewhere it dips; that is, it makes an angle with the horizontal. The angle of dip reaches about 75° in the northern United States.

What produces the earth's field? Theoretical physicists are beginning to get a fairly good picture of where it comes from. The core of the earth is a solid mass of iron and nickel, but it is surrounded by hot liquid metal more than a thousand miles deep. Because the earth is hottest at the center, there are convection currents in this liquid metal. The spin of the earth distorts these currents so that they flow around the earth. A complicated feedback mechanism is set up in which small currents in the liquid metal produce magnetic fields, which amplify the currents, which adds to the fields, and so on. This theory has even begun to make some sense out of the fact that every few hundred thousand years the polarity of the earth's field reverses, taking about 1,000 years to change the direction of the field.

Core Concept

The earth's magnetic field has a poorly concentrated S-pole near the geographic north pole.

Try This At a place where the magnitude of the earth's magnetic field is 5.3×10^{-5} Ts and the dip is 76°, how much is the horizontal component of the field?

11-10. Magnetism Makes Electricity

Consider the metal rod shown in Figure 11-21, placed in a magnetic field directed into the page. Let's push the rod to the right and see what happens.

In the rod there are positive particles (nuclei) and negative particles (electrons). Both are moving to the right, since both are in the rod. If we apply the right-hand rule for positive particles moving through a magnetic field, we will find that the field exerts a force on them, pushing them upward. Use your left hand to find that the force on the electrons is down-

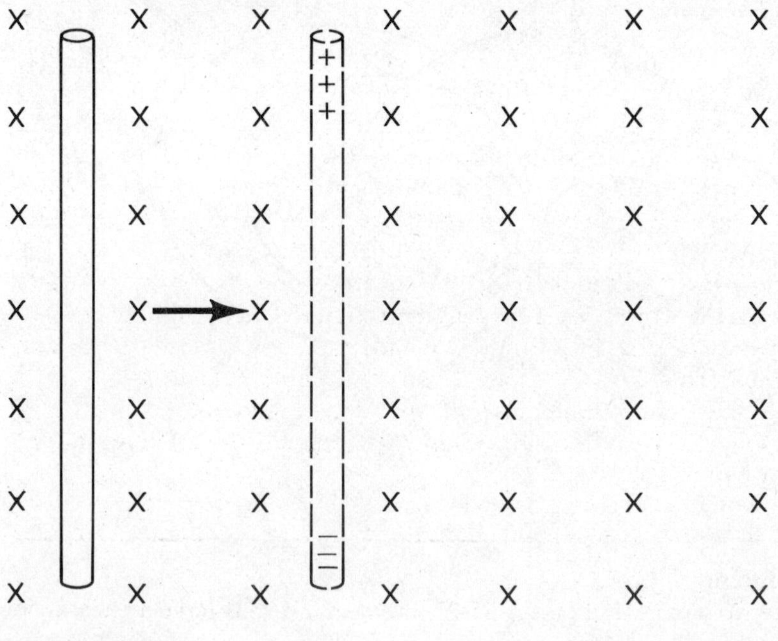

FIGURE 11-21

ward. The nuclei cannot move, but the electrons can, since a metal contains free electrons. The result is a separation of charge. As long as the rod keeps moving through the field at right angles to the field lines, there will be a separation of charge in the rod. The bottom becomes negative, the top positive.

As we have seen, separation of charge is a means of storing electric potential energy. A battery does this by chemical action, and an electric field by electrostatic induction. When charge is separated by moving an object through a magnetic field, we call the process *electromagnetic induction*.

The energy stored per unit of charge is the *induced emf* in the rod. We can measure the emf by means of the setup shown in Figure 11-22. The rod moves over a pair of rails connected to a voltmeter. The voltmeter will read the induced emf, which depends on the strength of the field, the length of the rod, and the speed with which it is moving. If the field, the rod, and its velocity are all perpendicular to each other, the induced emf is

$$\mathcal{E} = Blv \qquad \text{(Equation 11-10)}$$

See Sample Problem 11-14.

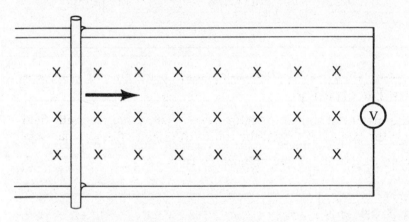

FIGURE 11-22

Sample Problem

11-14 What emf is induced between the ends of a rod 25 cm long that is moving perpendicularly to a magnetic field of 350 mTs at 5.0 m/s?

Solution $\mathcal{E} = Blv = (0.35 \text{ Ts})(0.25 \text{ m})(5.0 \text{ m/s}) = 0.44 \text{ V}$

Now suppose we remove the voltmeter and connect a wire between the rails, or a sensitive ammeter. Then the separated charges can reunite by flowing through the rails and the ammeter from one end of the rod to the other. We now have an induced *current*. As always, we get a current only when the source of emf is part of a complete circuit. The amount of current obeys Ohm's law, Equation 10-4, and its adaptation to complete circuits, Equation 10-6a. This is worked out in Sample Problem 11-15.

You have learned the principle of the electric generator, which supplies all the electricity that runs the country (except for the minute amount we get from batteries).

Sample Problem

11-15 In a setup like that of Figure 11-23, the rod and rails have a resistance of 0.15 Ω and the external resistance is 0.50 Ω. The rails are 20 cm apart, and the rod is to be moved at 8 m/s. How strong a field would be needed to produce 0.10 mA of current in the circuit?

Solution From Equation 10-6a,

$$\mathscr{E} = I(R_I + R_E) = (0.10 \times 10^{-3} \text{ A})(0.15 \ \Omega + 0.50 \ \Omega) = 6.5 \times 10^{-5} \text{ V}$$

Then, from Equation 11-10,

$$B = \frac{\mathscr{E}}{lv} = \frac{6.5 \times 10^{-5} \text{ V}}{(0.20 \text{ m})(8 \text{ m/s})} = 4 \times 10^{-5} \text{ Ts}$$

FIGURE 11-23

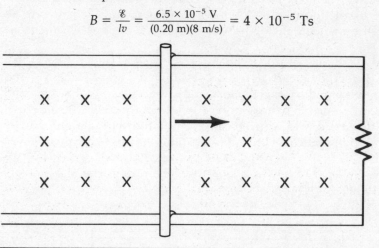

Core Concept

When a conductor moves through a magnetic field, an emf is induced across it:

$$\mathscr{E} = Blv$$

Try This The 15-cm rod of Figure 11-23 is moved to the left at 2.5 m/s in a field of 50 mTs. The complete circuit has a resistance of 10 Ω. Find the direction and magnitude of the induced current.

11-11. The Law of Induction

Aside from moving a rod, there are many ways to induce an emf by using a magnetic field. The most general rule that tells us how is called *Faraday's law of induction*.

Take another look at Figure 11-22. We can think of the arrangement as a loop consisting of the moving rod, the rails, and the wires running to the voltmeter. Magnetic field lines thread through the loop. If the field is strong, the lines are close together. The total number of field lines depends on the strength of the field and on the area of the loop. The product of these two quantities, which is represented by the number of field lines passing through the loop, is called the *magnetic flux* in the loop. If the plane of the loop is perpendicular to the field lines, the magnetic flux is given by the equation

$$\phi = BA \qquad \text{(Equation 11-11)}$$

where ϕ (the Greek letter phi) stands for the flux, measured in *webers* (Wb).

Moving the rod across the field changes the amount of flux passing through the loop. It is only while the rod is moving—while the amount of flux passing through the loop is changing—that an emf is induced and the voltmeter gives a reading. Faraday's law of induction tells us that *the emf induced in a loop is equal to the rate of change of the magnetic flux in the loop*. To see how this works out, see Sample Problem 11-16.

Sample Problem

11-16 A rectangular loop of wire consists of 50 turns 10 cm wide and 20 cm long. It is in a field of 80 mTs, with its plane perpendicular to the field. If the loop is yanked out of the field, taking 0.050 s to leave the field, how much emf is induced in it?

Solution The magnetic flux in the loop is

$$\phi = BA = (0.080 \text{ Ts})(0.10 \text{ m})(0.20 \text{ m})$$
$$\phi = 0.0016 \text{ weber}$$

This flux reduces to zero in 0.050 s, so the rate of change of flux is (0.0016 weber)/(0.050 s) = 0.032 V in each coil. With 50 coils in series, the total emf is (50)(0.032 V) = 1.6 V.

There are many ways of changing the amount of magnetic flux in a loop. For example, take a solenoid and connect its ends to a galvanometer, which is an instrument that can measure extremely small currents (Figure 11-24). Now insert a bar magnet into the solenoid. A current will register on the galvanometer while the magnet is moving. The current flows one way as the magnet moves in, and the other way as it comes out. When the magnet is at rest, there is no induced emf, and thus no current in the circuit.

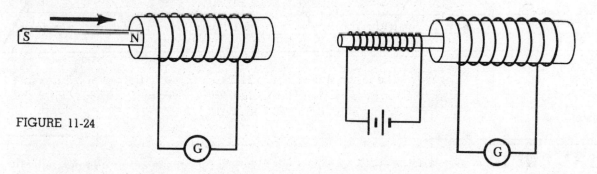

FIGURE 11-24

FIGURE 11-25

You can do the same thing with an electromagnet, as in Figure 11-25. An iron core will greatly strengthen the field of the electromagnet. As long as the electromagnet is moving in or out, the galvanometer will tell you that there is current in the circuit. But there are other things you can also do. You can weaken the field by removing the core; while this is going on, there is current. If you turn off the electromagnet current, the field suddenly collapses, and while it is diminishing, there is a sudden surge of current in the solenoid; then, nothing. Closing the electromagnet circuit induces a brief, sudden spurt of current in the other direction. The galvanometer pays no attention to a steady current in the electromagnet.

Yet another method is possible, the one used in electric generators. Figure 11-26 shows a coil of wire in a magnetic field. In position *A*, there is a lot of flux passing through the loop. At *B*, there is none; all the field lines go right by the loop without going through it. Rotating the loop thus changes the amount of flux in it and induces an emf. If the loop is rotated continuously, the field in it is constantly changing; every half-cycle, the field reverses the direction in which it passes through the loop. This causes the current in the loop to reverse, and it changes according to the sine-wave pattern of Figure 11-27.

This is alternating current, or AC.

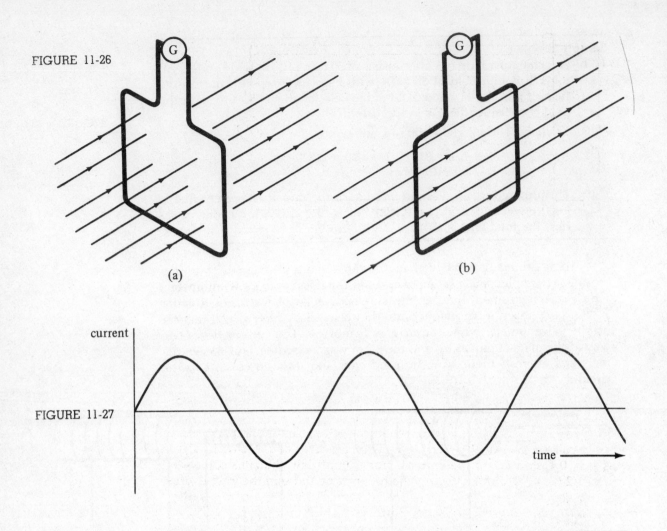

FIGURE 11-26

(a)

(b)

FIGURE 11-27

current

time

Core Concept

Emf is induced electromagnetically whenever the amount of magnetic flux passing through a loop of wire is changing:

$$\phi = BA$$

Try This A coil consisting of 50 turns of wire has an area of 150 cm^2 and is rotating in a field of 50 mTs. If it takes the coil 0.05 s to go from a position parallel to the field to a position perpendicular to it, what is the average emf induced in the coil?

11-12. Working to Make Electricity

Let's take another look at a simple electromagnetic induction process. Figure 11-28 shows a rectangular loop, 0.5 meter square, consisting of a rod that can ride on a pair of rails. The rails are connected to a voltmeter. A magnetic field of 800 mTs threads through the loop, directed into the page. You are going to push the rod the full 0.5-meter length of the loop at constant speed, doing the job in 0.10 s.

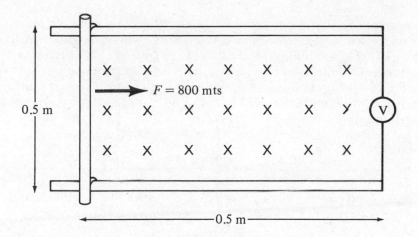

0.5 m

0.5 m

FIGURE 11-28

First question: How much emf is induced? The flux originally in the loop (from Equation 11-11) is $BA = (0.800 \text{ Ts})(0.5 \text{ m})^2 = 0.20$ weber. This changes to zero during the 0.10 s in which the rod is moving, so the rate of change of the flux is $(0.20 \text{ weber})/(0.10 \text{ s}) = 2.0$ webers/s. Faraday's law tells us that the induced emf is 2.0 volts, and that is what the voltmeter will read.

If you wish, you can calculate the induced emf from Equation 11-10 instead. It must be equal to $Blv = (0.80 \text{ Ts})(0.5 \text{ m})(5 \text{ m/s}) = 2.0$ volts. Same result.

Now suppose we replace the voltmeter by a piece of wire, completing the circuit so that current can flow. The resistance of the circuit is, let us say, 0.4 ohm. Then the current, from Ohm's law (Equation 10-6a) is $\mathcal{E}/R = (2.0 \text{ V})/(0.4 \text{ ohm}) = 5.0$ A. The rod moving through the field is producing electric energy; from Equations 5-10 and 10-3, the amount of energy produced is $I \Delta V \Delta t = (5.0 \text{ A})(2.0 \text{ V})(0.10 \text{ s}) = 1.0$ J.

Second question: Where does this energy come from? We can get a clue by considering the force that the magnetic field exerts on the current in the moving rod. Equation 11-2 tells us that, when a current flows across a magnetic field, the force on it is $IlB = (5.0 \text{ A})(0.5 \text{ m})(0.80 \text{ Ts}) = 2.0$ N. A familiar exercise with the right hand will inform us that this force is directed to the left, against the motion of the rod.

This should have been expected, surely. If you expect to get any energy out of the system, you have to put work into it. As you push the rod to the right, the magnetic field pushes it to the left, and you have to exert a force on it equal to the force exerted by the field if you want to keep it moving at constant speed. The work you do, from Equation 5-3, is $F \Delta s = (2.0 \text{ N})(0.5 \text{ m}) = 1.0$ J. And this is just the amount of electric energy you got out of the system. Once again, the law of conservation of energy is vindicated. See Sample Problem 11-17.

Sample Problem

11-17. In a setup like that of Figure 11-28, the strength of the magnetic field and the resistance of the circuit can be adjusted so that an emf of 3.5 V feeds 2.0 A into the circuit while the rod moves 20 cm in 0.50 s. How hard do you have to push on the rod?

Solution The electric energy that is produced is $I \Delta V \Delta t = (2.0 \text{ A})(3.5 \text{ V})(0.50 \text{ s}) = 3.5$ J. This is the amount of work you have to do, so $3.5 \text{ J} = F \Delta s = F(0.20 \text{ m})$, and $F = 18$ N.

When the circuit was open, there was no current, no energy output, no force on the rod, and no work input. Your utility company knows this; it can keep a generator spinning, ready for use, by burning very little oil to keep the steam turbines fired. The turbine does only enough work to compensate for the friction in the system. When the generator comes on line, sending current to your home, that current goes through the coils in the generator, and they have to be pushed much harder to keep them spinning. That's when they start to burn all that expensive oil.

Core Concept

When friction and other losses are neglected, the work done in producing electric energy is equal to the amount of energy produced.

Try This   A hand-cranked generator produces 50 V, which are used to operate a 6.0-A appliance for 30 s. If the handle of the crank is 15 cm long and must be turned 20 times to do the job, how much force must be exerted on the handle? (Neglect friction.)

11-13. The Direction of the Current

The moving rod in the last section is one example of an important principle known as *Lenz's law.* This law states that, whenever a current is induced in a loop, *the direction of the current must be such as to oppose the change that produces it.* In the example above, the current had the proper direction to push the rod to the left while you were pushing it to the right; the current opposed the motion of the rod. Only thus can the demand be met that you have to do work to produce energy.

Let's take another example. Look at the solenoid of Figure 11-29, which has its ends connected to a sensitive ammeter. If you push a bar magnet into the solenoid, you will induce a current in it. Problem: What is the direction of the current?

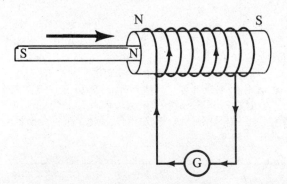

FIGURE 11-29

The current produces a magnetic field inside the solenoid. That field has to oppose the motion of the magnet. Since the magnet is being pushed in, the field of the solenoid must push it outward, to the left. This will happen only if the field of the solenoid is directed to the left, exerting a force to the left on the N-pole of the magnet. The right-hand rule for solenoids tells you that, if the current has the direction shown, it will generate a field directed to the left. See Sample Problem 11-18 for another try at this principle.

Sample Problem

11-18 In the solenoid of Figure 11-29, the S-pole of the magnet is being pulled out of the right-hand end of the solenoid. What is the direction of the current in the galvanometer?

Solution If the magnet is pulled out, the induced solenoid current must produce a field that pulls it back in. Therefore the field of the solenoid must be directed to the right so that it pushes the S-pole of the magnet to the left. The right-hand rule for currents tells us that the current must therefore be flowing down across the front face of the solenoid—and therefore from right to left through the galvanometer.

The method works as well if the magnet is an electromagnet rather than a permanent bar magnet. In Figure 11-30, the movement of the inner solenoid, which has a power supply to convert it into a magnet, induces a current in the outer solenoid. The right-hand rule for solenoids tells us that the right-hand end of the electromagnet is an S-pole. Since it is being moved *away* from the outer solenoid, the induced current must produce a field that pulls it back in. See Sample Problem 11-19.

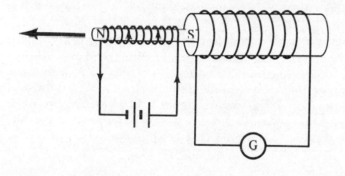

FIGURE 11-30

Sample Problem

11-19 What is the direction of the current in the solenoid of Figure 11-30 when the electromagnet is being pulled out of the solenoid, toward the left?

Solution Since the electromagnet is being pulled out, the solenoid field must pull it back in, toward the right, so the field must be diverted to the left. The way the solenoid is wound, this field will be produced by a current going from left to right through the galvanometer.

The example of Figure 11-31 is more difficult. In this case nothing moves. All we do is close the switch, sending current into the inner solenoid. As the current begins to flow, the magnetic field builds up, inducing a current in the outer solenoid. This must *oppose* the *change* in the field. Since the electromagnet field is increasing, the field of the solenoid must be in the opposite direction, to oppose the increase. The current must therefore be as shown.

When the switch is opened, the field of the electromagnet collapses. The field of the induced current must now *oppose* the *collapse* of the electro-

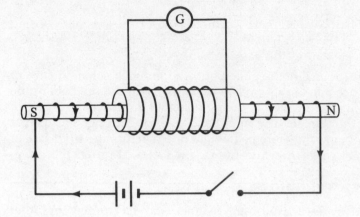

FIGURE 11-31

magnet field. The solenoid field is therefore in the same direction as the electromagnet field. See Sample Problem 11-20.

Sample Problem

11-20 There is a steady current in the inner solenoid, the electromagnet, of Figure 11-32. What will the galvanometer indicate when the switch is opened?

Solution The field inside the electromagnet is from left to right. When the switch is opened, the electromagnet field collapses, inducing a current in the solenoid. The direction of this current must be such as to oppose the collapse of the field; in other words, it must be in the same direction as the electromagnet field. A current from left to right through the solenoid and down across the front face of the solenoid will produce a right-directed field inside.

Core Concept

An induced current must be in such a direction as to oppose the change that produced it.

Try This In Figure 11-30, with the switch closed, the electromagnet is pulled out of the solenoid toward the left. Is the direction of the current in the ammeter to the right or to the left?

Summary Quiz

For each of the following, supply the missing word or phrase:

1. Parallel currents _____ each other.
2. The force between currents is called the _____ force.
3. The direction of a magnetic field is the direction pointed to by the end of a compass needle that points _____ on the earth.
4. The N-pole of a compass points to the _____ pole of a permanent magnet.
5. The force that a magnetic field exerts on a current is always perpendicular to the _____ and to the _____ .
6. A loop current is in equilibrium when its plane is _____ to the magnetic field in which it is placed.

7. A Weston galvanometer works by making use of the _____ that a magnetic field exerts on a loop current.

8. In a magnetic field pointing away from you, an electron traveling to the right will experience a force in the _____ direction.

9. To increase the kinetic energy of a charged particle, it is necessary to use a(n) _____ field.

10. When a charged particle travels perpendicularly to a magnetic field, freely in a vacuum, its path is a _____ .

11. The radius of the circle in which charged particles travel in magnetic fields is used to measure their _____ .

12. Magnetic fields are produced by _____ .

13. If a magnetic field is 10 mTs at a point 1 cm from the current that produces it, the field 5 cm from the current is _____ mTs.

14. You are looking into a solenoid, at its S-pole, along its axis. From your viewpoint, the direction of the current in the solenoid is _____ .

15. Crowding the wires of a solenoid more closely together will _____ the strength of the field inside it.

16. _____ materials can drastically increase the strength of a magnetic field.

17. The special magnetic properties of magnetic materials result from the existence in their atoms of spinning _____ .

18. Permanent magnets must be made of _____ of ferromagnetic elements.

19. The magnetic field of the earth points straight down at the _____ pole.

20. Separation of charge can be produced when a conductor is _____ in a magnetic field.

21. The amount of emf induced in a rod moving across a magnetic field depends on the _____ and the _____ of the rod.

22. An emf is induced in a loop of wire only while the _____ within the loop is changing.

23. In a generator, emf is induced in a loop by _____ the loop.

24. The electric energy produced by moving a rod, part of a circuit, through a magnetic field is equal to the _____ in moving the rod.

25. If a magnet induces a current in a solenoid by moving to the right, the direction of the force that the current exerts on the magnet is to the _____ .

26. An electromagnet induces currents in a solenoid by having its own current turned on and off. The field of the solenoid will be in the same direction as the field of the electromagnet when the switch of the electromagnet is _____ .

Problems

1. A vertical wire carries a current of 20 A straight down. At a point 15 cm north of the wire, what is (a) the magnitude and (b) the direction of the magnetic field produced by the current?

2. What current would have to flow in a wire to produce a field of 0.050 mTs at a distance of 2 cm from the wire?

3. What is the strength of a magnetic field that exerts a force of 2.5×10^{-4} N on each meter of wire carrying 5.0 A of current perpendicular to the field?

4. A #26 wire has a linear density of 1.50 g/m and can carry no more than 2.0 A of current. How strong a magnetic field would you need to support the entire weight of a length of this wire carrying maximum current?

5. Find (a) the direction and (b) the magnitude of the force exerted on a proton traveling perpendicular to a magnetic field of 1.5 Ts, if the proton is going 1.8×10^7 m/s. The proton is going north in an upward-directed field.

6. A charged particle going 2.6×10^7 m/s enters a magnetic field of 0.75 Ts and goes into a circular path of radius 2.4 m. If the particle has one elementary unit of charge (1.60×10^{-19} C), what is its mass?

7. How fast is a proton traveling if a magnetic field of 0.25 Ts can make it travel in a circular path of radius 40 cm? (See Appendix 3 for constants.)

8. What is the field inside a solenoid wound tightly in two layers of copper wire 0.10 mm in diameter if the current is 0.50 A?

9. A solenoid is made with 200 turns of wire wound on a core 15 cm long. The resistance of the wire is 30 Ω, and it is connected to a 12-V battery. (a) What is the field inside the solenoid? (b) If the same wire is wound in two layers 7.5 cm long and connected to the same battery, what is the field inside the solenoid?

10. In Chicago, the earth's magnetic field is 0.055 mTs, directly north, and dips down at 72°. Find the direction and magnitude of the force exerted on a wire 1.0 m long if it carries a current of 8.0 A directly eastward.

11. What emf is produced by moving a 20-cm rod perpendicularly to a 1.2-Ts magnetic field at 3.5 m/s?

12. A circular loop of wire 20 cm in diameter is set perpendicular to a magnetic field of 350 mTs. How much magnetic flux passes through the loop?

13. The loop of Problem 12 consists of 150 turns of wire, and its ends are connected to a potential-measuring device. If the magnetic field is turned off and drops to zero in 0.10 s, how much emf is induced in the loop?

14. A hand-cranked generator produces 65 V emf, and its internal resistance is 4.0 Ω. If the generator is connected to a 2.4-Ω resistance, how much power is needed to turn the crank? (Neglect friction.)

15. For the solenoid of Figure 11-32, tell whether the current in the galvanometer is to the right or to the left when (a) the switch is closed; (b) the electromagnet is moved to the right; (c) the switch is opened.

FIGURE 11-32

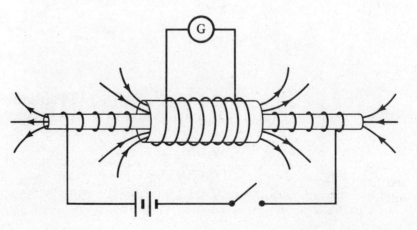

CHAPTER 12

Electricity in Use

12-1. Manufacturing Electricity

Few scientific discoveries have ever made an impact on society equal to that of Michael Faraday's observation that, when a wire is moved through a magnetic field, the electric charges in it can be separated (Chapter 11). It was this principle that made possible the production of electric energy in large quantities.

The principle is simple. Take a loop of wire and place it in a magnetic field, as in Figure 12-1. As the loop is turned, the amount of magnetic flux passing through it is constantly changing. The flux in the loop follows a

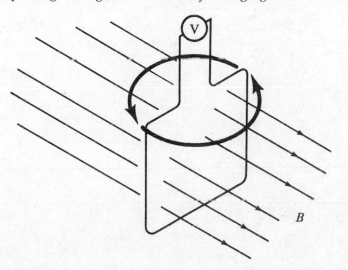

FIGURE 12-1

sine-wave pattern, as in Figure 12-2, reaching its maximum when the plane of the loop is perpendicular to the field. The rate of change of the flux in the loop (Figure 12-2a) is the slope of this graph and is plotted in Figure 12-2b. According to Faraday's law, this rate of change is equal to the emf induced in the loop. That sine-wave emf produces an alternating potential difference at the terminals of the loop.

FIGURE 12-2

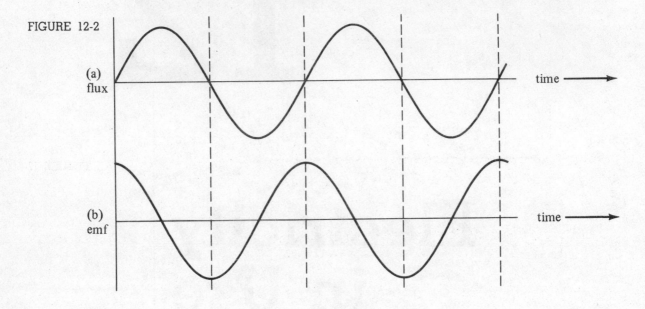

(a) flux   time ⟶

(b) emf   time ⟶

Attaching the ends of the loop to a split-ring commutator, as in Figure 12-3, brings this potential difference into an external circuit in the form shown in Figure 12-4. Every half-cycle, the external circuit is connected to a different end of the loop. The result is that the output potential difference is always in the same direction, although it fluctuates wildly.

In a real direct-current (DC) generator, there are many coils of wire in the rotating armature, connected to many segments of a split-ring commutator. Only that coil which, at any moment, has its plane parallel to the magnetic field feeds its output to the brushes. The result is an emf that fluctuates a great deal less, as in Figure 12-5.

FIGURE 12-3

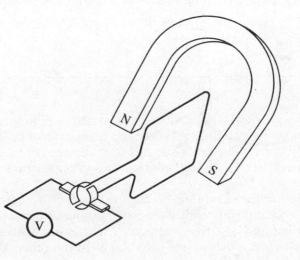

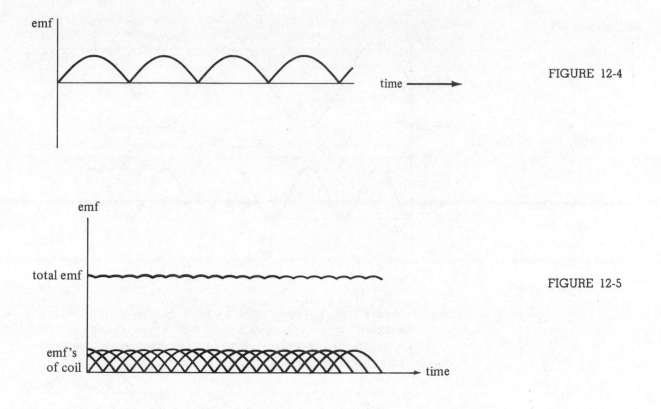

emf

time ——→

FIGURE 12-4

emf

total emf

emf's
of coil

time

FIGURE 12-5

Early in the electric power game, it was noticed that the alternating emf in the armature had some distinct advantages and that the split-ring commutator, which produced a DC output, was unnecessary. Today, all power-plant generators take the emf directly out of the armature. It is alternating current (AC) that is distributed to the consumer. Large generators are turned by steam turbines powered by burning coal, oil, or gas, or by steam produced in a nuclear reactor. The energy released in chemical or nuclear reactions is turned into heat and then into electrical energy to operate lights and machinery.

Core Concept

Electric energy is produced by generators which operate on the principle that a conductor moving through a magnetic field has its charges separated.

Try This   Explain why a rotating loop produces AC.

12-2. What Is AC?

If connected into a circuit, the alternating emf of a generator will produce an alternating current. As the potential reverses 120 times a second, the electrons throughout the circuit oscillate back and forth. Energy is transferred in a wavelike pattern, just as the oscillations of air molecules transfer energy in the form of a sound wave.

How shall we measure the size of an alternating current? No single number can express it, since it is constantly changing, varying sinusoidally with time as shown in Figure 12-6a. The average current is not a useful concept; since current flows as much in one direction as in the other, the

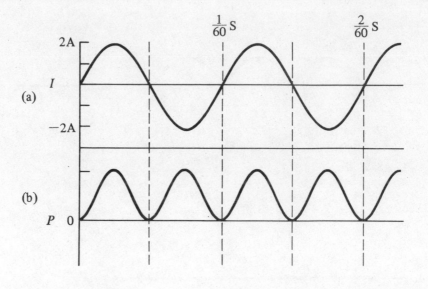

FIGURE 12-6

average is always zero. We need some measure of current that will tell us how to calculate the energy used in a light bulb when it is fed with AC.

We know that the power consumed in an electric circuit element, the rate at which it uses energy, is given by

$$P = I^2R$$ (Equation 10-10b)

The graph of Figure 12-6b shows how I^2R varies with time when the maximum current is 2 A (as in Figure 12-6a) and the resistance is 5 Ω. Whether I is positive or negative, P is always positive. The current delivers power to the bulb no matter which way it is flowing. The power goes from zero to 20 watts and back to zero 120 times a second. In special circumstances, it is possible to show that a light bulb grows dim and bright in just this rhythm.

The average power delivered to the bulb is a useful concept; it is, in fact, just half of the maximum power. To relate this power to the current, we can invent a quantity called the *effective value* of the current, which is defined by this equation:

$$P_{av} = I_{eff}^2R$$ (Equation 12-2a)

Now we can calculate power in ways familiar from our work with DC circuits, as in Sample Problem 12-1.

Sample Problem
12-1 How much is the effective AC current if it is delivering 120 W to a 30-Ω resistance?

Solution From Equation 12-2a,

$$I_{eff} = \sqrt{\frac{P_{av}}{R}} = \sqrt{\frac{120 \text{ W}}{30 \text{ }\Omega}} = 2.0 \text{ A}$$

From Figure 12-6, it is clear that

$$P_{max} = I_{max}^2R$$

and

$$P_{av} = \frac{P_{max}}{2}$$

so that

$$P_{av} = \frac{I_{max}^2 R}{2}$$

and combining this with Equation 12-2a gives the relationship between effective and maximum AC current:

$$I_{eff} = \frac{I_{max}}{\sqrt{2}}$$ **(Equation 12-2b)**

See Sample Problems 12-2 and 12-3.

Sample Problem

12-2 An AC ammeter, reading effective values, records a current of 2.5 A. How much is the maximum current?

Solution From Equation 12-2b,

$$I_{max} = I_{eff}\sqrt{2} = (2.5 \text{ A})(\sqrt{2}) = 3.5 \text{ A}$$

Sample Problem

12-3 What is the maximum power being delivered to a light bulb operating on 120 V AC and drawing 0.3 A AC?

Solution For a pure resistance, like a light bulb, Equation 10-3 applies, using effective values to get average power. Thus $P_{av} = I \Delta V = (120 \text{ V})(0.3 \text{ A}) = 36$ W. The maximum power is twice this, 72 W.

A similar line of argument gives the relationship between effective and maximum AC potential difference:

$$\Delta V_{eff} = \frac{\Delta V_{max}}{\sqrt{2}}$$ **(Equation 12-2c)**

When a household circuit is rated at "120 V," this means that the effective value of the potential difference is 120 volts. This calls for some caution. For example, if you were to use wire whose insulation is adequate protection at 150 V DC, you could not use it on 120 V AC, since there the potential rises to $120\sqrt{2} = 170$ V.

And a further warning: It is not always correct to calculate power in an AC circuit as the product $I_{eff} \Delta V_{eff}$. There are even cases in which a circuit element can have substantial AC current and potential difference without using any average power at all! Stick to $I_{eff}^2 R$; it always works.

Core Concept

Effective AC current, used in calculating power, is maximum current divided by $\sqrt{2}$:

$$P_{av} = I_{eff}^2 R; \quad I_{eff} = \frac{I_{max}}{\sqrt{2}}; \quad \Delta V_{eff} = \frac{\Delta V_{max}}{\sqrt{2}}$$

Try This 3.0 A AC (effective value) is flowing in a 20-Ω heater. Find (a) the maximum current; and (b) the power delivered.

12-3. The Great Transformation

AC has replaced DC chiefly for one reason: the transformer. This is a device that can use electric power at one potential to produce an equal amount of power at another potential. If you go to Europe, you can take along a transformer that will change 220 V AC to the 120 V AC you need to run your hair dryer. Your television set contains transformers that can change the 120 V AC you put into it into dozens of different potentials, all the way up to the 30,000 V you need to run the picture tube. Your electric utility company ships energy to you at 7,000 V and changes it down to the 120 V you use at home by means of a transformer near your home, either on a utility pole or underground. And transformers work only on AC.

The principle of the transformer is illustrated in Figure 12-7. A coil is wrapped around an iron core, and AC is fed into it. This sets up a magnetic field inside the iron, oscillating at 60 cycles or whatever the AC frequency happens to be. A second coil is also wrapped around the iron core and connected to an energy-using load of some sort. The second coil (called the *secondary*) surrounds a changing magnetic field; according to Faraday's law of induction, there must be a potential difference induced in the coil. If it becomes part of a complete circuit, there will be a current in the secondary.

FIGURE 12-7

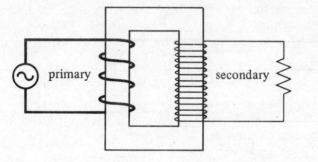

In the transformer, every turn of wire, whether in the input coil (the *primary*) or the secondary, encloses the same magnetic flux. Therefore, the potential emf induced in each turn of wire is the same in both coils. To make the emf in the secondary larger than the potential difference across the primary, all we have to do is to use more turns of wire in the secondary than in the primary. The two potential differences, in fact, will be in the same ratio as the number of turns of wire in the two coils:

$$\frac{\Delta V_s}{\Delta V_P} = \frac{N_s}{N_P}$$

(Equation 12-3a)

A transformer can increase (step up) or decrease (step down) potentials strictly according to the relative number of turns in the two coils. Some transformers have several secondaries, so that a number of different potentials can be obtained in the output. A variable transformer, like the one that feeds toy electric trains, has a movable tap that can contact a variable number of turns in the secondary. The possibilities are endless, once the principle has been established.

A transformer is a highly efficient device, and many of them can produce, in the secondary, well over 90 percent of the energy put into the primary. But we get no energy for nothing. However, if we neglect the small loss due to heating of the wires of the coils, the power output of a transformer will be equal to the power input. In a transformer, $I \Delta V$ is a valid expression for power, so we can write

$$I_s \Delta V_s = I_p \Delta V_p$$

Combining this with Equation 12-3a tells us that

$$\frac{I_p}{I_s} = \frac{N_s}{N_p} \qquad \textbf{(Equation 12-3b)}$$

Thus, if you want to step up potentials to higher values, you must expect to put correspondingly more current into the primary than you get out of the secondary. See Sample Problem 12-4.

Sample Problem

12-4 A transformer has 500 turns in the primary and 3,000 turns in the secondary. In the primary, the potential difference is 120 V AC and the current is 150 mA. Find (a) the secondary potential difference; and (b) the secondary current.

Solution (a) The potentials are in the same ratio as the number of turns:

$$\frac{\Delta V_s}{\Delta V_p} = \frac{N_s}{N_p}$$

$$\frac{\Delta V_s}{120 \text{ V}} = \frac{3,000}{500}$$

$$\Delta V_s = 720 \text{ V}$$

(b) The currents are in an inverse ratio to the number of turns:

$$\frac{I_p}{I_s} = \frac{N_s}{N_p}$$

$$\frac{150 \text{ mA}}{I_s} = \frac{3,000}{500}$$

$$I_s = 25 \text{ mA}$$

If you have a radio or any other electrical device that is marked "AC only," it is because the current you send into it is being transformed to use the power at different potentials. If you plug this appliance into a DC source, feeding power into a transformer, the current will be enormous and something will burn out.

Core Concept

A transformer converts potential differences and currents in an inverse ratio:

$$\frac{N_s}{N_p} = \frac{\Delta V_s}{\Delta V_p} = \frac{I_p}{I_s}$$

Try This A radio works on 15 V, obtained from a transformer that plugs into a 120-V wall outlet. It uses 3.0 W. Find (a) the turns ratio in the transformer; and (b) the current in the primary and secondary.

253

12-4. Power Lines

It was the transformer that made possible the power grid that carries electricity to the most remote parts of the country. Without the transformer, and the AC generator that feeds it, we would still need a powerhouse no more than $\frac{1}{2}$ mile from our homes, and we would still not get very much power.

To see why this is so, consider a modest home that needs, from time to time, 10 kW of power to keep its inhabitants happy and comfortable. To keep the numbers simple, let's assume that we will supply this power at a potential difference of 100 V. Then, from Equation 10-3,

$$I = \frac{P}{\Delta V} = \frac{10,000 \text{ W}}{100 \text{ V}} = 100 \text{ A}$$

so we have to send 100 A to the house.

Let's send this current through some good-sized aluminum wires. If we assume it is about 1 cm thick and 1 mi long, each wire will have a resistance of about 0.5 Ω. The two transmission wires and the house now constitute a series circuit connected to the powerhouse, as illustrated in Figure 12-8. This current of 100 A has to flow through the generator and the wires, as well as through the house.

As the current flows through the aluminum wires, it loses power; the wires have resistance, which converts electric energy into heat. The power loss in the wires is

$$P = I^2 R \qquad \text{(Equation 12-2a)}$$

$$P = (100 \text{ A})^2 (2 \times 0.5 \ \Omega) = 10,000 \text{ W}$$

We are using just as much energy to heat the transmission line as the house is using, and this is terribly wasteful. For another example, see Sample Problem 12-5.

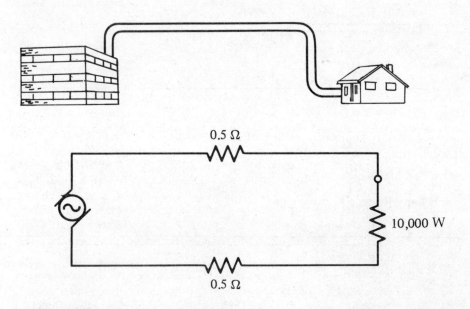

FIGURE 12-8

Sample Problem

12-5 How much power is wasted in supplying 30,000 W to a small factory at a potential difference of 240 V if the transmission lines have a resistance of 0.2 Ω in each direction?

Solution   From Equation 10-3, the current is

$$I = \frac{P}{\Delta V} = \frac{30,000 \text{ W}}{240 \text{ V}} = 125 \text{ A}$$

The waste occurs when this current passes through 0.4 Ω in the transmission lines; from Equation 12-2a,

$$P = I_{eff}^2 R = (125 \text{ A})^2 (0.4 \ \Omega) = 6,300 \text{ W}$$

There is a way out of this problem. Suppose we supply the 10 kW to the house at 1,000 V instead of 100 V. Then the house will get the same power by drawing only 10 A of current from the generator. The power loss in the lines will now be $(10 \text{ A})^2 (1 \ \Omega) = 100$ W. We have saved 99 percent of the wasted power.

The catch to this scheme is that no one wants a potential difference of 1,000 V in his electrical outlets. It is too dangerous. The transformer must come to the rescue. Power companies use very high potentials—and thus small currents and small line losses—in sending electric power over long distances. Locally, transformers reduce the potentials and increase the current, for safer conveyance of the power for short distances, where the wire resistance is small.

Power losses in transmission lines are minimized by the use of high potentials and small currents.

Try This   A school uses 140 kW of power, supplied at 120 V from a transformer nearby. The line from the transformer has a resistance of 0.002 Ω in each direction. How much power is lost in the lines?

12-5. Distributing Power

Let's trace your electricity from the generator where it is made to the fuse box where it enters your home.

In principle, the generator is no more than a loop of wire rotating in a magnetic field. But a generator that produces large amounts of energy with high efficiency is a large and complex machine. Every detail must be carefully designed so that the last joule of energy can be squeezed out of every drop of precious oil.

The field magnets of a typical large generator are electromagnets, energized at about 250 volts by some of the current produced by the generator itself. Instead of spinning the armature, the steam turbine is attached to the field magnets. It is more convenient to keep the armature stationary because the potentials in it are much higher. A field magnet usually has four poles—N S N S—so each time it goes around once, it induces two AC cycles in the armature.

There is no reason why the armature should contain only one loop of wire. There are three—or perhaps several groups of three. In each group, there are three coils set at angles of 120° to each other, as diagramed in Figure 12-9. Each loop produces its own AC, and the potentials of the three are 120° out of phase with each other. This three-phase AC is indicated in the graph of Figure 12-10. AC is carried out of the armature on three wires.

Typically, the generator produces AC of three phases, at 18,000 volts

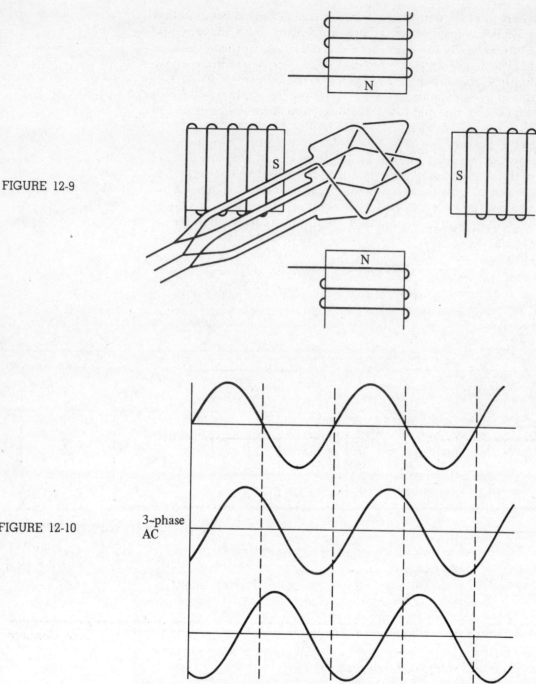

FIGURE 12-9

FIGURE 12-10

3-phase
AC

each phase. This current passes to a large transformer, where it is boosted to 345,000 volts. This is the electricity that is sent over those giant steel skeletons you see marching across the countryside, bearing wires on their arms. Note that the wires are in groups of three—identical currents differing in phase by 120°. The wires are not insulated. No conceivable covering could confine potentials that large. The wires are suspended from the towers by means of ceramic insulators 1 meter or so long. Another wire runs along the tops of the towers, connected to the steel framework. This is a "neutral" wire, electrically grounded through the steel to the earth. If lightning strikes the system, it will hit the neutral wire and pass harmlessly to the earth.

Somewhere in your community there is a transformer substation that receives the 345,000-volt power and steps it down to about 13,800 volts. This current travels locally for short distances, either on overhead poles or underground. It may wind up in a factory, where the whole three-phase AC is used to run a large motor, either at 13,800 volts or transformed down to something a little more modest. Or it may end up in a small transformer on your block. There, one phase is taken off, transformed down to 240 volts, and sent over a short wire into your home.

Figure 12-11 is a diagram of part of the transformer that supplies your home. The three-phase AC comes in on four wires, including the neutral wire. Each of the three primary coils, arranged in an electrical triangle, has 13,800 volts which are transformed down for you and your neighbors. One of the secondaries is shown; it has 240 volts between its ends, and a wire at its center is connected to the neutral and to ground. Between hot wire A and the neutral ground, there is 120 V AC. Between hot wire B and the neutral wire, there is also 120 V AC, but this potential is 180° out of phase with A. The result is that the AC potential between A and B is 240 volts. In your home, you can use this system to get two separate 120-volt circuits and one 240-volt circuit.

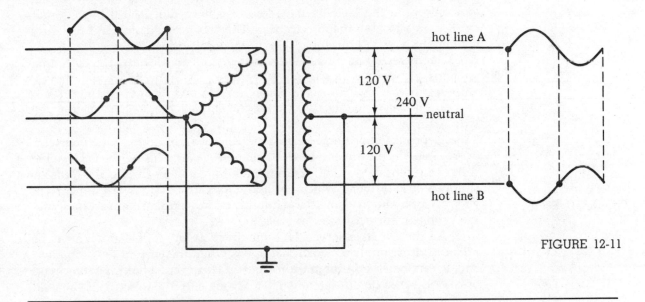

FIGURE 12-11

12-6. The Wires in Your Home

Electricity can be dangerous. Precautions must be taken against two kinds of hazard: fire and shock. Proper installation of all electrical equipment is essential. Standards are set by the *National Electrical Code*, a fat book that tells, in the greatest detail, what is and is not acceptable. Local communities may give this code the force of law, and they often make their own codes even more restrictive.

Unfortunately, every wire has resistance and converts some of the energy it carries into heat. When current flows, the wire warms up and keeps getting hotter until it loses heat to its surroundings at the same rate at which it is being made. Three factors determine the rate at which the wire can give up heat: (1) thickness—a thick wire has more surface area than a thin one, so there is a larger region over which heat can escape; (2) insulation—thick insulation tends to inhibit the escape of heat; (3) temperature—the hotter the wire is, the greater the rate at which it can radiate away its heat.

For any given wire, the controlling factor is temperature. The temperature of the wire will rise until it is in a thermal steady state, with heat escaping as fast as it is produced. If there is too much current, the wire may not reach this condition until it is dangerously hot. Then, the wire may melt, or it may set fire to the house.

The current that any given wire can safely carry is limited by its thickness and its insulation. Since insulation is absolutely necessary, it comes down to this: a fat wire can carry more current safely than a skinny one. Not only does the fat wire have less resistance, but it also has a larger surface area to help dispose of heat. The #14 wire (diameter 1.628 mm) commonly used in household wiring, made of copper and insulated, carries 15 amperes safely. Some household circuits, particularly those feeding kitchens, are wired with #12 wire (diameter 2.053 mm), which can carry 20 amperes. If you have an electric oven, it will need its own circuit; it operates on 240 volts, may use as much as 30 amperes, and thus calls for #10 wire (diameter 2.588 mm).

To protect your home against any circuit carrying too much current, every circuit incorporates a fuse or a circuit breaker. This device is placed in series with the line, so all the current in the circuit goes through it. A fuse is simply a strip of metal which has a low melting point. When the current gets too high, the fuse melts and interrupts the circuit. You must then find the problem, correct it, and replace the fuse.

A circuit breaker serves the same function, but it does not have to be replaced. It is simply a switch that opens when the current gets too high, and it can be reset after the problem is corrected. One kind of circuit breaker is based on an electromagnet. Strong current strengthens the field until it can pull on an iron core strongly enough to open the switch.

A fuse or circuit breaker must be matched to the wire it is designed to protect: On a circuit wired with #14 wire, for example, you must use a 15-ampere fuse.

Core Concept

Electric circuits produce heat and must be protected against overload.

Try This Considering the values given above, can you guess what gauge wire is protected by a 40-A fuse?

12-7. Shock Protection

The shock from a 120-volt line is dangerous and can be fatal. There are two lines of defense against shock: (1) Keep the electricity in the wires where it belongs; and (2) arrange the circuits so that in case the electricity escapes it does no harm.

A person receives a dangerous shock when a substantial current passes through his or her body. The current causes spastic contraction of muscles. If it goes through the heart, it can kill. This situation occurs when a hand touches a high-potential wire while the feet are at ground potential. Bare feet, especially if wet, are particularly dangerous; rubber shoes insulate the body from ground and prevent the formation of a complete circuit from high potential to ground. It is only the current passing through the body that counts. The white (neutral) wire of your household wiring may be carrying a lot of current, but it is at nearly ground potential, and it is not dangerous to touch it. Careful, though! Not every house is wired correctly!

Good insulation keeps the potentials inside the wire where they belong. But insulation may crack with age and may also wear off if it happens to touch a moving part. So even if wiring is properly installed at first, it may go bad and let the electricity leak out to a metal part. Allowance must be made for this eventuality.

The key to protection against shock is grounding. This means that circuits must be planned in such a way that a failure of the insulation of the high potential will result in a direct, low-resistance path from the high potential to the earth.

Suppose, for example, that the insulation fails inside your electric mixer, so that its metal casing makes contact with the high-potential wire. Then you will be in danger of a severe shock if you touch the casing, for your body provides a path from high potential to ground. However, if the shell of the mixer is connected to ground through a wire as in Figure 12-12, this pathway will have a much lower resistance than your body, and that is the path the current will follow.

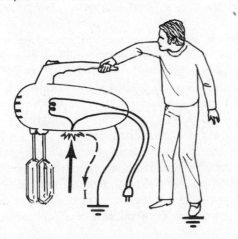

FIGURE 12-12

All electric codes call for careful grounding. Connections from one wire to another must be enclosed in a metal box. Every switch, every outlet, and every lamp connection must be surrounded by metal. All these metal boxes should be connected to each other by grounding wires in the cables that run through the walls of the house. Some codes even insist that all the cables be enclosed in metal sheathing. The entire electric circuitry is thus enclosed in metal, and the system grounded somewhere, either by connecting it to a water pipe or to a separate rod driven 10 feet into the earth.

Modern installations call for three prongs in every electric outlet. The third, connected to the green wire of the electric cable, is the grounding terminal. The cord of your mixer contains a third wire, running from the metal shell of the mixer to the third terminal of the outlet.

Accidental shock is prevented by enclosing all circuitry in grounded metal.

Try This   Explain why no grounding wire is needed in an electric drill made with a plastic case.

12-8. Appliances

Electrical appliances do many things for you, from brushing your teeth to waking you up in the morning. Whatever the gadget, however, the electrical part of it falls into one of four categories: (1) resistance heating, (2) operation of a motor, (3) fluorescent lighting, or (4) electronic control. The last of these, which includes your radio, television set, computer, and pocket calculator, is a whole world of its own, and we will not go into it.

Resistance heating devices include your toaster, electric iron, incandescent lamps, and electric heating system, if you have one. The principle of operation is simple: Just send a great deal of current through a wire, and it will convert the electric energy to heat at a rate equal to I^2R. The choice of metal for the heating element is crucial. You need something that will neither melt nor break at very high temperatures, and that has only moderate resistivity. For light bulbs, tungsten is used; for toasters and electric coffee pots, the choice is likely to be an alloy such as Nichrome.

If you stop to count the electric motors in your home, you will be surprised to discover how many there are. Many of them are of the universal type, which work on either AC or DC. They feed the current into the armature through brushes and a commutator, as described earlier. However, there are other kinds. An induction motor, which is probably the kind in your dishwasher, works on AC only and has neither brushes nor commutator. Its armature has no electrical connections at all. The current in the armature is induced by the alternating field produced in the field coils. It is like a transformer, with the secondary allowed to rotate when torque is applied to it by the field of the primary.

Electric clocks contain a device called a synchronous motor. This is a small permanent magnet placed inside an alternating field. The magnet spins in keeping up with the rapid changes in the field. On a 60-cycle AC, therefore, the permanent magnet spins at 60 revolutions per second. Since this frequency is accurately controlled by the utility company, the electric clock keeps excellent time. The newer clocks, battery-operated, work on an entirely different principle. Their motor is controlled by an oscillating quartz crystal.

The fluorescent light bulb is unique. It operates on the special electrical properties of an ionized gas, or plasma, which it contains. The bulb is largely evacuated, but it contains a small amount of mercury vapor. Hot filaments at the ends of the tube release electrons, which are accelerated toward the other end of the tube by an applied electric field. The electrons strike mercury atoms. The energy they add to these atoms increases the electron energy within the atom. These electrons within the mercury atom then given up their excess energy in the form of ultraviolet rays. The rays strike the coating on the inside of the tube, which is a special material that converts ultraviolet rays into visible light.

The special circuit of a hot-start fluorescent tube, shown in Figure 12-13, has to do three things: (1) heat the little filaments to make them release electrons; (2) apply a jolt of high potential across the tube to start the plasma current; and (3) apply a much lower, continuous potential dif-

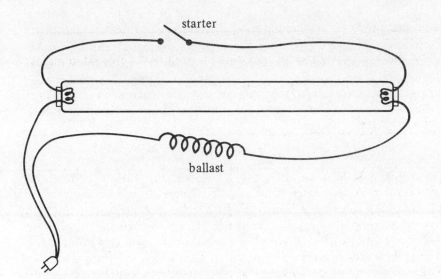

FIGURE 12-13

starter

ballast

ference between the ends of the tube to keep the plasma current flowing. To accomplish all this, the circuit needs two elements in addition to the tube itself: a starter switch and a ballast.

The starter switch is normally closed so, when you turn on the main switch, current flows through the filaments. This heats up the filaments and makes them emit electrons. After a second or so, the starter automatically opens. This interrupts the current in the ballast, which is a coil of wire wrapped around an iron core. The collapse of the magnetic field inside the core induces a sudden high potential pulse in the ballast, which is applied across the ends of the tube. This is the starting pulse.

Now the ballast has another function. It reduces the potential across the tube by developing an AC potential of its own. A short tube may need only about 25 volts, and the ballast drops the circuit potential to that level. A series resistor could do the same thing, but the unique nature of an iron-core coil enables it to do this without using any energy. In the next section you will see just how this is possible.

Core Concept

Electricity provides resistance heating, kinetic energy from motors, and other valuable services.

Try This   Count the electric motors in your home.

12-9. Inductors

The ballast of a fluorescent light bulb is only one example of a large class of devices that operate on AC by the principle of electromagnetic induction. Included in this category are transformers, induction motors, loudspeakers, electronic filter circuits, and many others.

As we saw in Chapter 11, the emf induced in a coil is equal to the rate of change of the field within the coil. If we put an AC into a coil, the alternating current produces an alternating field. And the alternating field, in turn, will induce an alternating emf in the coil. This *back emf* will be expressed as a potential difference between the ends of the coil. This is a self-induced potential difference.

This self-induced potential difference must oppose the change in current; it is greatest when the current is decreasing most rapidly, and greatest in the opposite direction when the current is increasing most rapidly. This relationship, for a sine-wave current, is shown in Figure 12-14. At maximum current, the emf is zero; at zero current, the current is increasing rapidly; then the emf is a maximum. The emf is a sine-wave function, 90° out of phase with the current.

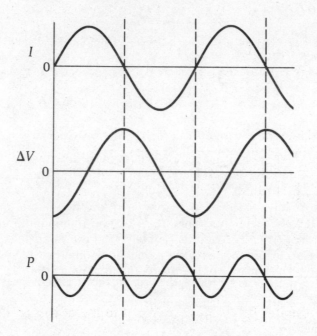

FIGURE 12-14

Now look at the power function. At any moment, the power is $I \, \Delta V$. Because of the 90° phase difference, the power is alternately positive and negative. Positive power means electric energy is being used; negative power means electric energy is being returned to the circuit. When the current is increasing, *in either direction*, it is losing electric energy to the magnetic field. As the current decreases, the field collapses and returns the energy to the circuit. The average power consumption of an inductor is zero, even though there is current in it and a potential difference across it.

Self-inductances of this sort have a wide use in all kinds of electronic circuitry. For example, they form an important part of electronic filters, whose function is to separate high frequencies from low ones. When the frequency is high, the current in an inductor is changing rapidly, so the back emf is large; this makes the current through the inductor small. Low-frequency currents pass through, but high frequencies are blocked. An inductor is part of the filter in a hi-fi system that sends the high frequencies to the tweeters and the lows to the woofer.

Core Concept

In an inductor, AC current and potential difference are 90° out of phase with each other, so that no power is consumed.

Try This In a transformer whose secondary is open, there can be no current, and therefore no power delivered. Yet, there is current in the primary, and potential differences in both the primary and the secondary. Explain why it is possible for there to be both current and potential difference in the primary coil, yet no conversion of energy.

Summary Quiz

For each of the following, supply the missing word or phrase:

1. To produce DC, the output of a generator must be fed through a _____ .
2. In any generator, the current in the armature is of the _____ type.
3. The average value of an alternating current is always _____ .
4. AC power is equal to I^2R if the value of I used in the equation is the _____ current.
5. If a transformer is used to step up potentials, the secondary current is _____ than the primary current.
6. In a transformer, energy is transferred from primary to secondary by means of a changing _____ .
7. Power loss in transmission lines is minimized by the use of high _____ .
8. Long-distance transmission of electric power is possible only because potentials can be changed by using a _____ .
9. The 120-V AC that comes into your home has _____ phases.
10. A typical value for the potential of a long-distance electric transmission line is _____ volts.
11. Wiring is protected from overload by a _____ or a _____ .
12. As the temperature of a wire rises, the rate at which it loses heat _____ .
13. Electric circuits are grounded to protect against accidental _____ .
14. The grounding wire of an electric installation can be recognized because it is _____ .
15. A(n) _____ motor operates by inducing an AC into the armature through a magnetic field.
16. The iron-core coil in the circuit of a fluorescent light bulb is called a _____ .
17. AC electric clocks operate on _____ motors.
18. As the frequency of an AC increases, the back emf it induces in a coil _____ .
19. The average power delivered to an inductance is _____ .

Problems

1. An alternating current in a 60-Ω resistor has a peak value of 0.50 A. Find (a) the peak power; (b) the effective potential difference.
2. On a 120-V AC line, a 300-W bulb is burning. Find (a) the effective current; (b) the peak power.
3. A transformer is to be used to step 120 V AC down to 6 V AC. If it has 2,400 turns of wire in the primary, how many must there be in the secondary?
4. A 60-W bulb is designed to operate on 12 V and is to be connected through a transformer into a 120-V circuit. Find (a) the secondary current; (b) the primary current; (c) the maximum potential difference the insulation in the primary circuit must withstand.
5. A large motor is needed to provide energy at the rate of 6.5 kW, and it is to be supplied with current through a transmission line with 0.20 Ω of resistance. How much power would be saved by designing the motor to operate at 1,500 V instead of 120 V?
6. Determine the amount of power wasted in providing a factory with 45 kW of electric power at 560 V through a transmission line with a resistance of 1.5 Ω.

7. What is the frequency of rotation of the field coils of a four-pole generator that produces 60-cycle, three-phase AC?

8. What is the greatest amount of power that can be delivered through a #14 wire operated at 120 V?

9. A large power transformer converts 325,000 V to 7,500 V, at the rate of 2.5 million W. Find (a) the primary current; (b) the secondary current.

10. What potential difference must be withstood by the insulation around the wire that delivers current at 30,000 V to the picture tube of your television set?

CHAPTER 13

Energy in Space

13-1. Field Interactions

Electricity and magnetism are intimately related phenomena. Magnetic fields are produced by electric currents. A changing magnetic field induces an electric potential difference in a loop.

The interrelationship between electric and magnetic phenomena is described mathematically by *Maxwell's laws of electromagnetism*. Two of these laws, applied together, produce a remarkable prediction.

One law is this: A changing magnetic field produces an electric field. This is a very general statement, one corollary of which is Faraday's law of induction.

A second law is: A changing electric field produces a magnetic field. This law leads to the familiar knowledge that an electric current is surrounded by a magnetic field.

A third law of Maxwell tells us that an electric charge is surrounded by an electric field. If the charge is in motion, it will produce a changing electric field, which generates a magnetic field. This is in accordance with the common knowledge that a moving charge constitutes a current, which produces a magnetic field.

A new feature enters if the charge is accelerated. Since the charge is in motion, it creates a changing electric field, and thus a magnetic field as well. However, since its velocity is changing, the magnetic field it produces is also changing. Therefore, it produces an electric field also. As this field changes, it generates a changing magnetic field, which produces an electric field, and so on. The two fields reinforce each other, traveling through space away from the accelerated charge that produced them. The accelerated charge produces an *electromagnetic wave*.

The theory of the electromagnetic wave tells us that it must travel with a speed that depends on the electric and magnetic properties of empty space. These are the well-known constants of Coulomb's law of static electricity (Equation 9-4; $k = 9.00 \times 10^9$ N·m²/C²) and the law of production of a magnetic field (Equation 11-6; $k' = 10^{-7}$ N/A²). The speed of an electromagnetic wave turns out to be

$$c = \sqrt{\frac{k}{k'}} = \sqrt{\frac{9.00 \times 10^9 \text{ N·m}^2/\text{C}^2}{10^{-7} \text{ N/A}^2}}$$

$$c = 3.00 \times 10^8 \text{ m/s}$$

See Sample Problem 13-1.

Sample Problem

13-1 The moon is 3.8×10^8 m from the earth. How long does it take a radio signal to get to the moon and back to a receiver on earth?

Solution Going both ways, the signal travels 7.6×10^8 m at speed c, so it takes

$$\Delta t = \frac{\Delta s}{v} = \frac{7.6 \times 10^8 \text{ m}}{3.0 \times 10^8 \text{ m/s}} = 2.5 \text{ s}$$

This value agrees very well with measured values of the speed of light in a vacuum. Light is apparently an electromagnetic wave. The constant c turns out to have many meanings, and it appears in most of the equations of the physics of fundamental particles.

Core Concept

Light is an electromagnetic wave traveling through empty space at a constant speed c.

Try This What is the frequency of a light wave whose wavelength is 5.5×10^{-7} m?

13-2. Electromagnetic Waves

The simplest kind of electromagnetic wave has a sine-wave form, and it is produced by sending an alternating current into an antenna. As the charge in the antenna oscillates back and forth, it is accelerated, and therefore it generates an electromagnetic wave. For best results, the antenna should be a half-wavelength long.

Suppose, for example, that an FM station wants to send a signal into space at a frequency of 98 MHz (megahertz, or 10^6 Hz). This is 98 on the dial of your FM radio. The station generates an AC at that frequency and sends it down a transmission line consisting of a pair of parallel wires. The wires end in an antenna; in the simplest case, this is simply the same wires bent at right angles to the transmission line, as shown in Figure 13-1. At a given moment, the current in the system might be as shown in Figure 13-1a. Looked at in the antenna only, it appears as a current flowing

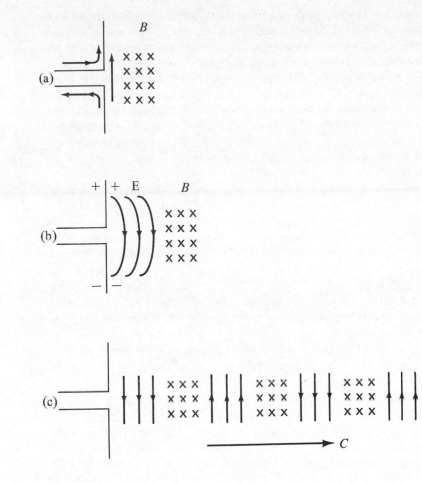

FIGURE 13-1

from one end to the other. When it reaches the end, it can go no further. A half-cycle later, the current piles up charge at the other end of the antenna. Remember that an AC in a line is a wave. Since there can be no current at the ends of the antenna, current nodes form there. If we make the antenna just a half-wavelength long, we can set up a stable standing wave of current in the antenna. We know that the wave travels in the wire at almost the speed of light, so we can easily figure out how long the antenna should be. See Sample Problem 13-2.

Sample Problem
13-2 How long should the antenna be to broadcast at 98 MHz?

Solution   The wavelength, from Equation 8-2, is

$$\lambda = \frac{c}{f} = \frac{3.0 \times 10^8 \text{ m/s}}{98 \times 10^6 \text{ Hz}} = 3.06 \text{ m}$$

The most efficient antenna is half this length, or 1.53 m.

It is the accelerated charges, moving back and forth in the antenna in a standing wave pattern, that produce the electromagnetic wave. In Figure 13-1a we have caught the current at a moment when it is going upward in the antenna. Using our familiar right-hand rule, we can see that, at that moment, there is a magnetic field around the antenna. In the space to the right of the antenna, the direction of the field is into the page.

Figure 13-1b shows the situation a quarter-cycle later. The current has dropped to zero. However, charge has piled up at the upper end of the antenna, leaving the lower end negative. This means that there must be an *electric* field around the antenna, its field lines running out at the top and into the antenna at the bottom. In the space to the right, the electric field lines point downward.

Both the electric and the magnetic fields are constantly changing, so each one reinforces the other. The final result is shown in Figure 13-1c. Magnetic fields and electric fields, both alternating in direction, chase each other through space, moving away from the antenna in all directions, at the speed of light. This is an electromagnetic wave.

(To set the record straight, it should be noted that the fields alternate at first, 90° out of phase with each other. As they travel, they gradually come into phase.)

Your receiving antenna, set parallel to the transmitting antenna, detects these alternating electric and magnetic fields and responds by setting up an alternating current, a standing wave of its own. This antenna is also a half-wavelength long, so that it resonates at its natural frequency. The AC signal is amplified electronically, and other electronic circuits extract the information from the signal and convert it to audio-frequency AC signals that operate the loudspeaker.

Electromagnetic waves of many different wavelengths are produced by a number of different processes. Here is a breakdown of the whole electromagnetic spectrum:

Radio waves, produced by electronic circuitry: wavelengths from 1,000 m to 10^{-3} m (1 mm). The longest waves are used in the AM broadcast band; then come short waves, the very high-frequency (VHF) waves used in FM and television, UHF waves, and microwaves.

Infrared rays, emitted by all warm objects, are produced by the random thermal motions of molecules. Wavelengths are from 10^{-3} m down to about 7×10^{-7} m.

Visible light occupies the tiny part of the spectrum between 7×10^{-7} m (700 *nanometers*, or 700×10^{-9} m) to about 400 nanometers. It is produced by changes in the energy levels of the outer electrons in atoms.

Ultraviolet rays are produced in much the same way, but their wavelength is too short to enter the eye; they cannot pass through the lens of the eye. Their wavelengths range from 400 nm to about 5 nm.

X-rays and *gamma rays* are produced by energy transitions in the innermost electrons of large atoms, or in nuclei. Wavelengths are from 5×10^{-9} m down to about 10^{-13} m, smaller than an atom. See Sample Problem 13-3.

Sample Problem

13-3 Compare the frequency of the longest radio waves with that of the shortest gamma rays.

Solution   Longest radio waves:

$$f = \frac{c}{\lambda} = \frac{3.0 \times 10^8 \text{ m/s}}{1,000 \text{ m}} = 3.0 \times 10^5 \text{ Hz, or 300 kHz}$$

Shortest gamma rays:

$$\frac{3.0 \times 10^8 \text{ m/s}}{10^{-13} \text{ m}} = 3 \times 10^{21} \text{ Hz}$$

which is a hundred million billion times as great.

Electromagnetic waves are produced by accelerated charges in a wide range of wavelengths.

Try This What is the best length for a microwave antenna for use on a frequency of 4,500 MHz?

13-3. Straight Lines or Curves?

Basically, the difference between one electromagnetic wave and another lies in their frequencies. The frequency difference produces profound differences between one kind of wave and another.

For example, we learned in Chapter 8 that waves tend to bend around corners (diffract) and that the amount of bending increases with wavelength. The waves of AM radio, from 60 m to 100 m long, easily bend around buildings and other barriers and find their way to your receiving antenna. They can travel for a thousand miles or more, following the curvature of the earth. In contrast, the range of shortwave radio, in straight-line transmission, is 100 miles or less. Short waves, however, bounce off the ionosphere, an ionized layer of air that lies above the stratosphere. That is why, especially at night, short waves may reach remote parts of the world.

The range of FM and television signals, at wavelengths of a few meters, can be received only up to 20 or 25 miles from the transmitter and are likely to be blocked by buildings or mountains. And the propagation of microwaves at wavelengths of a few centimeters is essentially line of sight. Microwave relay stations, now a familiar sight in the country, are placed on hilltops in view of each other. Each antenna is a half-wave long, no more than a couple of centimeters. It is backed by a parabolic dish that reflects the transmitted waves into a narrow beam aimed at the next relay tower. This is how long-distance telephone signals travel across the country.

At the extremely short wavelengths of visible light, diffraction is usually negligible. That is why we find that light travels in straight lines, and that is why an object placed near a point source of light casts a sharp-edged shadow, as in Figure 13-2. Even with visible light, however, effects due to diffraction can be found. Pierce a piece of aluminum foil with a fine needle and pass light through the hole. If you examine the spot of light that falls on a screen after passing through such a hole, you will see that it is surrounded by a series of bright and dark rings. Light is spilling out beyond

FIGURE 13-2

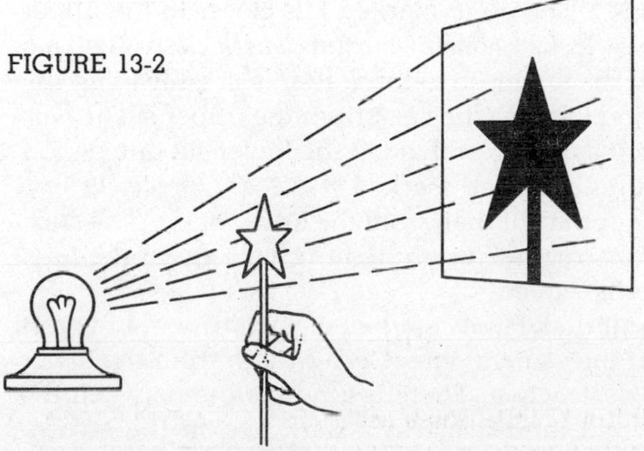

FIGURE 13-3

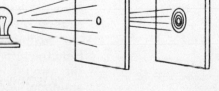

the spot that would form according to the straight-line propagation rule. The dark bands are produced by destructive interference where light from different parts of the hole arrives out of phase. See Figure 13-3.

Core Concept

Electromagnetic waves diffract, and the effect is greater at longer wavelengths.

Try This Explain why a microscope is unable to show the existence of objects no bigger than the wavelength of the light.

13-4. Measuring Wavelength

The wavelengths of light, small as they are, can be measured with extremely high accuracy. The method takes advantage of the phenomena of diffraction and interference.

As we saw in Chapter 8, a small opening, not much bigger than a wavelength, acts as a point source of waves. It is more convenient to use a thin slit rather than a point. If light is passed through such a slit, as in Figure 13-4, and then through two slits parallel to each other and close together, the light emerging from the double slit and falling on a screen forms a series of alternate light and dark bars, as shown. The bright bars are regions of constructive interference, and the dark bars are regions of destructive interference. From the geometry of the system, the wavelength of the light can be found.

FIGURE 13-4

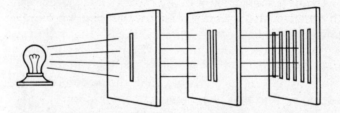

The geometry is diagramed in Figure 13-5. The two slits serve as point sources of light, oscillating in phase with each other. Any point that is the same distance from both slits will therefore receive crests from both slits in phase. Such a point is an *antinode*. All these points lie on the antinodal line marked $n = 0$ on the diagram; this line is perpendicular to the line joining the slits, and halfway between them. Bright light travels down this antinodal line, forming a bright bar on the screen.

There are other antinodes. The point marked P is closer to the upper slit than to the lower slit. It is, in fact, *exactly one wavelength closer*. Thus, a crest leaving the lower slit takes longer to get to P than a crest leaving the upper one—exactly one period longer. The crest from the upper slit arrives at point P at the same moment as the crest that left the lower slit one period earlier. All points on the antinodal line marked $n = 1$ are precisely one wavelength further from the upper slit than from the lower one. At all such points, the interference is constructive. Bright light travels down the line, lighting up another bar on the screen.

Halfway between the antinodal lines marked $n = 0$ and $n = 1$, a crest from the upper slit arrives at the same moment as a trough from the lower slit, and the interference is destructive. There is a nodal line between the antinodal lines $n = 0$ and $n = 1$, and it marks a dark bar on the screen.

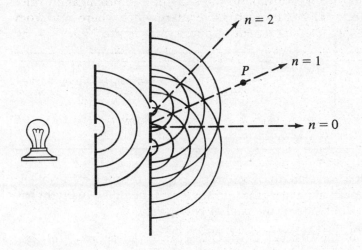

FIGURE 13-5

There are yet other antinodes. All of them must meet this condition: The path difference to any antinode, that is, the difference between the distances to the two slits, must be a whole number of wavelengths:

$$s_2 - s_1 = n\lambda \qquad \textbf{(Equation 13-4a)}$$

where n is any whole number and λ is the wavelength. Each of the bright bars is the end of an antinodal line, and each can be numbered, positive and negative, from 0 (the central line) to some higher value. See Sample Problem 13-4.

Sample Problem

13-4 In the interference pattern produced by a pair of slits acting on red light of wavelength 720 nm, how much farther is the fifth bright bar from one slit than from the other?

Solution At the fifth bright bar, a crest from one slit meets a crest that left the other slit 5 periods earlier and so traveled 5 wavelengths farther. The path difference is therefore 5×720 nm = 3,600 nm.

In practice, the device that makes use of this principle is a *diffraction grating*. It consists of a set of scratch marks made on a sheet of glass—a great many of them, accurately drawn with a diamond, straight and parallel. There may be 700 or 800 lines to a millimeter in such a grating. When light is passed through such a grating, it forms a series of nodes and antinodes, exactly like those of a double slit.

With a little geometry, it is possible to arrive at a useful expression that will make it possible to find the wavelength of the light:

$$\sin \theta_n = \frac{n\lambda}{d} \qquad \textbf{(Equation 13-4b)}$$

where d is the distance between the slits, n is a whole number, λ is the wavelength, and θ is the angle that any antinodal line makes with the central $n = 0$ line, as shown in Figure 13-6. See Sample Problems 13-5 and 13-6. The equation works well either for a pair of slits or for the multiple slits of a diffraction grating.

FIGURE 13-6

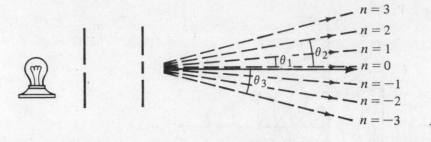

Sample Problem

13-5 If light with a wavelength of 580 nm is passed through a pair of slits 0.004 mm apart, at what diffraction angle will the first two bright bars appear on either side of the central bar?

Solution   For the first bar, $n = 1$, so

$$\sin \theta = \frac{\lambda}{d} = \frac{580 \times 10^{-9} \text{ m}}{0.004 \times 10^{-3} \text{ m}}$$

which gives $\theta = 8.3°$. For the next bar, $n = 2$ and

$$\sin \theta = \frac{2(580 \times 10^{-9} \text{ m})}{0.004 \times 10^{-3} \text{ m}}$$

and $\theta = 16.9°$.

Sample Problem

13-6 What is the wavelength of monochromatic light that makes a diffraction angle of 26.5° with a grating ruled 640 lines per millimeter?

Solution   This is presumably the first maximum, so $n = 1$ and, from Equation 13-4b,

$$\lambda = d \sin \theta$$

$$\lambda = \left(\frac{1}{640} \text{ mm}\right)(\sin 26.5°)$$

$$\lambda = 6.97 \times 10^{-4} \text{ mm, or 697 nm}$$

If you pass white light through the slits, the central $n = 0$ line will be white. But the $n = 1$ line, and any others will be spread out in a rainbow of colors, as in Figure 13-7. Many hues can be distinguished in this spectrum, but the most obvious colors form a sequence of six that looks like this: red, orange, yellow, green, blue, violet. Each color shades gradually into the next. Red appears farthest from the center line, and violet is closest.

Why are there colors? Since each color has been diffracted to a different angle, we need only look at Equation 13-4b to find the reason. Red is diffracted to a greater angle because it has a longer wavelength than violet. By measuring these angles, we can find out which wavelength corresponds to which color of the spectrum. See Sample Problem 13-7.

FIGURE 13-7

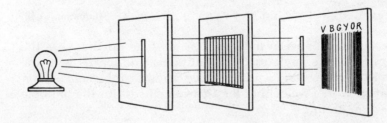

Sample Problem

13-7 As formed by a grating with 530 lines per millimeter, the entire visible spectrum is spread out from 11.6° at the violet end to 23.4° at the red end. What is the range of wavelengths of the visible spectrum?

Solution In practice, only the first diffraction spectrum is used, so $n = 1$. Therefore, from Equation 13-4b, $\lambda = d \sin \theta$. The value of d is 1/530 mm or 1/530,000 m; $d = 1.89 \times 10^{-6}$ m. Then, at the violet end, $\lambda = (1.89 \times 10^{-6}$ m$)(\sin 11.6°) = 3.8 \times 10^{-7}$ m. At the red end, $\lambda = (1.89 \times 10^{-6}$ m$)(\sin 23.4°) = 7.5 \times 10^{-7}$ m. In the customary units, the visible spectrum runs from 380 nm to 750 nm.

The interference pattern produced by a diffraction grating provides a means of measuring the wavelength of light:

$$\sin \theta_n = \frac{n\lambda}{d}$$

Try This What is the wavelength of the monochromatic (single-frequency) light of a laser if it is refracted to an angle of 37° by a grating ruled 640 lines per millimeter?

13-5. Coherence

There are many differences between the radio wave produced in an antenna and the light wave produced in a bulb, aside from wavelength. For one thing, the radio wave is *monochromatic;* that is, it consists of waves of a single frequency, while a light bulb produces a mixture of frequencies.

Another difference is that the radio wave is *coherent.* Just what this means is illustrated in Figure 13-8. If you put a receiving antenna at any point near a transmitting antenna, you can detect a regular oscillation of the electric field. It has a definite value at any moment. In contrast, the electric fields of the light from a bulb are all mixed up; every oscillating atom produces its own wave, and these waves arrive at any point in every conceivable phase. Some cancel others, and some reinforce others.

Another point is that the radio wave is continuous, and the light wave is pulsed. Each beam of light consists of enormous numbers of short bursts of waves, lasting perhaps 10^{-7} s each. It is like the difference between a stream of water flowing smoothly from a tap, and a shower.

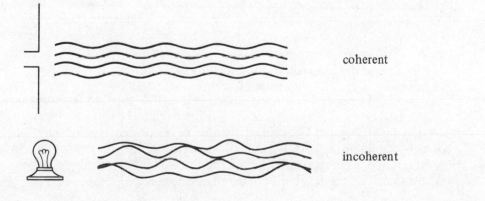

coherent

incoherent

FIGURE 13-8

A laser is a device that can produce light that is continuous, mono-chromatic, and coherent. The beam it produces is quite remarkable—a thin pencil of light that remains straight and spreads out very little. A lens can focus the beam down to a microscopic point. Some lasers are extremely powerful and can concentrate electromagnetic energy strongly and accurately.

Since their invention, lasers have found their way into a wide variety of technologies. Perhaps the earliest use of lasers was in drilling holes through industrial diamonds to be used as dies in drawing wire. A research biologist can direct a laser beam up into a microscope, where it can be focused on any tiny part of a cell to be destroyed. The laser technique is now the routine method of attaching retinas in the eye when they become separated from underlying tissue. The ability of laser beams to travel a long distance without spreading or bending has made them useful in surveying. Some lasers can be tuned to a desired frequency and have been used in chemical experiments to stimulate an experimental medium with a pin-point of light at a preselected frequency. A photographic film exposed in laser light forms an interference pattern which, when illuminated, gives the sensation of seeing an object in three dimensions; this is a hologram. Lasers also read the bar code on your supermarket purchases.

The ability to *amplitude-modulate* a laser beam gives it a new range of uses. Amplitude modulation (AM) is not new; this is how information—voice or music—is coded onto the AM radio signal. Figure 13-9 shows what this process involves. The amplitude of the radio frequency wave is made to vary according to the frequency of the sound the wave must carry. To transmit the note on a musical scale called "440-A", the radio signal, which might be at 1,200 kHz, increases and decreases in amplitude 440 times a second. In the receiver, the radio frequency is discarded and the modulation is detected, producing a 440-Hz signal that feeds into the loud-speaker.

modulated

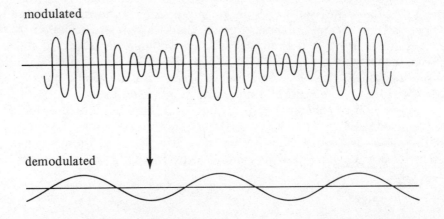

FIGURE 13-9

demodulated

The advantage of a laser in the transmission of information is that the light has such an extraordinarily high frequency. This makes it possible to transmit thousands of separate signals on a single modulated laser beam. The new telephone lines, now replacing copper wire, are tiny, transparent fibers fed by modulated laser light. A cable of a hundred such fibers can carry a half million conversations at the same time. Loss of energy in the fiber (damping) is compensated for by amplifiers at intervals. These amplifiers are lasers no bigger than a grain of salt.

Audio and video disks operated by laser are entering the market-place. A modulated laser signal is focused on the disk, where it burns a groove as the disk turns. In playing the disk, another laser focuses on these grooves and is modulated by them. Since no needle touches the record, there is no wear. Therefore, highest fidelity remains forever.

Core Concept

A laser, producing monochromatic, continuous, coherent light, has many practical uses.

Try This Determine the frequency of a red laser whose wavelength is 6.9×10^{-7} m.

13-6. Polarization

There is yet another difference between the electromagnetic wave produced by a radio antenna and that of a light bulb.

As Figure 13-10 shows, a vertical antenna produces a wave with a vertical electric field. This field varies both in time and in space, but it is always vertical. In contrast, the molecular vibrations that produce a light wave in a lamp are in every direction. At any given point in space, each member of this melange of waves oscillates in its own direction—always, to be sure, perpendicular to the direction of travel, but oriented at random in the plane perpendicular to that direction. A wave like that of the radio antenna, which has all its electric vibrations in the same direction, is said to be *polarized*.

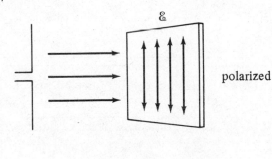

polarized

FIGURE 13-10

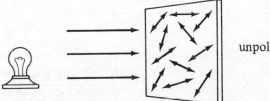

unpolarized

Light waves can be polarized. The simplest method is to use a polarizing filter; this filter looks like a sheet of gray plastic, but it is made with a thin layer of long, thin molecules, all lined up in the same direction. If light passes through such a filter, as in Figure 13-11, the electric vibrations parallel to the molecules of the filter will get through, but the electric vibrations perpendicular to the axis of the filter will be blocked. Thus the light that passes through the filter, about half of the total, will be polarized along the axis of the filter.

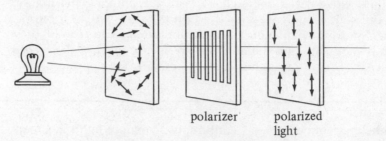

FIGURE 13-11

polarizer   polarized
             light

To the human eye (but not to that of a bee!) polarized light looks like any other light. You can tell whether or not light is polarized by passing it through a polarizing filter. Rotate the filter, as in Figure 13-12. If the light is polarized, it will be blocked when the axis of the filter is perpendicular to the electric field of the light. Rotating the filter through 90° makes the light entering your eye go from bright to dark, and another 90° restores the original brightness. Unpolarized light entering the filter looks equally bright no matter how the filter is oriented.

There is polarized light all around us, although we are never aware of it. The highlights produced when light bounces off a horizontal shiny surface is always more or less polarized, horizontally. Polarizing sunglasses take advantage of this; their lenses contain vertical polarizing filters, so that the horizontally polarized glare cannot get through.

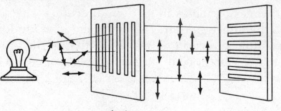

FIGURE 13-12

polarizer

Core Concept

Light can be polarized, and the polarization can be detected by rotating a polarizing filter.

Try This Explain how you would test the blue light of the sky to determine its direction of polarization.

13-7. Color

We tend to think of color as somehow being a property of an object: The book is red. But it is impossible to understand what that simple statement means unless we take into account the instrument that is making that decision: the human eye.

What is the meaning of the statement that the book is red? This: In ordinary sunlight, or any other source of light that provides a mixture of all visible wavelengths, the book absorbs certain wavelengths of light and reflects others. Those that are reflected enter the eye, where they stimulate certain cells in the retina, but not others. This particular mixture of wavelengths sets off a set of nerve impulses going to the brain, and the brain sees the color red.

The perception of "red" does not correspond to a unique set of wavelengths. Your repertory of visual perceptions is much more limited than the wavelength mixes that reach your eye. The human eye can distinguish several hundred colors and shades, but all these perceptions result from combinations of only three separate kinds of visual signals.

The nerve signals that tell the brain what color you are seeing originate in just three kinds of color-sensitive cells in the retina. These three kinds of cells are distinguished by their possession of one of three different visual pigments. Each of these pigments is sensitive in varying degree to different wavelengths of light. Figure 13-13 shows the sensitivities of these pigments. One, called "red," responds to light at the longest visible wavelength, about 760 nm, if the light is very bright. This pigment is most sensitive at about 600 nm. The "blue" pigment reaches, weakly, down to about 380 nm and is most sensitive at 430 nm. The "green" pigment is most sensitive in the middle of the spectrum at 520 nm and reaches nearly to both ends.

wavelength, nm

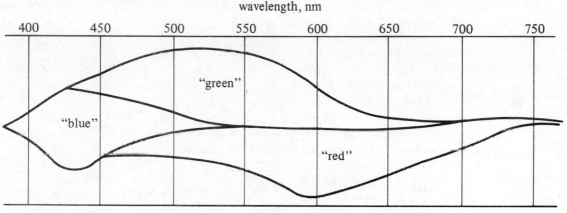

FIGURE 13-13

Under ordinary conditions, the color you see depends on the degree to which each of these pigments is stimulated. From 690 to 760 nm, things look red, since only the "red" pigment reacts in this range. The monochromatic light of a sodium vapor lamp, at 570 nm, acts on both the "red" and the "green" pigment almost equally; this combination produces the sensation we call yellow. Any combination of wavelengths that hits the red and green pigments equally looks yellow. Butter looks yellow because it absorbs light at the short-wavelength end, up to about 500 nm, and reflects the rest. When the reflected light hits the retina, it is thus received by the green and red pigments, but not the blue.

Wavelengths at three different parts of the spectrum can stimulate each of the three pigments selectively and thus can produce any sensation of color whatever. The screen of a color television set contains a mosaic of just three kinds of dots: red (long wave), blue (short wave), and green (medium wave). All the colors you see are produced in your eyes by proper combinations of these three wavelength ranges.

Understanding of the three pigments is only the beginning of a knowledge of color vision. Perception of color may often depend on the wavelengths reflected by nearby objects, as well as those reflected by the object itself; your eye reacts to contrast. Also, there is a mechanism, now partially understood, by which the retina can make allowance for differences in illumination. In the early morning, sunlight is quite red, and a color picture taken at that time will look red. But your retina has a mechanism to compensate for the imbalance in illumination, and objects will appear to you in their normal colors.

Perception of color depends on the relative stimulation of three kinds of color-sensitive cells in the retina.

Try This A magenta filter passes into the eye only the light that affects the "red" and "blue" pigments. What wavelengths are absorbed by the filter as white light passes through it?

13-8. Red Hot, White Hot

Every warm object emits infrared electromagnetic radiation. It is produced, as usual, by accelerated charges. Atoms are in constant oscillation, as we saw earlier. The energy of the oscillation is proportional to the temperature. It is these oscillations that constitute the accelerated charge producing the radiation.

Figure 13-14 shows the rate of radiation at all frequencies, for several different temperatures. At 500 K, the radiation is all infrared, ranging from a frequency of about 3×10^{12} Hz ($\lambda = 10^5$ nm) to 2×10^{14} Hz ($\lambda = 1,500$ nm). At 1,000 K, the total amount of energy radiated is much more—in fact, 16 times as much. Also, both the top frequency and the frequency of maximum radiation are higher. There is even a little at 4×10^{14} Hz, in the visible range. An object begins to glow red. When the temperature is raised to 4,000 K, a lot of energy is released over the entire visible range, and the object becomes white hot. There is even a little ultraviolet light. This is the temperature at which electric light bulbs operate.

These curves apply to what is called a *blackbody*—an ideal object that absorbs all the light that falls on it, reflecting nothing. A blackbody is also a

FIGURE 13-14

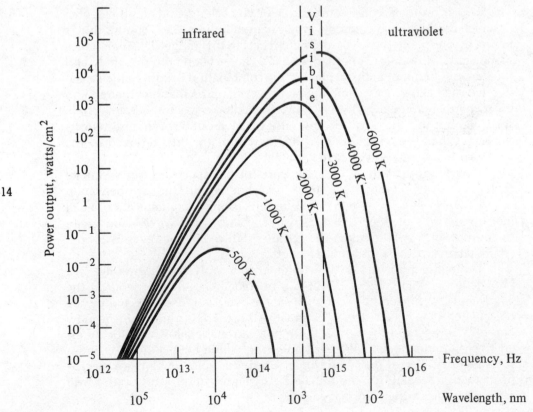

perfect radiator; any electromagnetic waves generated within it pass out through its surface freely. No real object is perfectly black, but some things come quite close. Analysis of the radiation of a blackbody is especially simple because it does not matter what the object is made of.

Each curve, at any temperature, has the same shape. Starting at the left end and moving to the right, we find that the amount of radiation is greater at the higher frequencies, until the curve reaches a peak at a certain frequency level. Above this point, the amount of radiation becomes smaller at higher frequencies, until there is none at all above a certain upper limit.

There is an old theoretical explanation for the fact that more energy is radiated at higher frequencies. It is assumed that the atomic oscillators set up standing electromagnetic waves within an object, something like the standing waves that an AC sets up in a radio antenna. It is these waves that produce the traveling waves that escape into the outside world. Now, within any given space, a lot more high-frequency (short-wavelength) waves can fit than low-frequency waves. In a guitar string 1 meter long, for example, there are 4 overtones with wavelengths between 10 and 15 centimeters, but 11 overtones with wavelengths between 5 and 10 centimeters. If the energy of the atomic oscillators is equally distributed among all possible wavelengths, there will be a lot more high frequencies than low ones, and the energy will concentrate at the upper end of the frequency spectrum.

This theory explains the left-hand side of the curve nicely, but it does not explain why the curve does not continue to rise toward the right forever. Clearly, there is something preventing energy from feeding into the higher frequencies.

The reason the higher frequencies do not get their fair share of the energy is this: *The energy of an oscillator cannot increase gradually; it goes up in steps.* An atomic oscillator cannot pick up just a little energy at a time; to jump up to the next step, it has to get enough energy in a package to move up a step. Furthermore, the steps are greater at higher frequencies. At very high frequencies, an oscillator is stuck at a low energy level unless it can get an enormous chunk of energy all at once. The higher the frequency, the more unlikely it is that the oscillator will get enough energy in one piece to jump up to the next level. It is easier at higher temperatures, when more energy is available, but at any temperature, there is an upper limit.

Mathematical analysis of measured blackbody radiation curves led to an equation for determining the size of the steps at any given frequency:

$$\Delta E = hf \qquad \text{(Equation 13-8)}$$

where ΔE is the size of the *quantum* of energy at a frequency f. The constant h is Planck's constant, which has a fundamental place in every equation of subatomic physics. Its value is

$$h = 6.62 \times 10^{-34} \text{ joule-second}$$

Sample Problems 13-8 and 13-9 show how this constant is used.

Sample Problem

13-8 An atomic oscillator is vibrating with a frequency of 3.0×10^{14} Hz. What is the amount of energy that will boost its energy by one step?

Solution The quantum of energy at that frequency is

$$\Delta E = hf$$
$$\Delta E = (6.62 \times 10^{-34} \text{ J·s})(3.0 \times 10^{14} \text{ Hz})$$
$$\Delta E = 1.99 \times 10^{-19} \text{ J}$$

Sample Problem

13-9 What is the size of the energy step by which an atomic oscillator drops when it emits a package of red light whose wavelength is 710 nm?

Solution Since the frequency of the light is $\frac{c}{\lambda}$, Equation 13-8 can be written as

$$E = \frac{hc}{\lambda} = \frac{(6.62 \times 10^{-34} \text{ J·s})(3.00 \times 10^8 \text{ m/s})}{710 \times 10^{-9} \text{ m}}$$

$$E = 2.8 \times 10^{-19} \text{ J}$$

In an electric light bulb, we put in electric energy that sets the molecules oscillating. They are going fast enough to generate electromagnetic waves in every part of the visible spectrum. The oscillators radiate out energy by dropping to a lower energy level, only to be soon pushed back up by the electric current. Every time an oscillator drops down one energy step, it gives off a tiny packet of light. If the size of the step is ΔE, the frequency of the light can be determined from Equation 13-8.

Core Concept

A radiating hot object gives off energy in packets, each with one quantum of energy, which is proportional to the frequency of the oscillation.

Try This How much energy is there in each packet of light emitted by a hot filament in the ultraviolet range with a wavelength of 240 nm?

13-9. Photoelectricity

Blackbody radiation studies showed that light is emitted in the form of tiny packages. Other studies have shown that, when light is absorbed, it must be taken into a material as tiny packages and not in a continuous stream. This model of light, as consisting of discrete packages of energy, explains so many observable phenomena that it must be taken as a useful way to look at light. The packages of light are called *photons*. Each has an amount of energy that depends only on the frequency of the light, according to the equation

$$E_{photon} = hf \qquad \text{(Equation 13-9a)}$$

See Sample Problem 13-10 for the use of this equation.

Sample Problem

13-10 What is the energy of a violet photon if the wavelength of the light is 470 nm?

Solution The frequency of the light, from Equation 8-2, is

$$f = \frac{c}{\lambda}$$

$$f = \frac{3.0 \times 10^8 \text{ m/s}}{470 \times 10^{-9} \text{ m}} = 6.38 \times 10^{14} \text{ Hz}$$

Therefore the energy of the photon is

$$hf = (6.62 \times 10^{-34} \text{ J·s})(6.38 \times 10^{14} \text{ Hz}) = 4.23 \times 10^{-19} \text{ J}$$

The nature of photons can be explored in the process of *photoelectric emission*. When the proper kind of light strikes the right kind of metal, electrons are given off. In a device called a *photoelectric cell*, or "electric eye," the emitted electrons come off a metal plate shaped like a half-cylinder, as shown in Figure 13-15. They are collected by a metal rod. If the rod and the plate are connected through an external circuit, current will flow in the circuit and can be measured by a sensitive ammeter. A photoelectric cell can be used to control other circuits, such as the one that opens a door when you interrupt a beam of light.

If we think of light as a wave reaching a surface and making electrons oscillate, we are led to expect a certain kind of behavior. The wave makes the electrons oscillate in resonance, like a tuning fork being stimulated by a sound wave. Electrons build up energy gradually until they have enough to get out of the metal. The brighter the light, the sooner the electrons escape and the more energy they have when they leave.

But that is not the way it works. If you shine visible light on a silver surface, no electrons leave, and the ammeter remains at zero no matter how bright the light is or how long you leave it shining. To get a photoelectric current, you have to use ultraviolet light. This works because each ultraviolet photon has enough energy to release one electron. If the kind of light you use produces photoelectric emission, making it brighter produces a larger photoelectric current because the metal surface releases more electrons per second. But to get any photoelectric emission at all, you must bombard the surface with photons that have enough energy to release the electrons.

Every metal has its characteristic electron binding energy. This is the amount of energy that has to be added to a surface electron to release it from the surface. Deeper electrons need more energy. The minimum energy needed to release an electron is called the *work function* of the metal. Metals such as sodium and potassium, which do not bind their electrons strongly, have smaller work functions. Orange light, at a wavelength of 530 nm, will produce a photoelectric current in potassium. Any shorter wavelength will do the same. But to get electron emission from lithium, you need a wavelength no longer than 430 nm, in the blue region. Most metals will not emit at all in the visible range, but need ultraviolet light. See Sample Problem 13-11.

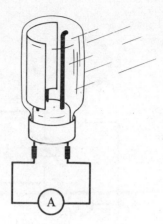

FIGURE 13-15

Sample Problem

13-11 Lithium can emit electrons using light of any wavelength less than 430 nm. What is the work function of lithium?

Solution The work function is the energy of the least energetic photon that produces photoemission:

$$E_{photon} = hf = \frac{hc}{\lambda}$$

$$E_{photon} = \frac{(6.62 \times 10^{-34} \text{ J·s})(3.00 \times 10^8 \text{ m/s})}{430 \times 10^{-9} \text{ m}} = 4.6 \times 10^{-19} \text{ J}$$

If the incident photon has more than enough energy to release electrons, the excess will become the kinetic energy of the electrons. This energy can be measured by adding a variable power supply and a voltmeter.

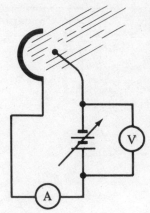

FIGURE 13-16

to the circuit, as shown in Figure 13-16. Note that the *negative* terminal of the battery is facing the collecting rod. To get back to the emitting plate, the electrons have to go through the battery in the wrong direction, thus losing energy as they travel. If the battery potential is turned up high enough, it will allow the battery to absorb completely the energy of the electrons, and the current will stop. This *cutoff potential* is the energy per unit charge of the electron as it leaves the plate, or, from Equation 9-9,

$$E_{kin} = q \, \Delta V_{cutoff}$$

This energy of each electron is the energy of the photon that released the electron, less the work function of the metal:

$$q \, \Delta V = hf - W \qquad \text{(Equation 13-9b)}$$

See Sample Problems 13-12, 13-13, and 13-14.

Sample Problem

13-12 An ultraviolet photon of wavelength 140 nm strikes the surface of a metal with a work function of 2.2×10^{-19} J. What is the maximum energy of the emitted electron?

Solution   The energy of the photon is

$$hf = \frac{hc}{\lambda} = \frac{(6.62 \times 10^{-34} \text{ J·s})(3.00 \times 10^8 \text{ m/s})}{140 \times 10^{-9} \text{ m}} = 1.42 \times 10^{-18} \text{ J}$$

The maximum electron energy is this value minus the work function:

$$14.2 \times 10^{-19} \text{ J} - 2.2 \times 10^{-19} \text{ J} = 12.0 \times 10^{-19} \text{ J}$$

Sample Problem

13-13 What is the cutoff potential for stopping the photoelectric current produced by the electrons of Sample Problem 13-12?

Solution   To lose all their kinetic energy, the electrons must gain electric potential energy, which is measured by $q \, \Delta V$. Then, $12.0 \times 10^{-19} \text{ J} = (1.60 \times 10^{-19} \text{ J})\Delta V$, from which $\Delta V = 7.5$ V.

Sample Problem

13-14 How much reverse potential will cut off the photoelectric current from a lithium surface (work function 4.63×10^{-19} J) illuminated by ultraviolet light at 2.2×10^{15} Hz?

Solution   The energy of each ultraviolet photon is

$$hf = (6.62 \times 10^{-34} \text{ J·s})(2.2 \times 10^{15} \text{ Hz}) = 1.46 \times 10^{-18} \text{ J}$$

Subtract the work function to get the electron energy:

$$(14.6 \times 10^{-19} \text{ J}) - (4.63 \times 10^{-19} \text{ J}) = 10.0 \times 10^{-19} \text{ J}$$

The electron energy is $q \, \Delta V$, so

$$\Delta V = \frac{E}{q} = \frac{10.0 \times 10^{-19} \text{ J}}{1.60 \times 10^{-19} \text{ J}} = 6.3 \text{ V}$$

Note that the cutoff potential does not depend on how bright the light is. This potential depends only on the photon energy, that is, on the frequency of the incident light. Both the work function of the metal and Planck's constant can be determined by measuring the cutoff potential at various frequencies.

A photoelectron is released by absorbing the energy of a single photon, which must have more energy than the work function of the metal:

$$E_{photon} = hf; \qquad q \, \Delta V = hf - W$$

Try This What is the work function of a metal if ultraviolet light at a frequency of 3.51×10^{15} Hz produces a photoelectric current that can be cut off with a reverse potential of 9.2 V?

13-10. The Energy of Photons

There are times when it is convenient to deviate from the rule of making all measurements in SI units. The joule is an inconveniently large unit for dealing with the energy of atoms and photons.

A more convenient unit is the *electron volt* (ev). This unit is defined as the amount that the energy of an electron changes in passing through a potential difference of 1 volt. To find its value, consider Equation 9-9:

$$\Delta E_{el} = q \, \Delta V$$

Then the value of an electron volt is found by setting q equal to the charge on an electron and ΔV at 1 volt:

$$1 \text{ ev} = (1.60 \times 10^{-19} \text{ C})(1 \text{ J/C})$$
$$1 \text{ ev} = 1.60 \times 10^{-19} \text{ J}$$

We can use this value to find an expression for Planck's constant that is often more convenient than the one expressed in joules:

$$h = (6.62 \times 10^{-34} \text{ J·s})\left(\frac{1 \text{ ev}}{1.60 \times 10^{-19} \text{ J}}\right)$$

$$h = 4.14 \times 10^{-15} \text{ ev·s}$$

To see how this simplifies problems, take a look at Sample Problem 13-15.

Sample Problem

13-15 What is the work function of a metal if the photoelectric current it produces is cut off at 1.5 V when it is illuminated with violet light at 460 nm?

Solution The energy of the photon is

$$E = hf = \frac{hc}{\lambda} = \frac{(4.14 \times 10^{-15} \text{ ev·s})(3.00 \times 10^8 \text{ m/s})}{460 \times 10^{-9} \text{ m}}$$

$$E = 2.70 \text{ ev}$$

Since a 2.7-ev photon releases electrons with no more than 1.5 ev of energy, the work function of the metal is 2.7 ev − 1.5 ev = 1.2 ev.

Note that the energy of the visible light photon in the problem is on the order of a few electron volts. This is no coincidence. In ordinary chemical reactions, energy changes per molecule are about that much. That is why electric batteries produce one or two volts per cell. The reason you cannot see infrared rays is that the energy of each infrared photon is too small to produce a chemical change in your retina. Neither can it produce photosynthesis or affect a photographic plate (with one minor exception).

You cannot see in the ultraviolet region because the energy of these photons—3 ev and more—is so great that it could damage your retina. The lens of your eye filters them out.

We have become more cautious, in recent years, about taking X-ray pictures, because we understand that the photon energies of these rays, which are thousands of ev can be extremely damaging to living tissue. And the photon energies of the gamma rays emitted by radioactive materials reach into the millions of ev.

Depending on the nature of the experiment, we elect to use either a wave model or a particle model for light. The wave model is necessary when we look at interference effects, or diffraction. In studying the photo-electric effect, or blackbody radiation, we must resort to the language of photons. It is futile to ask which model is "correct." We cannot settle the question by examining light in transit, for light can be studied only by letting it strike something. Then, it may show interference effects, or it may produce photoelectric emission. It may do both. Whatever it does, it no longer exists once it lands. Examine it and it is gone.

The models we use are free creations of the human mind, allowing us to think about light in familiar terms derived from experience with large objects. Which model is best might well depend on the amount of energy in the photons. It is quite impossible to detect photon effects in radio waves, but the gamma rays emerging from radioactive nuclei are never treated as waves. We count the separate photons.

You tend to think of a barrel of water as a continuous material. It can be poured, and it is measured in liters. You probably know that it consists of separate molecules, but they are so small that they rarely enter into our consideration. When you fill the same barrel with cannonballs, you treat them differently. The separate packages are so large that you cannot avoid dealing with them individually. Radio waves are like the water; X-rays are like the cannonballs.

Visible light comes in between, like a barrel of sand. You can pour it and measure it in liters, or in cubic yards. Yet you will surely notice that it consists of separate grains if you get one in your eye.

Core Concept

Particle properties of electromagnetic waves become noticeable at about 1 ev per photon and become more important as the photon energy increases.

Try This What is the wavelength of a 22,000-ev X-ray photon?

Summary Quiz

For each of the following, supply the missing word or phrase:

1. An electromagnetic wave is produced by a(n) _____ charge.
2. One piece of evidence suggesting that light is an electromagnetic wave is that its _____ agrees with a calculated value.
3. The ideal length for a broadcasting antenna is a _____ .
4. In an electromagnetic wave, the magnetic field is _____ to the electric field.

5. The shortest electromagnetic waves are the _____ .
6. Ultraviolet waves are _____ in wavelength than visible light.
7. The _____ wavelengths of electromagnetic waves travel in straighter lines.
8. The phenomenon of _____ can be noticed when a wave passes through a hole not much larger than a wavelength.
9. In the diffraction pattern of a double slit, the path difference at the $n = 3$ line is 3 _____ .
10. A bright bar on a diffraction pattern is the part of a standing wave called a(n) _____ .
11. A diffraction grating pattern separates light into colors because the colors differ in _____ .
12. The spectral color that is diffracted to the largest angle is _____ .
13. Coherent light is produced by a(n) _____ .
14. An information signal can be imposed on a beam by _____ the beam.
15. _____ light has all its electric field vectors oriented in the same direction.
16. No light passes through two polarizing filters if their axes are _____ to each other.
17. All color perception is produced by combinations of stimuli affecting just _____ kinds of nerve cells.
18. Equal stimulation of the red and green pigment cells produces a sensation of _____ color.
19. The "blue" pigment responds mostly to the _____ wavelength end of the spectrum.
20. A warm object produces _____ radiation.
21. As a blackbody increases its temperature, both the _____ and the _____ of the radiation increase.
22. There are more waves that can fit into any space at _____ frequencies.
23. The higher frequencies of a blackbody radiator do not receive as much energy as the low frequencies because the energy is _____ .
24. The size of energy quanta is proportional to _____ .
25. In the photoelectric effect, electrons are emitted from a metal under the influence of _____ .
26. Photoelectric emission works only if the light has a high enough _____ .
27. The minimum energy needed to release an electron from a metal is the _____ of the metal.
28. The cutoff potential of a photoelectric current is a measure of the _____ of the electrons.
29. The photon model of light is most appropriate at _____ wavelengths.

Problems

1. State whether each of the following will produce (A) an electric field only; (B) both electric and magnetic fields; (C) an electromagnetic wave; (D) none of these: (a) a proton falling in a vacuum; (b) an electron at rest; (c) a neutron traveling at constant speed in a straight line; (d) an electron traveling in a circle at constant speed; (e) a proton moving at constant speed in a straight line.
2. (a) What is the wavelength of a television station broadcasting on a frequency of 106 MHz? (b) What is the best length for an antenna for receiving this station?

3. A horizontal half-wave antenna is 2.2 m long. At a given moment, there is an electric field due to the radiation from this antenna 0.5 m from it. (a) How far from the antenna is the nearest maximum magnetic field? (b) What is the direction of this field?

4. What color is light whose frequency is 7.1×10^{14} Hz?

5. (a) What is the difference in path length to two slits for the tenth antinode if the slits are acting on green light of wavelength 530 nm? (b) If the diffraction angle for this antinode is 12.0°, how far apart are the slits?

6. A diffraction grating ruled 580 lines per mm is illuminated with white light. What is the range of all diffraction angles for the visible spectrum, with wavelengths from 380 to 760 nm? (Only the first-order spectrum is used.)

7. Show that there is no red light in the third-order spectrum produced by a diffraction grating ruled 530 lines per mm.

8. Referring to Figure 13-13, determine what color will be perceived if the eye receives an equal mixture of monochromatic light at 650 nm and 540 nm.

9. What wavelengths of light are blocked by a red filter?

10. What is the wavelength of the light at which energy radiated by a blackbody is greatest at a temperature of 4000 K? (See Figure 13-14.)

11. What is the size of the quantum step of energy at a frequency of 10^{14} Hz?

12. What is the photon energy of an ultraviolet photon if the wavelength of the light is 140 nm?

13. What is the longest wavelength of light that will release an electron from the surface of a metal whose work function is 3.1×10^{-19} J?

14. What is the work function of a metal if a retarding potential of 22.0 V is just enough to cut off the photoelectric current when the surface is illuminated by ultraviolet light whose frequency is 6.8×10^{15} Hz?

15. What is the maximum energy of an electron released from a surface whose work function is 1.6 ev by a yellow photon whose wavelength is 580 nm?

16. What is the cutoff potential of a photoelectric current produced by shining a blue light at 470 nm onto a surface whose work function is 2.2 ev?

CHAPTER 14

Rays and Images

14-1. What's a Ray?

Long before photons were dreamed of, even before there was an acceptable wave model for light, there were useful rules governing the behavior of light. These rules were based on a model that is still highly useful under many conditions. Light was visualized as consisting of rays, emerging from a lamp and traveling in straight lines unless deflected in some way.

A light ray may be thought of as a line drawn along the path of a photon. In the wave model, a ray is a line drawn in the direction of travel of the wave, perpendicular to the wave front. A light ray in empty space is a straight line, emanating from a lamp, the sun, or a star. Its path may be changed if it interacts with a material object.

The ray model helps in thinking about many of the properties of light. For example, the density of the rays, the number of rays per unit area, can be used to represent the intensity or brightness of light. In bright light, rays are crowded close together; they spread apart from each other as they move farther from the light source.

The ray model, however, is not always useful. When we consider the production or absorption of light, only the photon model provides meaningful explanations. And when light is diffracted through narrow slits or around small objects, its behavior can be accounted for only on the basis of a wave model. With these limitations, we can use a ray model effectively without concerning ourselves with the question of whether light is "really" a wave or a stream of particles.

A ray model is useful in dealing with light except where diffraction is important, or where light is absorbed or emitted.

Try This Using a ray model, explain the difference between the light coming from a light bulb and the light of a laser.

14-2. Illumination

Some lamps are brighter than others. We can think of the difference this way, if we like: There are more rays emerging from a bright lamp than from a dim one.

The total amount of light emerging from a lamp is called the *luminous flux* that the lamp produces. Luminous flux is measured in *lumens*, a unit that is carefully defined in the SI in terms of the light radiated under carefully controlled conditions by platinum at its melting point. An ordinary 100-watt light bulb produces about 1,700 lumens of luminous flux.

The lumen is not strictly a physical unit, since it depends on the sensitivity of the human eye. The eye is most sensitive to light at the center of the spectrum, to green light. The lumen measures how bright the light appears to the eye. In terms of the actual electromagnetic wave, or photon energy of the light, if the light is red or violet, it takes many more watts of power to constitute a lumen of luminous flux than if the light is green. A 1-watt green electromagnetic wave has 680 lumens of luminous flux; for a red or violet wave to appear as bright, and thus have the same amount of luminous flux, it would need more than ten times as much actual power. The luminous flux of ultraviolet and infrared rays is always zero, regardless of how much energy they deliver.

We are often concerned with the amount of light that falls on a surface. How much light must shine on a book, for example, to make it possible to read in comfort? How much light is needed to take a picture? This quantity is the *illumination* of the surface, and it is measured in lumens per square meter. A lumen per square meter is called a *lux*.

A simple application of the ray model enables us to determine the relationship between the amount of luminous flux produced by a lamp and the illumination falling on a surface any distance from the source. For simplicity, we will consider only a source that is considerably smaller than the distance from the source to the illuminated surface—a point source.

The relationship is illustrated in Figure 14-1. Think of the source as being at the center of a large sphere of radius R. Then all the luminous flux produced by the lamp spreads out radially, in straight light rays, to the inner surface of the sphere. The area of the sphere is $4\pi R^2$, so all this flux spreads out over that surface uniformly. The amount of flux per unit area, the illumination I, is therefore

$$I = \frac{F}{4\pi R^2}$$

(Equation 14-2)

This equation gives the illumination on any surface a distance R from a source emitting luminous flux F, provided that the surface is perpendicular to the rays reaching it.

An electric light bulb is extremely inefficient; about 97 percent of the electric energy put into it becomes invisible infrared rays. A 100-watt light bulb produces about 1,600 lumens, representing only about 3 watts of electromagnetic power. See Sample Problems 14-1 and 14-2.

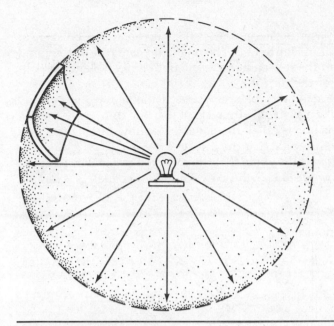

FIGURE 14-1

Sample Problem

14-1 What is the illumination on a book that is 2.5 m from a lamp that produces 1,200 lumens of flux?

Solution From Equation 14-2,

$$I = \frac{F}{4\pi R^2}$$

$$I = \frac{1,200 \text{ lumens}}{4\pi(2.5 \text{ m})^2} = 15 \text{ lumens/m}^2 = 15 \text{ lux}$$

Sample Problem

14-2 How much luminous flux must a light bulb produce if it is to be placed 3.7 m from a surface where the illumination needed is 20 lux?

Solution From Equation 14-2,

$$F = 4\pi R^2 I = 4\pi(3.7 \text{ m})^2(20 \text{ lux}) = 3,400 \text{ lumens}$$

Core Concept

Illumination, in luxes, is inversely proportional to the square of the distance from the source:

$$I = \frac{F}{4\pi R^2}$$

Try This A fluorescent bulb is much more efficient than an incandescent lamp. What is the illumination at a distance of 3.5 m from a 40-W fluorescent lamp that operates at 16% efficiency? (Figure that, on the average, 1 W of electromagnetic energy spread out over the visible spectrum produces about 450 lumens of luminous flux.)

14-3. Rays That Bounce

The physical system that comes closest to the mathematical line known as a light ray is the fine beam produced by a laser. It can be used to study the behavior of light rays. The beam can be made visible in space by providing lots of dust in the space.

If the beam of a laser is shone onto a flat mirror, it can be seen to reflect off the surface, as in Figure 14-2. The incident beam (the one coming toward the surface) and the reflected beam will both lie in a plane perpendicular to the surface of the mirror.

In discussing the geometry of rays, it is customary to measure all angles, where rays strike a surface, with respect to a line drawn perpendicular to the surface. This line is called the *normal*. As shown in Figure 14-2, the *angle of incidence* is the angle that the incident ray makes with the normal. The *angle of reflection* is the angle that the reflected ray makes with the normal. On a flat, mirrored surface, these two angles are equal. See Sample Problem 14-3.

FIGURE 14-2

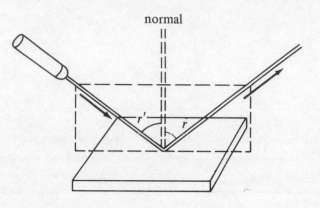

Sample Problem

14-3 A beam of light strikes a mirror at an angle of incidence of 40°. What is the angle between the incident and reflected beams?

Solution The normal bisects the required angle, and each half is 40°. The angle is thus 80°.

When light strikes an ordinary object, three things may happen to it:

1. It may be transmitted through the object; this is what happens when light strikes a pane of glass. If the glass is frosted, the light will pass through, but will be scattered, or diffused, in all directions instead of passing straight through.
2. It may be reflected off the surface; when the surface is mirrorlike and flat, the reflection obeys the rules discussed above and is said to be *specular*. Most surfaces are not of this type, and they reflect the light diffusely, in all directions, as illustrated in Figure 14-3.

FIGURE 14-3

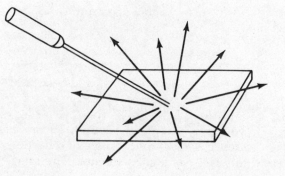

3. The light may be *absorbed* by the material, its energy being converted to heat. A perfectly black object is one that absorbs all the light that comes to it.

Typically, objects transmit, reflect, and absorb light selectively. A red filter is one that transmits long-wavelength rays (red) and absorbs the rest of the visible spectrum. A green book *reflects* the central part of the spectrum and absorbs both ends, so that the reflected light that reaches your eye appears green.

14-4. Seeing Things

When you look at a book, as in Figure 14-4, you can see it because light diffusely reflected from it enters your eye. *Each point* on the book scatters light in all directions, acting much like a point source of light. Some of the scattered light enters your eye. The rays coming into your eye from any single point on the book form a cone. These rays are divergent; that is, they spread apart from each other as they travel. When all these cones of divergent rays, from every point on the book, come into your eye, you see a book.

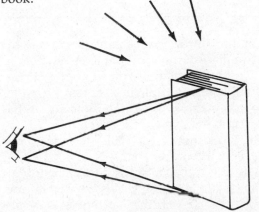

FIGURE 14-4

You can also see the book by looking at its reflection in a mirror. In this case, the cone of rays that passes from a point on the book to your eye is folded at the surface of the mirror, as shown in Figure 14-5. Your eye cannot detect the folding; it tells you that the rays are diverging from some point behind the mirror, as shown by the dashed lines in the figure. The point from which the light appears to be coming is called an *image point*. Every point on the object (the book) has a corresponding image point, so there is an *image* of the book behind the mirror. Since there is no light actually coming from the image to your eye, that image is said to be *virtual*.

For each ray, the angle of incidence is equal to the angle of reflection. A little geometry will show that this fact has a consequence: The image

FIGURE 14-5

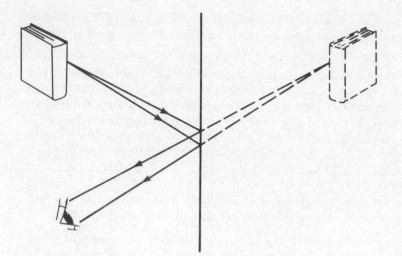

point is just as far behind the mirror as its corresponding object point is in front of it, and it lies on the normal drawn from the object point. This means that the virtual image in a plane mirror is the same size as the object and is just as far behind the mirror as the object is in front of it. The image has these properties regardless of the position from which it is viewed. See Sample Problem 14-4.

Sample Problem

14-4 You are taking a picture of yourself in a mirror while standing 12 ft from the mirror. At what distance should you focus your camera?

Solution Set it at 24 ft; your image is 12 ft behind the mirror, and thus 24 ft from you. If your eye cannot tell the reflected rays from rays coming directly from an object, neither can a camera.

Contrary to common opinion, the mirror image is not reversed right to left. Try this: Write something on a piece of transparent paper and hold it in front of you as if to read it. Do this while standing in front of a mirror and look in the mirror. You will find that the writing is still oriented in the usual way, and that you can read it in the mirror without difficulty. If the paper is not transparent, the only way you can read it in the mirror is to turn it around so that the writing faces the mirror. Then it appears reversed. But the mirror did not reverse it; you did!

The mirror image is reversed front to back, not right to left. If you look at yourself in a mirror, the image of your nose is closer to the mirror than the image of your ears. If you are facing north, your image is facing south. But if you raise your eastern hand, the eastern hand of your image waves back at you.

Core Concept

The image in a plane mirror is virtual, the same size as the object, as far behind the mirror as the object is in front of it, and reversed front to back.

Try This You are watching a friend walking toward a vertical plane mirror. What happens to the size of his image as he approaches the mirror?

The experienced fish archer of Figure 14-6 knows that, if he aims his arrow at the place where the fish appears to be, he will miss it. The light from the fish, coming to his eye, bends as it emerges from the water. This phenomenon, called *refraction*, is due to the fact that the light travels faster in air than it does in water.

Exploring this phenomenon with a laser beam, as shown in Figure 14-7, yields *Snell's law*, a rule that has been known for two centuries. When the beam enters a transparent medium from the air, making an angle of incidence i on the way in, it bends toward the normal, so that the angle of refraction r is smaller than the angle of incidence. Note that, as always, the angles are measured from the normal. As the angle of incidence is changed, the ratio between the sines of the two angles remains constant:

$$\frac{\sin i}{\sin r} = n \qquad \text{(Equation 14-5a)}$$

where n is a characteristic of the medium known as its *index of refraction*. See Sample Problems 14-5 and 14-6.

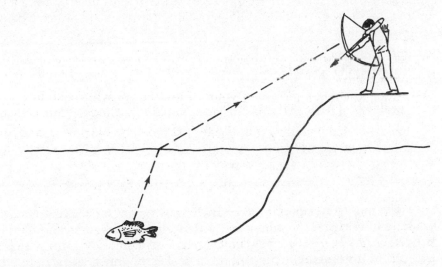

FIGURE 14-6

Sample Problem

14-5 A light ray enters a piece of crown glass at an angle of 57° and is refracted to 31° inside the glass. What is the index of refraction of the glass?

Solution From Equation 14-5a,

$$n = \frac{\sin i}{\sin r} = \frac{\sin 57°}{\sin 31°} = 1.63$$

Sample Problem

14-6 If a ray enters the glass in Sample Problem 14-5 at an angle of 26°, what will the angle of refraction be?

Solution From Equation 14-5a,

$$\sin r = \frac{\sin i}{n} = \frac{\sin 26°}{1.63}$$

from which $r = 16°$.

Geometric analysis of the passage of the light into a new medium, coming from a vacuum, shows that the index of refraction of the medium is an expression of the way the light slows down as it enters. It is easily shown that

$$n = \frac{c}{v}$$

(Equation 14-5b)

where c is the speed of light in a vacuum ($=2.998 \times 10^8$ m/s) and v is the speed of light in the medium whose index of refraction is n. See Sample Problem 14-7.

FIGURE 14-7

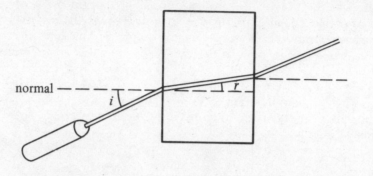

<hr />

Sample Problem

14-7 What is the speed of light in water, which has an index of refraction of 1.33?

Solution From Equation 14-5b,

$$v = \frac{c}{n} = \frac{2.998 \times 10^8 \text{ m/s}}{1.33}$$

$$v = 2.25 \times 10^8 \text{ m/s}$$

<hr />

The most convenient form of Snell's law would be one that makes it possible to calculate the angles of incidence or refraction at the interface between *any* two media—glass and water, Lucite and glycerine, benzene and alcohol, whatever. This rule even makes it unnecessary to specify which angle is the angle of incidence and which the angle of refraction, since the rays are always reversible. If the two media that form the interface are designated A and B as in Figure 14-8, and their indices of refraction

FIGURE 14-8

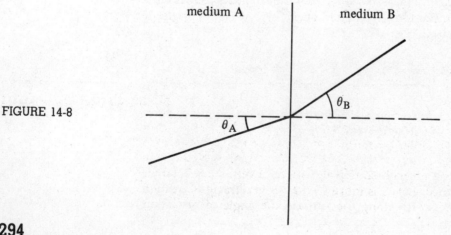

are n_A and n_B, we can just call the angles in the two media θ_A and θ_B, and Snell's law takes this form:

$$\frac{\sin \theta_A}{\sin \theta_B} = \frac{n_B}{n_A}$$

(Equation 14-5c)

See Sample Problems 14-8 and 14-9 for applications of this rule.

Sample Problem

14-8 A ray of light passes from water ($n = 1.33$) into a sheet of flint glass ($n = 1.61$), making an angle of incidence of 65°. What is the angle of refraction?

Solution Apply Equation 14-5c, letting the water be medium A and the glass be medium B,

$$\sin \theta_B = \sin \theta_A \left(\frac{n_A}{n_B}\right)$$

$$\sin \theta_B = \sin 65° \left(\frac{1.33}{1.61}\right)$$

$$\theta_B = 48°$$

Sample Problem

14-9 A block of crown glass ($n = 1.63$) is immersed in an unknown liquid. A ray of light is measured to make an angle of incidence within the unknown liquid of 48° as it approaches the glass. The angle in the glass is 36°. What is the index of refraction of the liquid?

Solution Using subscript g for glass and subscript l for liquid, we have

$$\frac{\sin \theta_g}{\sin \theta_l} = \frac{n_l}{n_g}$$

$$\frac{\sin 36°}{\sin 48°} = \frac{n_l}{1.63}$$

which gives $n_l = 1.29$.

Core Concept

When a ray strikes an interface, the sines of the angles of incidence and refraction are inversely proportional to the indices of refractions of the two media:

$$n = \frac{c}{v}; \qquad \frac{\sin \theta_A}{\sin \theta_B} = \frac{n_B}{n_A}$$

Try This A diamond ($n = 2.42$) is in water ($n = 1.33$), and a ray of light shines on it, making an angle of incidence of 55°. What is the angle of refraction inside the diamond?

14-6. Trapped Light

Consider the light rays emerging from a lamp that is under water, as in Figure 14-9. Ray A strikes the upper surface of the water along the normal; it speeds up as it emerges, but it does not bend. Ray B strikes at an angle and bends away from the normal, so that the angle of refraction is larger than the angle of incidence. Ray C is more so. Ray D is refracted so much that it emerges from the water along the surface; the angle of refraction is 90°.

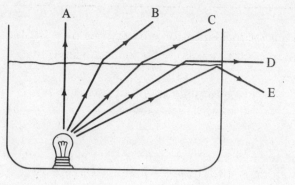

FIGURE 14-9

Now look at ray *E*, which makes an angle of incidence of 50°. Since light travels in air practically as fast as in a vacuum, $n_{air} = 1$. We calculate the angle of refraction in the usual way:

$$\frac{\sin 50°}{\sin r} = \frac{n_{air}}{n_{water}} = \frac{1}{1.33}$$

$$\sin r = (1.33)(0.766) = 1.019$$

Don't try to find *r* in your trig tables; there is no such angle. The largest possible angle of refraction is 90°.

Ray *E* undergoes *total internal reflection*. All rays that strike the surface at an angle larger than the *critical angle of incidence* are trapped inside the water. Reflection of this sort at an interior surface is more nearly complete and involves less loss of light than the best mirror can provide. That is why prisms like those in Figure 14-10 are used, rather than mirrors, in good binoculars, spotting scopes, and the finders of single-lens reflex cameras.

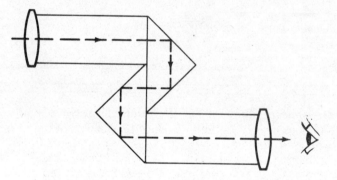

FIGURE 14-10

This kind of total reflection is always *internal;* that is, it occurs only when light is emerging into a medium of lower index of refraction. Only then is the ray bent away from the normal, and only then is there an angle of incidence that produces refraction at 90°. Total internal reflection, when it occurs, follows the usual rule of reflection: The angle of incidence equals the angle of reflection.

It is easy to find the critical angle of incidence; it is the angle at which the angle of refraction is 90°. For water, for example, using Equation 14-5c,

$$\frac{\sin i_c}{\sin 90°} = \frac{n_{air}}{n_{water}} = \frac{1}{1.33}$$

$$i_c = 49°$$

Thus, any ray inside the water that strikes the surface at more than 49° will be reflected back. See Sample Problems 14-10 and 14-11.

Sample Problem

14-10 What is the critical angle of incidence for crown glass ($n = 1.63$)?

Solution Apply Equation 14-5c to glass and air, with the angle in air taken as 90°:

$$\frac{\sin \theta_g}{\sin \theta_a} = \frac{n_a}{n_g}; \qquad \frac{\sin i_c}{\sin 90°} = \frac{1}{1.63}$$

which gives $i_c = 38°$.

Sample Problem

14-11 What index of refraction must a transparent material have in order for the total internal reflection to take place at all angles larger than 55°?

Solution Again,

$$\frac{\sin i_c}{\sin 90°} = \frac{1}{n}$$

Therefore

$$n = \frac{1}{\sin 55°} = 1.22$$

Core Concept

Total internal reflection occurs when the angle of incidence on an internal surface exceeds the angle at which the angle of refraction is 90°.

Try This What is the critical angle of incidence of diamond ($n = 2.42$)? (Can you see why a diamond sparkles?)

14-7. Colors by Prism

Figure 14-11 shows a beam of white light entering a triangular prism at a high angle of incidence. It is refracted downward toward the normal on entering the glass. It leaves at a surface slanted in the opposite direction and is refracted away from the normal—again downward, because of the different orientation of the surface. The emerging ray will be separated into the colors of the spectrum. Red refracts least, violet most.

In passing through a glass prism, light is refracted because it travels more slowly through the glass. Since we observe that the high-frequency end of the spectrum is refracted more than the low-frequency end, it is obvious that violet light slows up more than red light in passing through

FIGURE 14-11

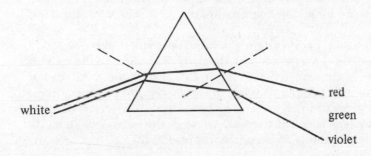

white

red

green

violet

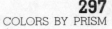

the glass. Glass *disperses* the light. Any medium in which the higher frequencies travel more slowly than the lower frequencies is said to be a *dispersive* medium.

Before the development of high-quality diffraction gratings, prisms were used in spectroscopes to analyze the wavelength content of light. They are by no means as good as gratings, however. Also note that the spectrum produced by a prism is the reverse of that made by a grating; in a grating, it is the red light that is deflected at the greatest angle from its original path.

Dispersion of light does not have much practical value, aside from producing the beautiful sparkling of a diamond or a crystal chandelier. It is a problem in optics. Until lens makers learned how to overcome it, everything seen in a microscope was surrounded by a rainbow-hued halo.

Core Concept

> Light is dispersed into colors by a prism because the different frequencies travel at different speeds in the glass.
>
> *Try This* In silicate flint glass, the index of refraction is 1.61 for red light and 1.66 for violet light. Find the angles of refraction for red and violet light if a beam of white light approaches the surface of such a piece of glass at an angle of 65°.

14-8. Focusing the Rays

Suppose we set a couple of prisms base to base, as shown in Figure 14-12, and send a couple of rays of light into them. The upper ray bends downward (toward the normal) on entering the prism, and downward (away from the normal) on leaving. The lower ray bends upward twice, and the two rays converge at a point.

FIGURE 14-12

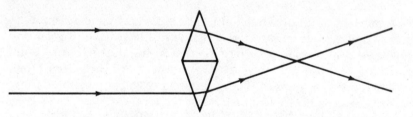

We can set up a system of this sort in which not just two but all the rays that may enter the system will converge at a point. A piece of glass or plastic that can do this is called a *converging lens.* Its surface (or both surfaces) is spherical—a small part of a large sphere. With such a *thin lens* entering light rays can be focused approximately at a point, as shown in Figure 14-13.

Lenses of this sort obey certain mathematical rules. First, we will define a line called the *principal axis* of the lens. This line is the axis of symmetry of the lens, drawn normal to its surface at the center. All rays of light that enter the lens parallel to this principal axis converge at a point called the *principal focus* of the lens. The distance from this principal focus to the center of the lens is called the *focal length* of the lens. This distance enters into all lens equations, where it is represented as *f*. At the same

16. Total internal reflection occurs when the angle of incidence is more than the _____ .
17. To get total internal reflection, the ray must approach an interface with a medium with a _____ index of refraction.
18. In a dispersive medium, the speed of light is a function of _____ .
19. When a spectrum is formed by a prism, it is the _____ light that is refracted most.
20. A _____ lens is thicker in the middle than at the edges.
21. Rays that enter a converging lens parallel to the principal axis will emerge and pass through the _____ of the lens.
22. To make a spotlight, the light source must be placed at the _____ of the lens.
23. If an image can be observed by casting it onto a screen, we know that the image is _____ .
24. In order for a converging lens to form a real image, the object distance must be more than _____ .
25. A ray diagram enables us to locate images by plotting exactly _____ rays emerging from a single object point.
26. A _____ lens can form only virtual images.
27. A converging lens forms a virtual image when the object distance is less than the _____ .
28. The ratio of object distance to image distance is the same as the ratio of object _____ to image _____ .
29. In an algebraic solution for the location of an image, a negative value indicates a _____ image.
30. In a camera, the distance from the center of the lens to the film is the focal length of the lens when the camera is focused at _____ .
31. Chromatic aberration results from _____ of light by the lens.
32. The light-gathering power of a lens is indicated by its _____ .
33. To form a large image of a distant object, a camera needs a lens with a _____ focal length.
34. The image on film is blurred when the _____ is large.

Problems

1. How much luminous flux must a lamp produce to provide 20 luxes of illumination at a distance of 1.5 m?
2. Approximately what is the luminous flux output of a 200-W lamp that operates at 96% efficiency?
3. Two mirrors are vertical and perpendicular to each other. Prove that, if any horizontal ray strikes one mirror and reflects from it to the other, the doubly reflected ray is parallel to the incident ray.
4. The image of a post in a plane mirror is 3 m high and 12 m behind the mirror. Find the size and position of the image if the object distance is halved.
5. What is the index of refraction of Lucite if light travels in it at 2.00×10^8 m/s?
6. Find the index of refraction of an unknown liquid if a ray of light enters it at an angle of 45° and is refracted to 32°.
7. A layer of benzene ($n = 1.50$) is floated on top of a mass of water ($n = 1.33$). If a light ray enters the benzene from above, making an angle of incidence of 55°, what is the angle of refraction inside the water?
8. A 45° prism is to be used to produce total internal reflection at two surfaces, as in Figure 14-10. What is the minimum index of refraction needed for the glass?

9. Determine the critical angle of incidence of a plastic if a ray entering it at 62° is refracted to 47°.

10. The index of refraction of borate flint glass is 1.590 for violet light and 1.566 for red. If a violet ray is refracted to 65.0° on entering the glass, what is the angle of refraction for a red ray?

11. Determine the size and location of the image in a magnifying glass of focal length +5.0 cm if it is being used to examine a beetle 1.2 cm long placed 2.0 cm in front of the lens.

12. What focal length lens would be needed to form a virtual image at a distance of 6.5 cm from the lens when the object is 9.0 cm from it?

13. Find the size and location of the image formed by a +20-cm lens when a vase 40 cm tall is placed 50 cm from the lens.

14. A lamp and a screen are 3.0 m apart, and an image of the lamp is to be formed on the screen by a +50-cm lens. What are the two possible distances from lamp to lens at which a real image will form on the screen?

15. A photographer, using a +80-mm lens, wants to take a portrait of the uppermost 1.00 m of a subject, to form an image on the film 6.5 cm high. How far from the lens must the subject be seated?

16. What is the diameter of a lens marked "f:3.5, focal length 80 mm"?

CHAPTER
15

Atoms
and Nuclei

15-1. Finding the Nucleus

Everything contains electrons; they can be transferred from one object to another; and they flow freely through metals. Electron charge has been measured, as well as electron mass.

Since electrons are negative, and ordinary objects have no electric charge, everything must also contain positive charges equal in magnitude to the charge on all their electrons. How is that positive charge distributed within a material? Just where is it located?

These questions are answered definitively by experiments on the scattering of alpha particles. These particles are given off by various kinds of radioactive materials and are known to bear positive charges. Their velocity and their energy can be measured, and they serve as useful bullets for probing the interior of atoms.

In the original experiment, a stream of alpha particles was obtained by enclosing radioactive polonium inside a small lead box with a hole in it (Figure 15-1). Alpha particles streamed out of the hole and struck an extremely thin sheet of gold foil. On the other side of the foil was a screen coated with a chemical salt that produced a flash of light every time it was struck by an alpha particle. The whole apparatus was enclosed in high vacuum.

FIGURE 15-1

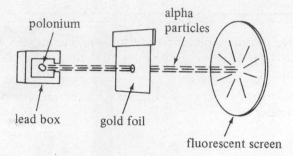

The first discovery was that the gold foil was mostly empty space; 99.99 percent of all the alpha particles went straight through the foil, 400 atoms thick, just as though there was nothing there. An occasional particle bounced back in the direction it had come from. Whatever it is that makes up gold, it is concentrated in many tiny lumps separated by distances about 1,000 times their own size—nuclei.

Most of the particles that did not go straight through the gold foil did not bounce straight back but were deflected to one side or another of their straight-line path. They struck the screen off to the side of the main, straight-line path. Mathematical analysis of this deflection, shown in Figure 15-2, made it possible to determine the nature of the force the nuclei exerted on the particles. It turned out to be an inverse-square force of repulsion and agreed perfectly with values calculated from Coulomb's law of electric force, Equation 9-4. It was evident that the positive charge in the gold foil was concentrated just where the mass is found, in the nucleus.

When this experiment was repeated with other kinds of foil, it was found that each chemical element had its own characteristic electric charge on the nucleus. Since each atom as a whole is electrically neutral, the charge on the nucleus must be equal in magnitude to the charge on some whole number of electrons. The nucleus contains an integral number of positive elementary charges, and that number, unique for each chemical element, is the *atomic number* of the element.

FIGURE 15-2

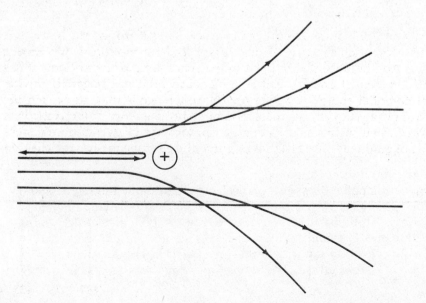

An atom consists of a nucleus, containing most of its mass and a quantized positive charge, surrounded by negative electrons whose number is the atomic number of the element.

Try This How much is the positive charge, in coulombs, in the nucleus of an atom of tin, atomic number 50?

15-2. The Electrons of the Atom

Positive nucleus, negative electrons. Why don't the electrons fall into the nucleus? In an early model of the atom the electrons were in orbit around the nucleus. They were supposed to be held in place by the electrical attraction, just as the planets are held in orbit around the sun by gravity. Unfortunately, this does not work out well. An electron in a circular orbit is accelerated, and any accelerated charge produces electromagnetic radiation, as we saw in Chapter 13. Losing its energy in this way, every electron would soon fall into the nucleus.

Many models of the atom have been proposed, but none of them has ever been able to explain all of its properties on the basis of the rules of classical physics alone. We run into the same problem we had in trying to describe the nature of light. The only models we have are those based on our experience with large objects. Like photons, electrons cannot be completely described by these rules.

The simplest atom is that of hydrogen—one electron and a nucleus with a single unit of charge. The first successful atomic models explained very nicely most of the observed properties of the hydrogen atom. There were two models that did well—one based on the electron as a particle, and the other treating it as a wave. Each works fairly well; neither tells the whole story.

Any model of the hydrogen atom must explain its spectrum. When hydrogen gas at low pressure is exposed to an electric arc, it glows. A spectrometer reveals that the light of this glow consists of a set of discrete spectral lines, as shown in Figure 15-3. Why does the gas produce light at wavelengths of 658.5 nm and 487.8 nm (among others), but nothing in between?

If you think of the electron as a particle in orbit around the nucleus, it has two forms of energy. Because it is a mass in motion, it has kinetic energy. And because it is trapped in the space around the nucleus, it has negative potential energy, just as a planet in orbit has negative gravitational potential energy (Chapter 6). The difference is that, to explain the spectrum of hydrogen, we must assume that the energy of the electron is *quantized*. That means that its energy can increase or decrease only stepwise, not gradually. The electric current in the gas tube pushes the electron

FIGURE 15-3

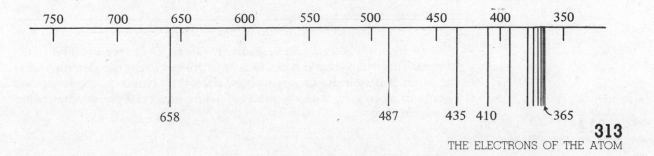

| 750 | 700 | 650 | 600 | 550 | 500 | 450 | 400 | 350 |

658   487   435 410   365

to a higher energy level; it then drops to a lower level, giving up the excess energy in the form of a photon. The photon has exactly the same amount of energy as the difference between the two step values of the electron in the atom.

When the nucleus and the electron are separated, their energy is zero. The energy decreases when they come together to form an atom with negative possible values that can be calculated from a simple formula:

$$E_{electron} = \frac{-13.6 \text{ ev}}{n^2}$$ **(Equation 15-2)**

where n is any whole number. It is called the *principal quantum number* of the atomic state. See Sample Problems 15-1 and 15-2 to find out how to calculate the wavelengths of the spectrum of atomic hydrogen.

Sample Problem

15-1 A hydrogen nucleus (a proton) comes near an electron, which is attracted to the nucleus and comes close to it, forming a hydrogen atom in its lowest energy, or ground, state. Find (a) the energy of the photon that is emitted; and (b) its wavelength.

Solution   (a) Since the energy of the hydrogen atom drops from 0 to −13.6 ev, the photon that is emitted must have 13.6 ev of energy. (b) Since $E_{photon} = hc/\lambda$,

$$\lambda = \frac{hc}{E} = \frac{(4.14 \times 10^{-15} \text{ ev·s})(3.00 \times 10^8 \text{ m/s})}{13.6 \text{ ev}} = 9.1 \times 10^{-8} \text{ m, or } 91 \text{ nm}$$

This is well into the ultraviolet range.

Sample Problem

15-2 What is the wavelength of the spectral line produced when the electron of a hydrogen atom drops from the fourth to the second quantum state?

Solution From Equation 15-2, the energy of the electron in the fourth state is

$$\frac{-13.6 \text{ ev}}{4^2} = -0.850 \text{ ev}$$

and in the second state it is

$$\frac{-13.6 \text{ ev}}{2^2} = -3.40 \text{ ev}$$

The photon energy is the difference, or 2.55 ev. Its wavelength, from Equations 8-2 and 13-9a, is

$$\lambda = \frac{hc}{E} = \frac{(4.14 \times 10^{-15} \text{ ev·s})(3.00 \times 10^8 \text{ m/s})}{2.55 \text{ ev}}$$

$$\lambda = 4.87 \times 10^{-7} \text{ m, or } 487 \text{ nm}$$

To free the electron entirely from the atom, its energy must be raised to zero. This means that it takes 13.6 ev of energy to get the electron out of the atom if it is in its lowest energy state. This value is the ionization energy of hydrogen. It agrees perfectly with values determined by chemical experiments.

Energy levels of the hydrogen electron are quantized; transitions to lower energies produce photons that carry away the excess energy.

Try This Find the wavelength of the ultraviolet photon produced when a hydrogen electron drops from its $n = 2$ to its $n = 1$ state.

15-3. Why Is the Energy Quantized?

The energy levels of hydrogen are amply demonstrated and measured by a variety of experimental methods. Is there any theoretical way of conceptualizing the reason for these levels?

The earliest attempt was based on the concept of the electron as a particle in orbit. The usual rules for calculating energy gave results consistent with the measured size of the atom, its ionization energy, and the wavelengths of its spectral lines. The calculation involved Coulomb's law, the kinetic energy formula, and the equation for centripetal force. It worked beautifully provided one additional assumption was made: *Angular momentum is quantized* in units of $h/2\pi$. In algebraic language,

$$mvr = \frac{nh}{2\pi} \qquad \text{(Equation 15-3a)}$$

where m is the mass of the electron, v is its speed, r is the radius of its orbit, and mvr is therefore its angular momentum. There was no theoretical basis for this assumption; it worked, that's all.

The basis for this equation can be firmed up a little by treating the electron as a wave. There is ample experimental basis for this; a beam of electrons exhibits such wave properties as diffraction and interference. To find the wavelength of such a wave, let's go back to the photon. Its energy is

$$E_{\text{photon}} = hf \qquad \text{(Equation 13-9a)}$$

However, relativity theory tells us that the total energy of anything can be related to its mass in this way:

$$E = mc^2 \qquad \text{(Equation 6-9)}$$

Consequently, we can write, for a photon,

$$mc^2 = hf \qquad \text{or} \qquad mc = \frac{hf}{c}$$

Now mc is the momentum of a photon (from Equation 4-4) and $\lambda = c/f$ (from Equation 8-2), so

$$\text{momentum} = \frac{h}{\lambda}$$

Assuming the same relationship holds for particles other than photons, whose momentum is mv, we can give an equation for the wavelength of a particle in motion:

$$\lambda = \frac{h}{mv} \qquad \text{(Equation 15-3b)}$$

This equation has been tested experimentally for many kinds of particles, and it works. Interference effects have been found with beams of many kinds of particles. This equation does not work with baseballs only because the wavelengths are too small to produce a detectable interference pattern.

In the hydrogen atom, we assume that the electron must form a standing wave around the nucleus. This implies that the circumference of its orbit is some whole number of wavelengths, so

$$2\pi r = n\lambda$$

or, since $\lambda = h/mv$,

$$2\pi r = \frac{nh}{mv}$$

which is exactly the same as Equation 15-3a.

Both of these assumptions lead to the same prediction; both produce correct values for the energy levels of a hydrogen electron. Sounds great. Unfortunately, neither one of them produces any useful predictions for other kinds of atoms. As soon as a second electron is introduced, these simple theories break down. More complex atoms also produce line spectra, and have detectable quantized energy levels, but they call for a much more complex theory to explain them.

Core Concept

Energy levels in a hydrogen atom can be explained by either a particle or a wave model of the electron; $mvr = nh/2\pi$.

Try This The radius of a hydrogen atom in its lowest energy state is about 5×10^{-11} m. Find (a) the wavelength of the electron; and (b) its angular momentum.

15-4. Within the Nucleus

Every nucleus possesses a positive charge that is an integral multiple of the elementary charge. This must be so, for the atom as a whole is electrically neutral, and the nuclear charge must exactly cancel out the negative charge on some whole number of electrons. This suggests that the nucleus contains some sort of particle possessing a single elementary unit of positive charge. This particle is called a *proton*.

The mass of the nucleus presents a problem. Chemists had long since discovered that the masses of most elements are very nearly whole-number multiples of the mass of the smallest atom, that of hydrogen. But this multiple is always larger than the atomic number. If the protons contain all the charge of the nucleus, there must be something else there to account for all the mass. A nucleus of gold, for example, has exactly 79 times the charge on a hydrogen nucleus (a proton). But its mass, on the average, is 195.4 times as much as the mass of a proton.

The problem was partially solved with the discovery of the *neutron*. This particle was found in one of the many alpha-particle bombardment experiments being done to investigate the atom. An alpha particle, on striking a beryllium nucleus, knocked out of it a neutral particle with a mass a little greater than that of a proton. Now it became clear that the extra mass of the nuclei was provided by these neutral particles. The neutron has no charge, but its mass turned out to be nearly the same as the mass of a proton.

There was still a problem. If the nucleus is made of some integral number of protons and neutrons, then the mass of each nucleus ought to be some whole-number multiple of the mass of one proton. Most nuclei came pretty close, but there were some disturbing deviations. How was it possible to account for the atomic mass of chlorine, for example, which is 35.5 times the mass of a proton?

The answer is that not all atoms of chlorine contain the same number of neutrons. In a mass spectrometer a beam of chlorine ions is divided into two parts with different masses. All have the same number of protons; that is what makes them chlorine atoms. But there are several *isotopes* of chlorine, containing different numbers of neutrons. The total number of particles in the nucleus, protons, and neutrons together, is called the *atomic mass number* of the particular atomic species.

The composition of a nucleus is thus represented by two numbers, written before the chemical symbol of the element as a subscript and superscript. Chlorine-35, for example, is symbolized in this way:

$$^{35}_{17}Cl$$

indicating that its atomic number (number of protons) is 17 and its atomic mass number (both kinds of particles) is 35. The other common isotope of chlorine is chlorine-37:

$$^{37}_{17}Cl$$

Still, there is a problem. The actual masses of the various isotopes are *still* not exact multiples of the mass of a proton, or of the separate masses of the protons and the neutrons in the nucleus! We measure the masses of nuclei *in daltons*; a dalton (Dl) is defined as $\frac{1}{12}$ of the mass of a nucleus of carbon-12. This, of course, is an extremely small unit of mass; it takes 6.02×10^{26} Dl to make up one kilogram. Isotope masses (except for carbon-12) are never exactly whole numbers of daltons. As the sample values of Table 1 show, they are always within a few hundredths of a dalton of the atomic mass number. The deviation from integral values lies at the heart of our knowledge of nuclear energy.

Core Concept

Nuclei are made of protons and neutrons; their masses are not quite whole numbers of daltons.

Try This The atomic number of iron is 26. State the number of protons and neutrons in a nucleus of iron-54.

15-5. Tracking the Particles

All of the heaviest metals—polonium, radium, uranium, thorium—seem to violate the law of energy conservation. They produce a steady stream of energetic particles, with no observable energy input. These elements are said to be *radioactive*.

This peculiar property was found when it was noted that an ore containing these metals fogged a photographic plate. A charged particle passing through a photographic emulsion gives up some of its energy in colliding with the atoms of the sensitive material. Its path is marked by a streak of ionized atoms, which are converted into visible black specks of metallic silver when the plate is developed. Using a stack of photographic plates, an investigator can trace the three-dimensional path of a charged particle. Since each speck of silver represents a definite amount of energy lost by the particle, it is possible to analyze the plates to study the energy of the particle.

In a *cloud chamber*, the entire track of a charged particle is visible as it is made. This device, shown in the picture, is a round box with a glass top. Inside, the air is kept saturated with water vapor. The ionization trail of a charged particle forms tiny droplets of water, which are visible against a black background as a thin line of fog. If the particle is moving fast enough, it can be seen that the trail is made of separate droplets, each representing the ionization of a water molecule, with energy transferred from the moving charged particle.

Modern equipment produces charged particles of extremely high energy, and much more elaborate equipment is needed to trace their history. Most of them disintegrate into other particles within times as short as 10^{-20} s. One tool for studying the path and the interactions of these particles is the *bubble chamber*. Such a chamber may contain 50 gallons of liquid hydrogen at its boiling point of about 4 K. A charged particle going through it leaves a trail of tiny hydrogen bubbles. A light flashes, and two cameras take pictures from different angles. The advantage of a bubble chamber is that its high density enables it to absorb the energy of the particle more effectively than water vapor; a high-energy particle could pass completely through a cloud chamber before anything interesting happened to it.

Most of the pictures taken of cloud chamber or bubble chamber tracks show the paths of the particles to be curved. This curvature is produced by a magnetic field, as we saw in Chapter 11. According to Equation 11-5, the radius of curvature of the path can be used to calculate the momentum of the particle. The spacing of the bubbles gives the kinetic energy of the particles. Knowing both mv and $\frac{1}{2}mv^2$, both the mass and the velocity of the particles can be found.

Core Concept

The properties of charged particles can be found by studying the tracks they leave in various kinds of detection devices.

Try This Explain why the path of an electron in a bubble chamber is a spiral rather than a straight line or a circle.

15-6. Radioactive Elements

A piece of uranium or polonium emits three kinds of radiation. In a cloud chamber in a magnetic field, alpha rays curve gently in one direction; beta rays curve sharply in the other; gamma rays leave no track but can be detected by various kinds of devices; they pass straight through without deflection.

Alpha rays consist of massive particles, each composed of two protons and two neutrons. With atomic number 2, they are considered to be nuclei of helium, ^4_2He.

Beta rays are electrons emitted with much higher energy than can be produced in any sort of chemical reaction.

Gamma rays are X-ray photons of extremely high frequency and energy. They give us an inkling of the difference between the energy of chemical reactions and of nuclear reactions. Electron transitions in the outer atom, as in ordinary chemical processes, produce visible photons with energies of a few electron volts; gamma photon energies are measured in hundreds of thousands of electron volts. Nuclear energies are related to the *strong nuclear force*. Electron transitions obey the rules of the much weaker electromagnetic force.

The stability of a nucleus depends on the balance between the electric repulsion between protons, which tends to make the nucleus blow apart, and the strong nuclear attraction of protons and neutrons. In very large nuclei especially, this balance is precarious. Sooner or later, the nucleus gives off its excess energy and converts itself into some other kind of nucleus. It may do this several times before it becomes stable at a low energy level. The bursts of emitted energy and particles are what we call *radioactivity*.

One form of radioactivity is alpha emission, in which an alpha particle is emitted, for example:

$$^{238}_{92}U \rightarrow {}^{234}_{90}Th + {}^{4}_{2}He$$

A nucleus of uranium-238 has emitted an alpha particle (helium-4 nucleus), thus reducing its atomic number by 2 and its atomic mass number by 4 and becoming thorium-90. For another example, see Sample Problem 15-3.

Sample Problem

15-3 Write the equation for the alpha decay of thorium-230.

Solution Consult Appendix 3; thorium-230 has an atomic number of 90, just like all the other forms of thorium. So

$$^{230}_{90}Th \rightarrow {}^{4}_{2}He + {}^{226}_{88}Ra$$

Element 88 is radium, according to the table.

Emitted alpha particles always have, in any particular reaction, certain definite energy values. Just like the definite energy values of the photons emitted in an electron transition, these values indicate that there are quantized energy levels within the nucleus. If the alpha particle energy is less than the maximum possible, the rest is released as a gamma photon.

Alpha emission decreases the charge on the nucleus. In other nuclei, stability is increased by *increasing* the nuclear charge. This is done by converting a neutron to a proton, with the negative charge being carried off in the form of an electron, a beta particle. A gamma photon may or may not come off also, but there is always another particle known as a neutrino, which carries off some of the excess energy. Thorium decay is an example:

$$^{234}_{90}Th \rightarrow {}^{234}_{91}Pa + {}^{0}_{-1}e + {}^{0}_{0}\nu$$

The neutrino, with no charge and negligible mass, appears in the equation as the Greek letter nu (ν). Note that, in beta decay, the atomic mass number does not change but the atomic number increases by 1. See Sample Problem 15-4. The neutrino, having no charge, makes no change in the atomic number. Whether it has mass is one of the still unsolved questions, but if it does, it is surely not much.

15-4 Write the equation for the beta decay of palladium-234.

Solution   Palladium is element 91:

$$^{234}_{91}\text{Pa} \rightarrow\ ^{0}_{-1}\text{e} +\ ^{234}_{92}\text{U} +\ ^{0}_{0}\nu$$

Palladium is still unstable and undergoes another beta decay. This is followed by a series of alpha and beta decays, until the nucleus finally becomes stable in the form of lead-206. The entire series is illustrated in Figure 15-4.

FIGURE 15-4

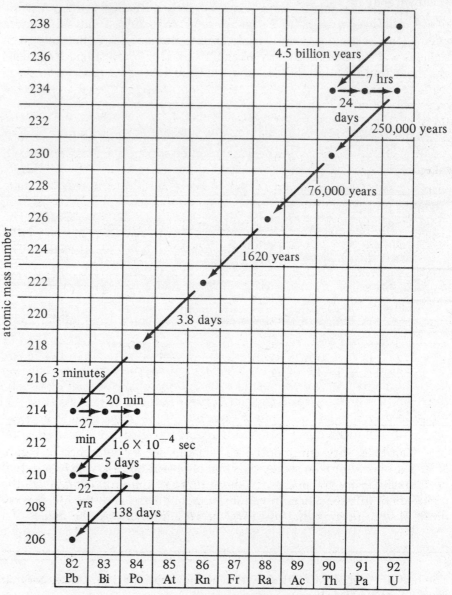

In natural radioactivity, nuclei become more stable by emitting alpha and beta particles.

Try This Write the equations for an alpha decay, followed by a beta decay, of polonium-214. (Atomic numbers are listed in Appendix 3.)

15-7. How Long Does It Take?

Each radioactivity event represents the transformation of a single nucleus. These nuclei have been resting in the earth ever since there was an earth. A nucleus, having been a uranium nucleus for billions of years, suddenly decides that now is the time to turn into palladium.

We know of no way to predict when any particular nucleus will explode. The event seems to be completely random. As with all random events, the best we can do is to determine the *probability* that any given nucleus will explode within any given time period. Some radioactive nuclei are more unstable than others, and the probability that any nucleus will explode is greater for them. This probability is a billion to one *against* the explosion of any given uranium-238 nucleus within the next year; for a thorium-234 nucleus, it is 40,000 to 1 *in favor*.

The stability of radioactive substances can be expressed in terms of a quantity called *half-life*. The half-life of a substance is the length of time it takes for half of any sample to undergo radioactive decay. The half-life of uranium-238, for example, is 4.5 billion years. Since this is approximately the age of the earth, it means that half of all the uranium-238 that was in the earth at the beginning is still there. The rest has gone through the decay sequence of Figure 15-4, or some part of it. For a typical calculation involving half-life, see Sample Problem 15-5.

Sample Problem

15-5 Thorium-234 has a half-life of 24 days. If you have a 1-kg block of thorium-234, how much of it is unchanged after 4 months?

Solution Four months is about 120 days, which is five half-lives of thorium. The mass of thorium then must drop to one-half five times, so it becomes

$$(1 \text{ kg})\left(\frac{1}{2}\right)^5 = 0.031 \text{ kg}$$

Half-lives vary enormously, as you can see by reference to Figure 15-4. The shortest in the sequence is that of polonium-214, which loses half of its nuclei every 160 microseconds. A lump of uranium contains all the members of this sequence, in various stages of decay, but the life of polonium-84 is so fleeting that there will be very little of it in the mix.

The half-life of a radioactive substance is the length of time it takes for half of its nuclei to disintegrate.

Try This If you start with 1 gram of polonium-214, how much will be left at the end of $\frac{1}{100}$ s?

15-8. The Energy of the Rays

What is the source of the energy of these emissions? The answer turns out to be intimately associated with a question we asked earlier: Why are all nuclear masses not equal to some integral multiple of the mass of a proton?

The answer lies in the famous Einstein mass–energy relationship:

$$E = mc^2 \qquad \text{(Equation 6-9)}$$

which tells us that mass and energy are simply two different ways of measuring the same thing.

This leads to a contradiction with the most basic rule of chemistry. Chemists tell us that 16 grams of oxygen and 2 grams of hydrogen form an explosive mixture. A spark ignites it, and the oxygen and hydrogen combine to form exactly 18 grams of water. Mass is conserved. But if mass is energy, we ought to expect that the mass of the water produced in the explosion will be *less* than 18 grams, since a lot of energy is given off in the explosion. In general, every molecule ought to have less mass than the separate atoms that make it up, since it always takes energy to separate the atoms from each other.

The reason that chemists have never noticed this missing mass is that it is far too small to measure. But with nuclei, the situation is different. The nuclear force is far stronger than the electromagnetic force that holds molecules together. Thus, the energy given up when protons and neutrons combine to form nuclei produces a *measurable* difference in mass. Every nucleus has a *mass deficit*, which is the difference between the mass of the nucleus and the masses of the separate particles of which it is composed. Consider, for example, iron-56, the most common isotope of iron:

mass of 26 protons	
26×1.007277	26.18920
mass of 30 neutrons	
30×1.008665	30.25995
total mass of separate particles	56.44915
mass of $^{56}_{26}$Fe	55.9349 Dl
mass deficit of $^{56}_{26}$Fe	0.5143 Dl

This mass deficit represents the energy that would be released if we could find a way to make 30 neutrons and 26 protons combine into a single nucleus. It is also the energy that would have to be added to separate the nucleus into its parts.

While energy and mass are completely interchangeable, we do not usually think of a dalton as a unit of energy. We can use Einstein's equation to express this energy in more familiar units. See Sample Problem 15-6. The result is

$$1 \text{ dalton (Dl)} = 931 \text{ million electron volts (Mev)}$$

Sample Problem

15-6 How many million electron volts correspond to a mass of 1 Dl?

Solution　From Equation 6-9,

$$E = mc^2 = (1 \text{ Dl})(2.998 \times 10^8 \text{ m/s})^2 \times \frac{\text{kg}}{6.025 \times 10^{26} \text{ Dl}}$$

$$E = (1.4918 \times 10^{-10} \text{ J}) \times \frac{\text{ev}}{1.602 \times 10^{-19} \text{ J}}$$

$$E = 9.31 \times 10^8 \text{ ev} = 931 \text{ Mev}$$

Thus the mass deficit of iron-56 (0.5143 Dl) corresponds to a nuclear binding energy of 479 Mev. Compare this with the chemical binding energy of the two atoms of hydrogen and one of oxygen that make up a water molecule. It is a mere 7 ev, corresponding to a few billionths of a dalton of mass. See Sample Problem 15-7.

Sample Problem

15-7 Find (a) the mass deficit and (b) the binding energy of the nucleus of boron-11.

Solution (a) From Table 1, the atomic number of boron is 5, so boron-11 contains five protons and six neutrons.

mass of five protons	
5 × 1.007277	5.03635 Dl
mass of six neutrons	
6 × 1.008665	6.051990
total separate masses	11.08834
mass of boron-11	11.00931
mass deficit	0.07903 Dl

(b) The binding energy is found by transforming mass units to energy units:

$$0.07903 \text{ Dl}\left(\frac{931 \text{ Mev}}{\text{Dl}}\right) = 73.6 \text{ Mev}$$

In any radioactive decay process, the mass deficit must increase; the total mass of the products must be less than the mass of the nucleus that exploded. The lost mass provides the energy of the emitted particles. Consider, for example, the alpha decay of the most stable isotope of radium, which has a half-life of 1,600 years:

$$^{226}_{88}\text{Ra} \rightarrow {}^{222}_{86}\text{Rn} + {}^{4}_{2}\text{He} \qquad (1,600 \text{ years})$$

It decays to form a nucleus of radon-222 and an alpha particle. Then here is the mass-energy budget:

mass of $^{226}_{88}\text{Ra}$	226.0254 Dl
mass of $^{222}_{86}\text{Rn}$	222.0175
mass of $^{4}_{2}\text{He}$	4.0026
total mass of product nuclei	226.0201
additional mass deficit	0.0053 Dl
corresponding to a binding energy of	4.93 Mev

Most of the alpha particles emitted have only 4.78 Mev of kinetic energy. The additional 0.15 Mev of energy is then given off as a gamma photon. It is an extremely high-energy photon, with a wavelength of only about 10^{-11} m. See Sample Problem 15-8.

Sample Problem

15-8 The alpha particle emitted in the decay of radium-226 sometimes has an energy of 4.340 ev. What is the wavelength of the gamma photon that follows?

Solution The total increase in binding energy is still 4.93 Mev, so the energy of the photon must be $4.93 - 4.34 = 0.59$ Mev. The wavelength, from Equations 8-2 and 13-9a, is

$$\lambda = \frac{hc}{E} = \frac{(4.14 \times 10^{-15} \text{ ev·s})(3.00 \times 10^8 \text{ m/s})}{0.59 \times 10^6 \text{ ev}}$$

$$\lambda = 2.1 \times 10^{-12} \text{ m}$$

Core Concept

A nuclear reaction always results in increased mass deficit, with the lost mass appearing in the form of the energy of the emitted particles.

Try This What is the binding energy of an alpha particle? (All needed data appear in Appendix 3.)

15-9. Increasing the Energy

The energies of alpha particles emitted in radioactive decay run to about 5 Mev. If they are used to bombard gold foil, they cannot collide head-on with a nucleus because it takes more energy than this for the particle to overcome the electric repulsion. However, if these alpha particles are sent into a metal with a smaller nucleus, the nuclear charge will be smaller and the alpha particle might actually enter the nucleus. Usually, it forms an unstable combination that breaks up immediately. In fact, it was this kind of nuclear reaction that first revealed the existence of the neutron. Alpha particles were sent into a piece of beryllium, with this result:

$$_4^9\text{Be} + _2^4\text{He} \rightarrow (_6^{13}\text{C}) \rightarrow _6^{12}\text{C} + _0^1\text{n}$$

The carbon-13 formed was in a highly excited state and settled down by emitting a neutron (n).

To make alpha particles, or any other charged particles, enter a large nucleus they must be given more energy. This is the function of a particle accelerator. Many types of accelerators have been designed, and the newest ones are the largest and most expensive pieces of research equipment ever produced.

A charged particle is accelerated by placing it in an electric field. Equation 9-9 tells us that, as a positive charge is allowed to fall in an electric field from high potential to low, it gains an amount of energy equal to $q\,\Delta V$. If the value of q is one elementary charge, the energy the particle gains, in electron volts, is equal to the difference of potential in volts.

The simplest particle accelerator is a Van de Graaff generator, such as that described in Chapter 9. A large Van de Graaff generator can produce potentials up to several million volts. To prevent sparking, they are made with smooth, round terminals and enclosed in a case filled with a special gas under high pressure. Protons have been accelerated up to 9 Mev with a device of this sort.

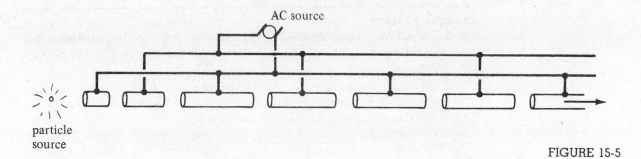

AC source

particle
source

FIGURE 15-5

The key to producing particles with much higher energies than that is to accelerate them in steps rather than all at once. This can be done in a *linear accelerator*, which consists of a series of *drift tubes* in a linear array, as in Figure 15-5. There is no force on a particle while it is inside a tube. Acceleration occurs in the gaps between tubes and is produced by a difference of potential between the two tubes. Thus, if the potential of the second tube is 10,000 volts lower than that of the first, a positive particle will gain 10,000 ev of energy as it crosses the gap. The tubes are connected to carefully synchronized AC sources, so that the polarity switches while the particles are passing through the tubes. Then, when the particle reaches the next gap, it finds an electric field that will boost its energy by another 10,000 ev.

The largest linear accelerator is in a tunnel 2 miles long at Stanford University. Its 960 drift tubes accelerate electrons up to 20 gigaelectron volts (1 Gev = 1,000 Mev). The lengths of the drift tubes must be carefully controlled; the first few must increase in length as the electrons speed up. Very soon, however, the electrons are going practically at the speed of light, and nothing can go faster. Therefore, all but the first few tubes are the same length. As the electrons gain kinetic energy, their speed cannot increase, but their mass does. See Sample Problems 15-9 and 15-10.

Sample Problem

15-9 What is the mass of a proton that has been accelerated to a kinetic energy of 20,000 Mev?

Solution   The kinetic energy has mass equal to

$$(20,000 \text{ Mev})\left(\frac{\text{Dl}}{931 \text{ Mev}}\right) = 21 \text{ Dl}$$

Added to the rest mass of 1 Dl, this gives a total of 22 Dl.

Sample Problem

15-10 The mass of an electron at rest is 9.1×10^{-31} kg. By what factor does its mass increase when its energy is raised to 20 Gev?

Solution   From Equation 6-9,

$$m = \frac{E}{c^2} = \frac{(20 \times 10^9 \text{ ev})(1.60 \times 10^{-19} \text{ J/ev})}{(3.00 \times 10^8 \text{ m/s})^2}$$

$$m = 3.56 \times 10^{-26} \text{ kg}$$

To get the factor of the increase, divide this by the rest mass:

$$\frac{3.56 \times 10^{-26} \text{ kg}}{9.1 \times 10^{-31} \text{ kg}} = 39,000$$

With protons, much higher energies can be obtained by passing them repeatedly through the same drift tubes. The tubes are arcs of a circle, each surrounded by a large electromagnet. The field of the magnet bends the path of the particles into a circular path that keeps them inside the tube. Drift tubes alternate with acceleration gaps around a circular path, and the protons make many circuits, gaining energy all the while, until another magnetic field releases them for use.

This device is called a *synchrotron*. The largest of them is the Fermilab machine, in Illinois. First, a simple transformer-rectifier circuit produces 750,000 volts and accelerates protons to this level. The protons are then injected into a linear accelerator that speeds them up to 200 Mev. Then they enter a small(!) synchrotron, 500 feet in diameter, for their final acceleration before being pushed into the main ring. The synchrotron has a diameter of *4 miles*, which can boost the protons up to 1,000 Gev—that's 1 million Mev! Still larger and more powerful accelerators are being planned.

These high-speed particles are used as bullets to be fired into other particles. As a result, many new kinds of particles are produced, whose properties can be studied in a bubble chamber or other detection device. Studying the inner structure of a proton or a neutron with such high-energy bullets is now routine. Such investigations give us ever-deeper insights into the nature of matter.

Core Concept

High-energy accelerators use electric fields to boost the energy of charged particles, and magnetic fields to control their paths.

Try This What is the mass of a proton with 500 Gev of kinetic energy? How does this compare with its rest mass?

15-10. Transforming the Nucleus

Firing energetic bullets into nuclei can transform one element into another. In one common type of reaction, an alpha particle is absorbed by a nucleus, which forms a new and highly unstable nucleus. This immediately breaks down with emission of a proton:

$$\,_2^4\text{He} + \,_{13}^{27}\text{Al} \rightarrow (\,_{15}^{31}\text{P}) \rightarrow \,_{14}^{30}\text{Si} + \,_1^1\text{H}$$

The alpha bullet has transformed aluminum-27 into silicon-30.

In a similar reaction, the bombarded nucleus emits not a proton, but a neutron. We saw such a reaction earlier. For another example, see Sample Problem 15-11.

Sample Problem

15-11 Fluorine-19 emits a neutron when bombarded by an alpha particle. Write the equation.

Solution

$$\,_9^{19}\text{F} + \,_2^4\text{He} \rightarrow (\,_{11}^{23}\text{Na}) \rightarrow \,_{11}^{22}\text{Na} + \,_0^1\text{n}$$

The product is sodium-22.

Often the unstable nucleus produced by alpha bombardment does not break down immediately. It becomes an artificially produced radioactive substance with a half-life that might be many years. Carbon-14 is one such artificially radioactive nucleus. It can be produced by alpha bombardment of boron:

$$\,_2^4\text{He} + \,_5^{11}\text{B} \rightarrow (\,_7^{15}\text{N}) \rightarrow \,_6^{14}\text{C} + \,_1^1\text{H}$$

Carbon-14 is also produced naturally in the atmosphere, when a nitrogen nucleus absorbs a neutron:

$$\,_7^{14}\text{N} + \,_0^1\text{n} \rightarrow (\,_7^{15}\text{N}) \rightarrow \,_6^{14}\text{C} + \,_1^1\text{H}$$

The carbon-14 decays by beta emission with a half-life of 5,730 years:

$$\,_6^{14}\text{C} \rightarrow \,_7^{14}\text{N} + \,_{-1}^0\text{e} + \,_0^0\nu \quad (5{,}730 \text{ years})$$

Artificially induced radioactivity takes some forms that are quite different from the alpha and beta decay of naturally radioactive substances. In one form, a proton in the nucleus converts itself into a neutron by emitting a *positron*. This is a positive electron; it has all the usual properties of an electron, except that its charge is positive. An example is the decay of phosphorus-30, produced by alpha bombardment of aluminum. It decays by positron emission:

$$\,_{15}^{30}\text{P} \rightarrow \,_{14}^{30}\text{Si} + \,_{+1}^0\text{e} + \,_0^0\nu \quad (2.5 \text{ minutes})$$

For another example, see Sample Problem 15-12.

Sample Problem

15-12 When aluminum-27 is bombarded with an alpha particle, a neutron is released and a new kind of nucleus is formed. (a) Write the equation for this reaction. (b) Write the equation for the decay of the new nucleus by positron emission.

Solution (a) $\,_{13}^{27}\text{Al} + \,_2^4\text{He} \rightarrow (\,_{15}^{31}\text{P}) \rightarrow \,_0^1\text{n} + \,_{15}^{30}\text{P}$.

(b) $\,_{15}^{30}\text{P} \rightarrow \,_{14}^{30}\text{Si} + \,_0^0\nu$.

Nuclear transformation produces a whole string of new elements larger than uranium-238 which is the largest naturally occurring nucleus. In this reaction uranium-238 is bombarded with neutrons. The neutrons do not need high energy; since they have no charge, they are not repelled by the uranium nucleus and can simply fall inward. This is what happens:

$$\,_{92}^{238}\text{U} + \,_0^1\text{n} \rightarrow \,_{92}^{239}\text{U}$$

The new isotope of uranium undergoes beta decay:

$$\,_{92}^{239}\text{U} \rightarrow \,_{93}^{239}\text{Np} + \,_{-1}^0\text{e} + \,_0^0\nu \quad (24 \text{ minutes})$$

Neptunium (Np) is a new element. It also decays:

$$\,_{93}^{239}\text{Np} \rightarrow \,_{94}^{239}\text{Pu} + \,_{-1}^0\text{e} + \,_0^0\nu \quad (2.4 \text{ days})$$

The other new element is plutonium (Pu). It is much more stable than the others:

$$\,_{94}^{239}\text{Pu} \rightarrow \,_{92}^{235}\text{U} + \,_2^4\text{He} \quad (24{,}360 \text{ years})$$

As we shall see, plutonium is now available in large quantities. It is an extremely useful material—and the most dangerous ever to come out of a laboratory.

Core Concept

Nuclei can be transformed by bombarding them with particles.

Try This Complete the following nuclear equation:

$$\ce{^4_2He + ^{39}_{19}K} \rightarrow (\ce{^?_?Sc}) \rightarrow \ce{^1_1H + ^?_?Ca}$$

15-11. Nuclei That Split

Certain nuclei, made unstable by absorption of a particle, will split nearly in half instead of just emitting a particle of some sort:

$$\ce{^{235}_{92}U + ^1_0n} \rightarrow (\ce{^{236}_{92}U}) \rightarrow \ce{^{141}_{56}Ba + ^{92}_{36}Kr} + 3\,\ce{^1_0n}$$

This is only one of a number of ways in which the uranium-235 nucleus can split when it is struck by a neutron. Fission produces a wide variety of fragments, always radioactive, medium-sized nuclei. Most very large nuclei can undergo fission this way, but many of them do not unless the incident neutron has a great deal of energy.

Medium-sized nuclei are more stable and more tightly bound than the real giants. This means that their mass deficits, for each particle, are much larger. That is why the fission of a big nucleus releases a lot of energy. If you calculate the additional binding energy produced in the fission of uranium-235 into barium-141 and krypton-92 (as in Sample Problem 15-13), you will find that it comes to 174 Mev, about 20 times as much as in an ordinary alpha decay.

Sample Problem

15-13 How much additional binding energy is generated when a uranium-235 nucleus is struck by a neutron and splits to form barium-141, krypton-92, and three neutrons?

Solution The total mass going into the reaction—one uranium-235 plus one neutron—is

$$235.0439 + 1.008665 = 236.0526 \text{ Dl}$$

and coming out of the reaction:

$$140.9140 + 91.9261 + 3(1.008665) = 235.8661 \text{ Dl}$$

The difference is 0.1865 Dl. Multiplying by 931 Mev per Dl gives 174 Mev as the energy released.

It is another feature of the fission process shown above that provides the key to usable atomic energy. The process uses 1 neutron and produces 3. Each extra neutron can strike another nucleus and cause it to split, releasing still more neutrons. This *chain reaction* can build up rapidly, until enormous numbers of nuclei are splitting. Uncontrolled, this produces an atomic explosion, which releases thousands of times more energy than the largest TNT bomb ever made.

To sustain a chain reaction, more neutrons must be produced than consumed. Only three kinds of nuclei can be used: uranium-235, uranium-233, and plutonium-239. Of these, only uranium-235 is found naturally, and it is scarce. Natural uranium has 140 times as much uranium-238 as uranium-235. Uranium-238 cannot sustain a chain reaction because it undergoes fission only with extremely energetic neutrons, and when it splits it does not release them. To obtain nuclear fuel, it is necessary to separate uranium-235 from uranium-238. This cannot be done by chemical means, since both are chemically the same. The physical separation process, based on the slight difference in mass, is difficult and expensive.

The most significant bomb fuel today is plutonium-239, which is produced from the abundant uranium-238 by bombardment with slow neutrons. The equations for this reaction were given earlier.

To make a bomb, you have to have fairly pure fuel. Impurities soak up neutrons, so that they become unavailable for producing additional fissions. The fast neutrons produced in fission can knock as many as five additional neutrons out of a nucleus, but they must not be lost. When the mass of fuel is too small, many of the nuclei are near its surface and can escape. On the other hand, if the size of the fuel package is big enough, all it needs to set it off is one neutron. A few kilograms of uranium-235 or plutonium-239 will explode spontaneously.

The trick in making an atom bomb is to bring the fuel together and keep it in one place long enough for the reaction to take place—say, a few millionths of a second. In a bomb, the fuel is at the center, loosely packed. It is surrounded by TNT, and the whole thing is packaged within a strong, thick metal wall. The explosion of the TNT cannot breach the wall, so it actually implodes; it pushes toward the center, compressing the fuel into a critical mass and holding it there just long enough. Boom!

Core Concept

A nuclear chain reaction can occur when nuclear fission releases more neutrons than are consumed or lost.

Try This   An atomic bomb contains 2.0 kg of fuel. What fraction of the fuel undergoes fission if the bomb produces energy equivalent to 25 kilotons of TNT? (A ton of TNT produces about 4×10^9 J of energy.)

15-12. The Reactor

The fission of uranium-235 does not have to be used to make bombs. The reaction can be controlled, and when operated slowly can produce electricity. This is done in a *nuclear reactor*.

There are many kinds of reactors. Some are designed for research purposes, some for the production of artificial radioisotopes, some to drive ships, and others to operate electric power stations. A *breeder reactor* exposes uranium-238, the commonest kind, to a flux of neutrons, converting it into fissionable plutonium.

Nuclear power plants in the United States are of the kind called boiling water reactors. The fuel is in the form of long rods containing uranium oxide pellets. A typical reactor has 560 fuel assemblies, each consisting of 64 rods several meters long. The rods are immersed in very pure water, and the whole assembly is housed inside a steel vessel about 10 m tall. This is the reactor core. The uranium used is "enriched"; that is, the proportion of fissionable uranium-235 has been increased from its natural value of 0.7 percent to about 3 percent. This, of course, is far below the 90 percent purity needed to make it explode like a bomb.

With the fuel in long, thin rods, most of the neutrons produced in the fission of uranium-235 escape from the rod into the water. The water acts as a *moderator*; it absorbs most of the energy of the neutrons, slowing them down. This is important, since slow neutrons are captured for fission by uranium-235 and are not as likely to get lost inside other nuclei, such as uranium-238. The slow neutrons reach another rod, where they do their job properly. As fission products accumulate, they trap some neutrons, but the reaction continues as long as one neutron from each fission, on the average, manages to reach another fissionable nucleus.

The large amount of uranium-238 in the rods can catch fast neutrons before they leave the rod, converting their nuclei into uranium-239. Two beta decays turn the uranium-239 into fissionable plutonium-239—more nuclear fuel. Usually, the fission of this plutonium is a rather small part of the energy produced in the reactor. A breeder reactor is designed to produce plutonium, which can be extracted chemically from the spent fuel rods. It can be fed back into a reactor as fuel, or used to make bombs.

To control the rate of the reaction, a number of *control rods* project into the reactor core. These are made of a boron–carbon alloy. Their function is to absorb neutrons. As the accumulation of fission products slows the reaction down, the control rods are gradually withdrawn to keep the flow of neutrons from one fuel rod to another at the desired level. Inserting the control rods quickly into the reactor core will shut down the whole operation promptly.

The highly purified water that serves as a moderator has another function as well. It is the working fluid of the power generation system. The energy of the neutrons and other particles generated in the reaction heats up the water. It is kept under pressure, so that it boils at a temperature far above its usual value. The steam produced flows into the turbines that turn the electric generators.

Once a year, the reactor is shut down and about 30 percent of the fuel rods are replaced. The spent rods are highly radioactive. This makes them extremely dangerous to the public health, so they must be stored in isolation for thousands of years. Now, they are considered waste, but they are undoubtedly the plutonium mines of the future. Surely, many uses will be found for the other radioactive substances they also contain. Meanwhile, disposal of radioactive waste remains the thorniest problem for the nuclear energy industry.

Core Concept

In a reactor, fission of uranium-235 is accomplished by means of neutrons slowed down by passage through a moderator.

Try This Like uranium-238, thorium-232 is "fertile"; that is, it can be converted into a fissionable substance by bombardment with neutrons. Write the three nuclear reaction equations showing the effect of the thorium absorbing a neutron, followed by two beta decays to produce fissionable uranium-233.

Energy is released by any nuclear reaction in which the binding energy, or mass deficit, of the products is more than that of the original nuclei. As we have seen, this situation exists whenever a véry large nucleus is split into medium-sized ones. It also exists, however, when very small nuclei are combined to form medium-sized ones. This is called *nuclear fusion*. Medium-sized nuclei have the largest mass deficit per particle.

The fusion of small nuclei provides the energy radiated out by the sun and other stars. Many different processes go on; one of the most important, in the sun at least, is a three-step process that converts hydrogen to helium. It starts with two protons (nuclei of hydrogen-1) combining to form hydrogen-2, also known as deuterium, with emission of a positron:

$$\mathrm{{}^1_1H + {}^1_1H \rightarrow {}^2_1H + {}^0_{+1}e + {}^0_0\nu}$$

Then deuterium reacts with another proton to form helium-3:

$$\mathrm{{}^2_1H + {}^1_1H \rightarrow {}^3_2He + {}^0_0\nu}$$

Finally, two of these helium-3 nuclei combine to form helium-4 and a couple of protons:

$$\mathrm{{}^3_2He + {}^3_2He \rightarrow {}^4_2He + 2\,{}^1_1H}$$

At each step, the total mass decreases and energy is released.

These reactions are called *thermonuclear* because they can occur only at extraordinarily high temperatures, about 40 million K. The reason is that all the particles involved have positive charges, so they repel each other. Violent vibration at high temperatures is necessary to bring the nuclei close enough for the nuclear force to take over and draw the two nuclei together.

On earth, we have mastered nuclear physics well enough to produce our own thermonuclear reactions—in hydrogen bombs. One type of fuel used for this sort of bomb is the chemical lithium deuteride, a solid composed of atoms of lithium-6 and hydrogen-2 (deuterium) in equal numbers. The explosion of a hydrogen bomb is triggered by an ordinary atom bomb, in which uranium-235 undergoes fission. This reaction produces not only the necessary temperature for fusion, but also a tremendous flux of fast neutrons. These neutrons react with the lithium nuclei:

$$\mathrm{{}^6_3Li + {}^1_0n \rightarrow {}^4_2He + {}^3_1H}$$

Now the nuclei of hydrogen-3 (tritium) combine in a thermonuclear reaction with the deuterium nuclei:

$$\mathrm{{}^3_1H + {}^2_1H \rightarrow {}^4_2He + {}^1_0n}$$

This reaction releases an enormous amount of energy, but this bomb has still another trick. All fast neutrons strike the uranium-238 shell that holds the bomb together. The resulting fission of the uranium adds its energy to the mix. The overall output is enormous. The largest hydrogen bombs produce about 25 megatons of energy, a thousand times as much as that produced by the atom bomb that destroyed Hiroshima in World War II.

Efforts are being made to find ways to use nuclear fusion in a controlled way to produce electricity. If this can be done economically, it will put an end to our shortage of energy. The fuel will be deuterium, which is found as 1 part in 6,000 parts of all natural hydrogen and is thus available in endless supply. There is no fission involved, so very little radioactive waste will be produced. The process will be nonpolluting. Controlled fusion has already been achieved on a laboratory scale, but there are great obstacles in the way of making it commercially useful.

Fusion of small nuclei produces great amounts of nuclear energy.

Try This The hydrogen bomb thermonuclear fusion results in an overall effect, which is the conversion of lithium-6 and deuterium into two nuclei of helium-4. How much energy is released?

Summary Quiz

For each of the following, supply the missing word or phrase:

1. The atomic nucleus was discovered by bombarding gold foil with _____ .
2. The nucleus of any atom contains some whole number of _____ charges.
3. The number of electrons surrounding the nucleus of an atom is equal to the _____ of the nucleus.
4. The _____ atom has an atomic number of 1.
5. In dropping from a high energy level to a lower one, the electron of a hydrogen atom emits a _____ .
6. When the principal quantum number of the electron energy of a hydrogen atom reaches infinity, the atom is _____ .
7. Quantized energy levels of the hydrogen electron can be derived on the assumption that its _____ is quantized.
8. Angular momentum is always an integral multiple of _____ .
9. In the wave model, electrons form _____ in their orbits.
10. The positively charged particle in the nucleus is called a _____ .
11. The number of protons plus the number of _____ equals the atomic mass number of the nucleus.
12. _____ of the same element differ only in the number of neutrons in their nuclei.
13. Nuclear masses are measured in _____ .
14. Charged particles can be tracked through detection devices because they lose energy in _____ the material through which they pass.
15. The tracks of charged particles are curved as a result of the influence of a _____ .
16. Charged particles passing through liquid hydrogen at its boiling point leave a trail of _____ .
17. The photons emitted by radioactive materials are called _____ rays.
18. Beta rays are _____ .
19. Alpha particles are nuclei of _____ .
20. The process of _____ decay increases the atomic number by 1.
21. In alpha decay, the atomic mass number decreases by _____ .
22. If a gram of a radioactive substance is reduced to $\frac{1}{2}$ g in 6 hr, then the _____ of that substance is 6 hr.
23. The mass of a nucleus is always _____ the mass of the separate particles of which it is composed.
24. Mass deficit is _____ energy.
25. In any spontaneous nuclear process, mass deficit always _____ .
26. Charged particles are increased in energy by the action of a(n) _____ field.

27. In a linear accelerator, energy is added to particles as they pass through the gap between the _____ .
28. In a synchrotron, a _____ field bends the path of the particles.
29. A positron is a positive _____ .
30. Small radioactive nuclei can be made artificially by bombardment with _____ .
31. _____ with very low energy can enter nuclei.
32. Plutonium is produced by reacting uranium with a _____ .
33. A chain reaction can be sustained only by a fuel that produces _____ when it undergoes fission.
34. The preferred fuel for atom bombs is _____ .
35. In a nuclear reactor, it is the function of the _____ to slow down the neutrons.
36. In a reactor core, neutrons are absorbed by the _____ .
37. Temperatures in the millions of kelvins are needed to produce nuclear _____ .
38. The largest mass deficit per particle is found in _____ nuclei.

Problems

1. An atomic species is represented as $^{190}_{77}$Ir. State the number of (a) electrons in the rings; (b) neutrons in the nucleus; and (c) protons in the nucleus.
2. What is the energy of a hydrogen electron in (a) its ground, or lowest energy, state; (b) its third quantum state; and (c) its ionized state?
3. A hydrogen electron drops from its third quantum state to the first (ground) state. Find (a) the energy of the photon emitted; and (b) its wavelength.
4. In a collision with another particle, a hydrogen atom is ionized, gaining 18 ev of energy in the process. What is the kinetic energy of the electron after it leaves the atom?
5. In an interference experiment, an electron beam shows a measured wavelength of 3.0×10^{-10} m. Find (a) the momentum of the electrons; (b) their velocity; and (c) their kinetic energy in electron volts.
6. A hydrogen electron in its third quantum state has an orbital radius of about 1.5×10^{-10} m. Find (a) its wavelength; and (b) its momentum.
7. Determine the mass deficit of phosphorus-30 and the corresponding binding energy.
8. (a) Write the equation for the alpha decay of polonium-218; and (b) calculate the combined energy of the alpha particle and the gamma photon released.
9. How long would it take for a sample of polonium-218 to be reduced to $\frac{1}{16}$ of its original mass?
10. (a) Write the equation for the beta decay of lead-210; and (b) calculate the combined energy of the electron and the neutrino produced.
11. What is the kinetic energy of an alpha particle that has been accelerated in an electric field by a potential difference of 650,000 V?
12. Through what potential difference would a proton have to be accelerated to double its mass?
13. An alpha particle strikes the nucleus of boron-10 and combines with it. The compound nucleus stabilizes by emitting a neutron. Write the equation.

14. Write the equation for the decay of sodium-22 by emission of a positron.

15. Plutonium-239 receives a slow neutron and splits, releasing 3 neutrons and a nucleus of xenon-142. Determine the other fission product and write the complete equation for the reaction.

16. Determine the energy of the gamma photon released in the following fusion reaction, which takes place in the sun:

$$^{1}_{1}H + {}^{12}_{6}C \rightarrow {}^{13}_{7}N + \text{gamma}$$

APPENDIX 1

The Mathematics You Need

A1-1. Constants and Variables

Physicists are always making measurements. Most of what they know about the universe has come from measurements of one kind or another.

When you measure something, you get a definite value for some particular quantity. A value of this sort is called a *constant*. Here are some examples of constants: the scale tells me I weigh 130 pounds; the clock says 2:45; I used 2 cups of flour; my desk is 30 inches high.

While particular facts of this sort are often interesting, and even important, scientists look for something deeper. They want to find out the general laws that govern the universe. They would like to make some sort of statement not about *my* weight or *your* weight, but about weights in general. They often start by inventing a code, such as, "Let w stand for the weight of any object." Then they can talk about weights without specifying *which* weight they have in mind. A quantity of this sort, which can take on many different values, is called a *variable*.

In the language of physics—algebra—a variable is always represented by a letter. There is some standardization of usage: w is generally used to stand for weight, t means time, s represents distance, and so on. We also use letters to stand for constants; we usually write π (the Greek letter pi) for convenience instead of its numerical value of 3.14159. . . .

The constant π is an example of a *universal* constant which has the same value everywhere. There are some physical constants that are universal. An example is the speed of light in a vacuum, which is always and

335

forever the same everywhere, so far as we know. It appears in many equations represented by the letter c. Another example is e, the electric charge of an electron.

Many of the constants we use are not universal, but apply only under a specifically defined set of conditions. These are *local* constants. For example, when you are discussing the traffic conditions on a particular road, an important constant is surely the speed limit. But when you are discussing the conditions on a great many roads, the speed limit is a variable, with a particular constant value for each road.

Core Concept

A *constant* is a quantity that has a definite value under a specified set of conditions; a *variable* is a letter representing a quantity that may take on many different values.

Try This Classify each of the following quantities as a universal constant, local constant, or variable: (a) the height of the ceiling in your room, (b) the heights of all ceilings in your town, (c) the number of inches in a foot, (d) the speeds of all cars on a particular highway, (e) the speeds of all the cars in a freight train.

A1-2. Proportionality

Physicists try to discover the laws of nature. These laws are expressed as relationships between variables. Physics deals with such questions as, What happens to the speed of an object if the force on it increases? As a nail is brought closer to a magnet, what happens to the force on the nail? Each of these relationships between two variables is expressed as a *proportionality*.

For example, suppose you have a spool of wire and you cut off many pieces from it. Each piece has a different length; length is a variable for the whole set of pieces. Each piece will also have a different weight; weight is another variable. But there is a consistent relationship between the lengths and the weights of the pieces. If one piece is twice as long as another, it will also weigh twice as much, and a piece six times as long as the first will have six times its weight. Algebraically, this relationship is expressed this way:

$$w \propto L$$

which is read as "weight is directly proportional to length."

If you wanted to express this relationship in terms of the measured values, it turns out to be very cumbersome. You could let w_1 stand for the weight of the first piece and L_1 stand for its length; then use w_2 and L_2 for the second piece, and so on. Then you would find that

$$\frac{w_1}{L_1} = \frac{w_2}{L_2} = \frac{w_3}{L_3} = \frac{w_4}{L_4} = \frac{w_5}{L_5} = \text{etc.}$$

In other words, the *ratio* between weight and length is the same for all the pieces of wire. For this wire, this ratio is a constant, even though the weight and the length are variables. The easy way to write this relationship is

$$\frac{w}{L} = d$$

where w and L are variables and d is the constant ratio between them. This constant is known as the linear density of the wire and can be expressed as a certain number of ounces per yard. If we know this constant for any particular kind of wire, we can easily calculate the length of any weight of this wire, or the weight of any length.

A direct proportionality is a statement that the ratio between two variables is a constant.

Try This Gas mileage is the constant ratio between the distance a car travels and the amount of fuel it uses. (a) What is this constant of proportionality if a car uses 2.5 gal of gasoline in traveling 70 mi? (b) How much gasoline would the car use in going 350 mi?

A1-3. Other Proportionalities

Unfortunately, not all relationships between variables are of the simple, direct proportionality type. There are many kinds, two of which occur commonly.

Suppose, for example, that you are making a scaffolding and you need an iron rod for a connection between two points. You have to know how much the rod weighs. Since its length and material are fixed (constant), the weight of the rod will depend only on how thick it is.

If you make the rod twice as thick, will it weigh twice as much? No, indeed; it will weigh four times as much. The weight of the rod depends on its cross-section area, which is equal to πr^2. Changing r to $2r$ changes the cross-section area to $\pi(2r)^2$, which is $4\pi r^2$. For a set of rods all of the same length and made of the same material, weight is proportional to the *square* of the radius, or

$$w \propto r^2$$

and the proportionality constant is

$$\frac{w}{r^2} = \text{constant}$$

Look at Sample Problem A1-1 to see how a proportionality of this kind can be used.

Sample Problem

A1-1 An iron rod 5.0 mm thick weighs 2.40 lb. What is the weight of an iron rod of the same length if it is 8.5 mm thick?

Solution Since w/T^2 is the same for both rods,

$$\frac{2.40 \text{ lb}}{(5.0 \text{ mm})^2} = \frac{w}{(8.5 \text{ mm})^2}$$

which is easily changed to

$$w = 2.40 \text{ lb} \left(\frac{8.5 \text{ mm}}{5.0 \text{ mm}}\right)^2$$

$$w = 6.9 \text{ lb}$$

We will also meet *inverse* ratios, in which one variable gets larger as the other gets smaller. For example, suppose you want to buy tiles to redo your kitchen floor. You will have two variables to deal with: the number of tiles you need, and the size of each tile. The bigger the tiles you decide on, the fewer you will need. This can be written as

$$N \propto \frac{1}{s}$$

which is read, "The number of tiles is inversely proportional to the tile size." In this kind of relationship, the constant of proportionality is the *product* of the two variables:

$$Ns = A$$

where A is the total area of the floor.

Study Sample Problem A1-2 to see how an inverse proportionality is used in calculation.

Sample Problem

A1-2 The time it takes to travel a given distance is inversely proportional to the speed. If it takes 12 min to make a trip at 20 mi/hr, how long will it take to make the same trip at 35 mi/hr?

Solution Since time is inversely proportional to speed, the product of speed and time is constant. Therefore

$$(20 \text{ mi/hr})(12 \text{ min}) = (35 \text{ mi/hr})t$$

therefore

$$t = (12 \text{ min})(\frac{20 \text{ mi/hr}}{35 \text{ mi/hr}})$$

$$t = 6.9 \text{ min}$$

You will meet other kinds of proportionalities as we go along, but these will do for now.

Core Concept

There are many kinds of proportionalities; the proportionality constant is the ratio or the product of two variables, which may be raised to a given power.

Try These 1. An iron rod 0.20 in. thick weighs 2.5 lb. How much will another rod weigh if it is made of the same material, is just as long, and is 0.60 in. thick?

2. A floor can be tiled with 1,500 tiles, each with an area of 12 in.^{2}. How many tiles would be needed if you decided to use tiles whose area is 30 in.^{2} instead?

A1-4. Right Triangles and Trigonometric Functions

Triangles in which one angle is 90° have certain special properties that make it especially simple to use them in calculations. The sum of the acute angles, for example, is exactly 90°. Also, thanks to Pythagoras, we know that the squares of the two legs add up to the square of the hypotenuse. In this analysis, we refer to the various parts of a right triangle, and they are labeled in Figure A1-1.

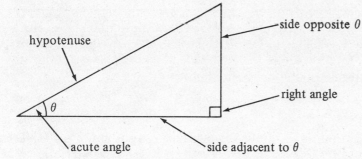

hypotenuse

acute angle

side opposite θ

right angle

side adjacent to θ

θ

FIG. A1-1

Here is a most useful theorem: In a group of right triangles all having the same acute angle, the sides are in proportion. This is illustrated in Figure A1-2. Since the angle θ is the same in all three triangles, it follows that

$$\frac{a}{b} = \frac{a'}{b'} = \frac{a''}{b''} \text{ ; also } \frac{c}{b} = \frac{c'}{b'} = \frac{c''}{b''} \text{ etc.}$$

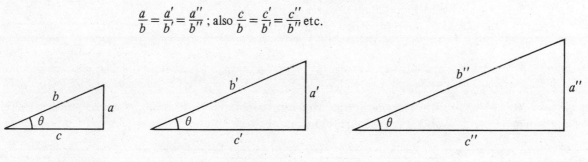

FIG. A1 2

The ratio $\frac{a}{b}$ is the same for all right triangles having the acute angle θ. It is called the sine of the angle θ, and it can be looked up in tables of trigonometric functions. It is the ratio of the side opposite θ to the hypotenuse of the triangle.

The similar triangle theorem can be used to measure the height of a tree, as in Figure A1-3. Put a vertical stick into the ground and sight past it from ground level to the top of the tree. Mark the stick where it intersects your line of sight. Then, since the triangles are similar:

$$\frac{a}{b} = \frac{A}{B}$$

And A can be found by measuring a, b and B, as in Sample Problem A1-3.

You do not really need the stick. You could use a protractor and measure the angle that your line of sight makes with the ground. The ratio a/b is the tangent of this angle.

FIG. A1-3

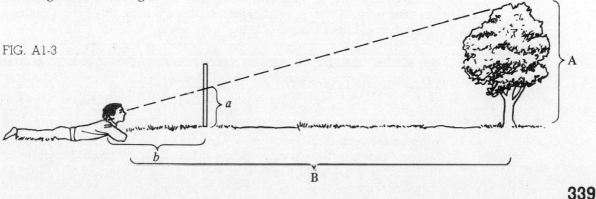

339
THE MATHEMATICS YOU NEED

Sample Problem

A1-3 In measuring a tree, it is found that the angle θ is 35° and the distance B, from eye to tree, is 52 ft. How high is the tree?

Solution By the definition of the tangent,

$$\tan 35° = \frac{A}{52 \text{ ft}}$$

so $A = (52 \text{ ft})(\tan 35°)$. From trigonometry tables, $\tan 35° = 0.7002$. Therefore, $A = 0.7002 \times 52 \text{ ft} = 36 \text{ ft}$.

Trigonometric tables and scientific calculators give values of three trigonometric functions. If θ is an acute angle of a right triangle,

$$\text{the sine of } \theta \ (\sin \theta) = \frac{\text{side opposite } \theta}{\text{hypotenuse}}$$

$$\text{the cosine of } \theta \ (\cos \theta) = \frac{\text{side adjacent to } \theta}{\text{hypotenuse}}$$

$$\text{the tangent of } \theta \ (\tan \theta) = \frac{\text{side opposite } \theta}{\text{side adjacent to } \theta}$$

The trigonometric tables in Appendix 2 will enable you to solve for sides and angles of triangles as needed.

Core Concept

The ratio of any two sides of a right triangle is a standard trigonometry function that can be looked up in trigonometry tables.

Try These 1. What is the angle in a right triangle if the side opposite it is 35 cm and the hypotenuse is 50 cm?

2. What is the hypotenuse of a right triangle if one angle is 65° and the side adjacent to it is 30 ft?

3. What angle in a right triangle has the side opposite it equal to 20 in and the side adjacent to it equal to 35 in?

A1-5. Manipulating Equations

Sometimes an algebraic equation can tell you what you want to know more conveniently if you first change it around a little. Follow one simple rule in altering equations: Do the same thing to both sides of the equation. You can add, subtract, multiply, divide, take the square root, or whatever, provided you treat both sides of the equation exactly alike.

Often, you want to arrange to have one constant or variable alone on one side of the equation. Getting v alone on the left is called "solving the equation for v." Suppose you have the equation

$$A + d = 6r$$

and you want to solve it for A. All you have to do is subtract d from both sides of the equation:

$$A + d - d = 6r - d$$

$$A = 6r - d$$

The quantity you want to solve for may be part of a term; that is, it may exist in the equation multiplied or divided by something else. For example, you want to solve this equation for P:

$$6P + 9m - R$$

The first step is to put the whole term all by itself on the left. Therefore, you start by subtracting $9m$ from both sides. This gives

$$6P = R - 9m$$

Now you are ready to isolate P. Do this by dividing both sides of the equation by 6:

$$\frac{6P}{6} = \frac{R - 9m}{6}$$

$$P = \frac{R - 9m}{6}$$

Which operations do you perform first? Follow this rule: *first* add or subtract terms; *then* multiply or divide; and *finally*, take roots or raise to powers.

Look at Sample Problems A1-4 and A1-5 for examples of how to solve equations for one of the given quantities.

Sample Problem

A1-4 Solve for P: $P/4 - 3t = q$.

Solution   First, add $3t$ to both sides. This gives

$$\frac{P}{4} = q + 3t$$

Then multiply all terms by 4, to get

$$P = 4q + 12t$$

Sample Problem

A1-5 Solve for r: $3r^2 + 9 = 12s$.

Solution   First, subtract 9 from both sides to get

$$3r^2 = 12s - 9$$

Then, divide all terms by 3:

$$r^2 = 4s - 3$$

Finally, take the square root of both sides:

$$r = \sqrt{4s - 3}$$

Core Concept

To transform an equation, perform the same process on both sides, in this sequence: add or subtract; multiply or divide; take roots or powers.

Try These   1. Solve for d: $d/4 - 6 = 22a$.
2. Solve for r: $3r + t = 5t + 7$.
3. Solve for Y: $aY + 3t^2 = 44$.
4. Solve for p: $p^2 + 6 = r - 3$.

A1-6. Dimensionality

Every measurement must be expressed in two parts: a number and a label. If you tell someone that your height is 160, she will know nothing unless you tell her whether you measured in inches, centimeters, feet, or miles.

If your height is 160 centimeters, are you taller or shorter than your friend, who tells you he is 64 inches tall? To find out, you must either convert your height to inches or your friend's to centimeters. You can do this if you happen to know that

$$2.54 \text{ centimeters} = 1 \text{ inch}$$

Now, let's see, do you multiply the inches or divide them to change them to centimeters? Making the wrong choice in this situation is a very common error, which can be avoided absolutely by following a simple rule: *Treat the labels exactly like algebraic quantities.*

Here's how it works: Since 2.54 centimeters = 1 inch, it follows that

$$\frac{2.54 \text{ cm}}{1 \text{ inch}} = 1 = \frac{1 \text{ inch}}{2.54 \text{ cm}}$$

Now everyone knows that you can multiply anything by 1 without changing it. So you can multiply your friend's height in inches by either of the fractions above without breaking any rules. Choose the one that eliminates the inches and leaves your answer in centimeters:

$$64 \text{ inches} \times \frac{2.54 \text{ cm}}{1 \text{ inch}} = 162.56 \text{ cm}$$

Don't antagonize the guy—he's bigger than you are. And it does not help to convert your height to inches instead:

$$160 \text{ cm} \times \frac{1 \text{ inch}}{2.54 \text{ cm}} = 63.0 \text{ inches}$$

Sample Problem A1-6 is another example of how this method is used.

Sample Problem

A1-6 How many pounds is 45 oz? There are 16 oz to a pound.

Solution

$$(45 \text{ oz})\left(\frac{1 \text{ lb}}{16 \text{ oz}}\right) = 2.8 \text{ lb}$$

You can add your height to your friend's. If he stands on your head, the distance from the top of his head to the floor is 160 cm + 163 cm = 323 cm, or 63 inches + 64 inches = 127 inches. You can also subtract the two values. He is 1 inch, or 2.54 centimeters taller than you are.

Which is more, 125 pounds or 6 feet? This question has no answer. Pounds measure weight, and feet measure length or distance. There is no way to convert pounds to feet, or vice versa. Furthermore, you cannot add or subtract pounds and feet. The only way you can add, subtract, or compare two quantities is by expressing them in the same units. And this can be done only if they measure the same kind of quantity. Then they are said to have the same *dimensionality*.

Only quantities with the same dimensionality can be compared, added, or subtracted by converting units to express the quantities in the same units.

Try These 1. There are 16 oz to a pound. What is the sum of 3 lb and 27 oz, in ounces?

2. A board is 57 in long. If you cut off 65 cm, how many inches are left?

A1-7. Creating New Labels

While it is impossible to add pounds to feet, you can multiply or divide them. In fact, *any* quantity can be multiplied or divided by any other. When you perform this kind of arithmetic, you create a quantity with a new label, a new dimensionality.

For example, suppose you have a rectangular table with a top measuring 30 inches long and 18 inches wide. What is its surface area? The area of a rectangle is the length times the width. Treating the labels like algebraic quantities, the solution looks like this:

$$A = (30 \text{ in})(18 \text{ in}) = 540 \text{ in}^2$$

The answer is read "five hundred forty square inches." The in^2 is a new label that cannot be compared with, added to, or subtracted from inches. It has the dimensionality of area and can be added to square centimeters, acres, or any other unit of area.

Similarly, the cubic inch (in^3) is a measure of volume and can be converted to cubic centimeters, quarts, and so on. Sample Problem A1-7 shows how this can be used.

Sample Problem

A1-7 How many gallons of water are there in a rectangular aquarium 20 in long and 12 in wide if the water is 9.5 in deep? There are 231 in^3 to a gallon.

Solution Volume is length × width × height:

$$(20 \text{ in})(12 \text{ in})(9.5 \text{ in}) = 2,280 \text{ in}^3$$

Now convert the units:

$$(2,280 \text{ in}^3)\left(\frac{1 \text{ gal}}{231 \text{ in}^3}\right) = 9.9 \text{ gal}$$

A ratio formed by dividing one quantity by another also generates a new label. What, for example, is the linear density of a wire if a piece 550 yards long weighs 210 pounds?

$$d = \frac{w}{L} = \frac{210 \text{ lb}}{550 \text{ yd}} = 0.38 \text{ lb/yd}$$

The answer is read "zero point three eight pounds per yard." The label of a ratio is always written as a fraction. You will find another example in Sample Problem A1-8.

Sample Problem

A1-8 What is the average speed of a car that travels 240 mi in 6.2 hr?

Solution   Average speed is distance divided by time:

$$\frac{240 \text{ mi}}{6.2 \text{ h}} = 38.7 \text{ mi/hr}$$

Multiplication also produces new labels. Your public utility, for example, decides how much to charge you for the electric energy you use by multiplying your power consumption by the length of time you have it turned on. Suppose you burn a 100-watt bulb for 6 hours. How much energy do you use? Solution:

$$(100 \text{ watts})(6 \text{ hours}) = 600 \text{ watt} \cdot \text{hours}$$

The answer is read as "six hundred watt-hours." You would be charged the same amount if you burned a 60-watt bulb for 10 hours, or a 200-watt bulb for 3 hours.

Core Concept

Labeled quantities can be multiplied and divided, and the labels follow the ordinary rules of algebra.

Try These   1. Divide 125 lb by 30 in^2.
2. Multiply 20 ft by 16 lb.
3. It takes 100 cm to make a meter. Take the cube of 100 cm to find the number of cubic centimeters in a cubic meter.

A1-8. Le Système Internationale d'Unités

The name of this system is in French because the International Bureau of Standards has its headquarters in Paris, where the metric system was invented. The *Système Internationale* (SI) used throughout this book, is a special simplification of the metric system.

Traditional systems of measurement have evolved through the centuries and are different in every country. The only thing they have in common is that they make no sense at all. No one in his or her right mind would sit down and invent the English system, in which there are 437.5 grains to the ounce, 16 ounces to the pound, 14 pounds to the stone, 8 stone to a hundredweight, and 20 hundredweight to a long ton.

In the metric system, which *was* invented, things are much simpler. All the multipliers are powers of 10. There are 100 cm to a meter, 1,000 cm^3 to a liter, 1,000,000 micrograms to a gram, and so on.

Even simpler is the SI, which uses no multipliers at all. Some are made available to those who feel they need them, but they are not necessary. The system is extremely simple because, as long as you stick to it, it is never necessary to convert units.

In the SI, there are exactly seven fundamental units. Dozens of other units are derived from these by multiplication and division of these seven, but without introducing any numbers at all. These are the fundamental units:

the meter (m) for distance
the kilogram (kg) for mass
the second (s) for time
the kelvin (K) for temperature
the ampere (A) for electric current
the candela (cd) for luminous intensity
the mole (mol) for number of particles

Here are a few examples of derived SI units:

the square meter (m^2) for area
the cubic meter (m^3) for volume
the kilogram per cubic meter (kg/m^3) for mass density
the coulomb (C) or ampere-second ($A \cdot s$) for charge
the newton (N) or kilogram-meter per second squared ($kg \cdot m/s^2$) for
force

The system allows certain multiples and submultiples to be used also. They are limited to powers of 10^3 only. Here are the ones we will use:

nano = 10^{-9}; example: 1 nanometer (nm) = 10^{-9} m
micro = 10^{-6}; example: 1 microcoulomb (μC) = 10^{-6} C
milli = 10^{-3}; example: 1 milliampere (mA) = 10^{-3} A
kilo = 10^3; example: 1 kilowatt (kW) = 10^3 W
mega = 10^6; example: 1 megajoule (MJ) = 10^6 J
giga = 10^9; example: 1 gigavolt (GV) $-$ 10^9 V

You will find that there is one metric unit that always refuses to fit this pattern, but is used all the time anyway from pure force of long habit:

1 centimeter (cm) = 10^{-2} m

While this system is simple, it is not always convenient. Some of the units are too large or too small for practical use, and we still like the old, familiar ones. Don't ask your grocer for 0.00091 m^3 of milk; tell him you want a quart. Soon enough you will be getting a slightly larger bottle when you ask for a liter.

Core Concept

The SI is a convenient system of units for scientific work in which each dimensionality is expressed in only one unit.

Try This The rate of flow of water in a pipe can be expressed as the amount of mass per unit of time. What would be the correct SI unit for this rate?

A1-9. Scientific Notation

If you want to express all distances in the SI unit, the meter, you will run into trouble when you write the distance to the nearest star or the diameter of an atomic nucleus. In either case, you will need 15 zeroes—at the end of the number for the star, and at the beginning of the number for the nucleus.

The only function of all these zeroes is to indicate where the decimal point belongs. But there is a better way to do this. It is called *scientific notation*. We write every number with the decimal point just after the first nonzero digit and then multiply by some power of 10 to show where it really belongs. For example, instead of writing a number like 15,200, we write

$$15,200 = 1.52 \times 10,000 = 1.52 \times 10^4$$

This may seem like a lot of trouble, but it is extremely convenient once you get used to it, and it is really the only way to deal with very large and very small numbers.

To put a number in scientific notation, just count the number of places to the left you have to move the decimal point to place it after the first digit. That number becomes the exponent of 10. In the example given above, the decimal point was moved four places to the left.

If the decimal point is somewhere else, and powers of 10 are already given, just move it to the left and add the number of places to the exponent of 10. For an example, see Sample Problem A1-9.

Sample Problem

A1-9 Put the number 351.7×10^4 into standard form.

Solution The decimal point must be moved two places to the left, which amounts to dividing by 100. Therefore, you must also multiply by 100 by increasing the exponent of 10 by 2; the answer is 3.517×10^6.

For small numbers—less than 1—you will have to move the decimal point to the right to get it where you want it. Then you *subtract* the number of places moved from the exponent of 10. For example:

$$0.0000162 = 1.62 \times 10^{-5}$$

If the number already has a power-of-ten multiplier, follow the same rule:

$$0.135 \times 10^4 = 1.35 \times 10^3$$

Core Concept

To put a number into scientific notation, write it with the decimal point after the first nonzero digit and multiply by the appropriate power of 10.

Try These Put the following numbers into the standard form of scientific notation:

1. 14 million
2. 234×10^{-3}
3. 4,190
4. 0.000000398.

A1-10. Arithmetic in Scientific Notation

One benefit of scientific notation is that it is easier to add exponents than to count zeroes. If you have to multiply 23,400 by 17,000,000 you can greatly simplify the process by first putting the numbers into scientific notation. Then the problem looks like this:

$$(2.34 \times 10^4)(1.7 \times 10^7)$$

Then all you have to do is multiply the numbers in the usual way and add the exponents of 10. The answer is therefore 3.978×10^{11}. See Sample Problem A1-10 for another example.

Sample Problem
A1-10 How much is $(5.6 \times 10^4)(7.9 \times 10^2)$?

Solution Multiply the numbers and add the exponents to get 44.2×10^6; shift the decimal point, and the answer is 4.42×10^7.

The same rule holds if the exponents are negative:

$$(2.91 \times 10^{-3})(5.0 \times 10^{-6}) = 14.55 \times 10^{-9}$$

Now shift the decimal point to its proper position, remembering that a move to the left requires you to increase the exponent of 10. Was your answer 1.455×10^{-8}?

To divide, just divide the numbers in the usual way and *subtract* the exponent of 10 in the denominator from that in the numerator:

$$\frac{4.5 \times 10^8}{1.7 \times 10^3} = 2.6 \times 10^5$$

When the exponents are negative, watch out! Remember that subtracting a negative number is like adding!

$$\frac{2.8 \times 10^{-2}}{1.1 \times 10^{-6}} = 2.5 \times 10^4$$

If that surprises you, perhaps you have forgotten that $-2 - (-6) = +4$.

You can add and subtract numbers in scientific notation only if they have the same exponent; the exponent of the sum or difference is then the same as for the numbers you are combining. For example, you know that 4.5 million plus 2.2 million is 6.7 million. In scientific notation, it looks like this:

$$(4.5 \times 10^6) + (2.2 \times 10^6) = 6.7 \times 10^6$$

If you have to add numbers that do not have the same exponent of 10, you first have to shift the decimal point. See Sample Problem A1-11.

Sample Problem
A1-11 Subtract 4.17×10^{-8} from 2.25×10^{-7}.

Solution Shift the decimal point of one of the numbers until both have the same exponent of 10; then subtract:

$$(2.25 \times 10^{-7}) - (0.417 \times 10^{-7}) = 1.83 \times 10^{-7}$$

When multiplying numbers in scientific notation, add the exponents of 10; when dividing, subtract the exponents of 10.

Try These 1. Multiply 3.5×10^2 by 4.6×10^4.
                2. Multiply 5.7×10^{-5} by 6.0×10^8.
                3. Divide 2.6×10^8 by 1.3×10^{12}.
                4. Add 3.75×10^4 to 1.91×10^3.
                5. Subtract 5.17×10^5 from 3.95×10^4.

A1-11. Graphs

An equation is one means of expressing the relationship between two variables. A graph is a second method. Graphs are used when the relationship is too complex to be expressed by an equation, or when it is desirable to see the relationship in visual terms.

In the most common kind of graph, values of the variables are plotted on a set of axes, as in Figure A1-4. The axes are perpendicular to each other, and each is marked off as a number line. The numbers represent the values of the variables.

In many relationships, one of the variables depends on the value of the other. In the relationship between the distance a car travels and the amount of gasoline used, for example, the amount of gasoline used *depends on* how far you drive. The gasoline used is called the *dependent variable*, and the distance traveled is the *independent variable*. On a graph, the independent variable is always plotted along the horizontal axis, and the dependent variable on the vertical axis.

In plotting a graph, each data point is represented as a little + mark. One data point might be obtained, for example, by determining that, in going 50 miles, the car uses 1.6 gallons of gasoline. The + mark farthest to the left in Figure A1-4 represents this pair of readings. It was plotted by placing the point at the 50-mile position to the right of the origin (the zero point), and 1.6 gallons up from the origin on the vertical axis.

FIG. A1-4

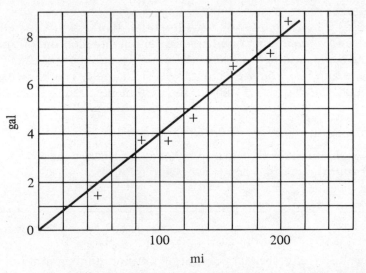

The other data points on the graph represent other readings. The second point shows that, in traveling 85 miles, the car used 3.5 gallons of gasoline, and so on. Because of uncertainties in measurement and unexplainable variations in the conditions of the experiment, the points will never lie exactly on a straight line, or on any smooth curve. For the data on this graph, we draw a straight line, coming as close as possible to all the data points. This line represents our best estimate of the way the amount of gasoline used depends on the distance the car is driven. If we wanted to know the amount of gasoline used in traveling 180 miles, for example, we could look at the graph at the point where the line crosses the 180-mile mark on the distance line. It corresponds to 7.1 gallons on the gasoline axis.

A graph is a useful method of expressing the relationship between an independent and a dependent variable.

Try This From the graph, how far can the car travel on 5 gal of gasoline?

APPENDIX

Values of the Trigonometric Functions

Angle	Sin	Cos	Tan	Angle	Sin	Cos	Tan
1°	.0175	.9998	.0175	21°	.3584	.9336	.3839
2°	.0349	.9994	.0349	22°	.3746	.9272	.4040
3°	.0523	.9986	.0524	23°	.3907	.9205	.4245
4°	.0698	.9976	.0699	24°	.4067	.9135	.4452
5°	.0872	.9962	.0875	25°	.4226	.9063	.4663
6°	.1045	.9945	.1051	26°	.4384	.8988	.4877
7°	.1219	.9925	.1228	27°	.4540	.8910	.5095
8°	.1392	.9903	.1405	28°	.4695	.8829	.5317
9°	.1564	.9877	.1584	29°	.4848	.8746	.5543
10°	.1736	.9848	.1763	30°	.5000	.8660	.5774
11°	.1908	.9816	.1944	31°	.5150	.8572	.6009
12°	.2079	.9781	.2126	32°	.5299	.8480	.6249
13°	.2250	.9744	.2309	33°	.5446	.8387	.6494
14°	.2419	.9703	.2493	34°	.5592	.8290	.6745
15°	.2588	.9659	.2679	35°	.5736	.8192	.7002
16°	.2756	.9613	.2867	36°	.5878	.8090	.7265
17°	.2924	.9563	.3057	37°	.6018	.7986	.7536
18°	.3090	.9511	.3249	38°	.6157	.7880	.7813
19°	.3256	.9455	.3443	39°	.6293	.7771	.8098
20°	.3420	.9397	.3640	40°	.6428	.7660	.8391

Angle	Sin	Cos	Tan	Angle	Sin	Cos	Tan
41°	.6561	.7547	.8693	66°	.9135	.4067	2.2460
42°	.6691	.7431	.9004	67°	.9205	.3907	2.3559
43°	.6820	.7314	.9325	68°	.9272	.3746	2.4751
44°	.6947	.7193	.9657	69°	.9336	.3584	2.6051
45°	.7071	.7071	1.0000	70°	.9397	.3420	2.7475
46°	.7193	.6947	1.0355	71°	.9455	.3256	2.9042
47°	.7314	.6820	1.0724	72°	.9511	.3090	3.0777
48°	.7431	.6691	1.1106	73°	.9563	.2924	3.2709
49°	.7547	.6561	1.1504	74°	.9613	.2756	3.4874
50°	.7660	.6428	1.1918	75°	.9659	.2588	3.7321
51°	.7771	.6293	1.2349	76°	.9703	.2419	4.0108
52°	.7880	.6157	1.2799	77°	.9744	.2250	4.3315
53°	.7986	.6018	1.3270	78°	.9781	.2079	4.7046
54°	.8090	.5878	1.3764	79°	.9816	.1908	5.1446
55°	.8192	.5736	1.4281	80°	.9848	.1736	5.6713
56°	.8290	.5592	1.4826	81°	.9877	.1564	6.3138
57°	.8387	.5446	1.5399	82°	.9903	.1392	7.1154
58°	.8480	.5299	1.6003	83°	.9925	.1219	8.1443
59°	.8572	.5150	1.6643	84°	.9945	.1045	9.5144
60°	.8660	.5000	1.7321	85°	.9962	.0872	11.4301
61°	.8746	.4848	1.8040	86°	.9976	.0698	14.3007
62°	.8829	.4695	1.8807	87°	.9986	.0523	19.0811
63°	.8910	.4540	1.9626	88°	.9994	.0349	28.6363
64°	.8988	.4384	2.0503	89°	.9998	.0175	57.2900
65°	.9063	.4226	2.1445	90°	1.0000	.0000	

APPENDIX 3

Properties of Selected Nuclides

Atomic number	Name	Symbol	Atomic mass number	Atomic mass (Dl)	Half-life
0	Electron	$_{-1}e$	0	5.49×10^{-4}	Stable
0	Neutron	n	1	1.008665	11 minutes
1	Hydrogen	H	1	1.007825	Stable
	(deuterium)		2	2.0140	Stable
	(tritium)		3	3.01605	12 years
2	Helium	He	3	3.01603	Stable
			4	4.00260	Stable
3	Lithium	Li	6	6.01512	Stable
			7	7.01600	Stable
4	Beryllium	Be	9	9.01218	Stable
5	Boron	B	10	10.0129	Stable
			11	11.00931	Stable
6	Carbon	C	12	12 exactly	Stable
			14	14.0032	5730 years
7	Nitrogen	N	13	13.0057	10 minutes
			14	14.00307	Stable
9	Fluorine	F	19	18.99840	Stable
10	Neon	Ne	22	21.99138	Stable
11	Sodium	Na	22	21.9944	2.6 years
13	Aluminum	Al	27	26.98153	Stable
14	Silicon	Si	30	29.97376	Stable

Atomic number	Name	Symbol	Atomic mass number	Atomic mass (Dl)	Half-life
15	Phosphorus	P	30	29.9783	2.5 minutes
17	Chlorine	Cl	35	34.96885	Stable
			37	36.9659	Stable
19	Potassium	K	39	38.96371	Stable
20	Calcium	Ca	42	41.95863	Stable
26	Iron	Fe	54	53.9396	Stable
			56	55.9349	Stable
36	Krypton	Kr	92	?	3.0 seconds
40	Zirconium	Zr	95	?	65 days
54	Xenon	Xe	142	?	1.5 seconds
56	Barium	Ba	141	?	18 minutes
82	Lead	Pb	206	205.9745	Stable
			210	209.9848	21 years
			214	213.9998	27 minutes
83	Bismuth	Bi	210	209.9840	5 days
84	Polonium	Po	214	213.9952	1.6×10^{-4} second
			218	218.0089	3 minutes
86	Radon	Rn	222	222.0154	38 seconds
88	Radium	Ra	226	226.0254	1,600 years
90	Thorium	Th	230	230.0331	80,000 years
			234	234.0436	24 days
91	Palladium	Pa	234	234.0433	6.7 hours
92	Uranium	U	233	233.0395	160,000 years
			235	235.0439	710 million years
			238	238.0508	4.5 billion years
			239	239.05433	24 minutes
93	Neptunium	Np	239	239.05295	2.4 days
94	Plutonium	Pu	239	239.0522	24,000 years

APPENDIX

4

Physical Constants

Acceleration due to gravity on earth (average): $g_o = 9.81$ m/s^2

Atomic mass unit: 6.023×10^{26} daltons = 1 kg

Boltzmann gas constant: $k = 1.380 \times 10^{-23}$ J/mlc·K

Electric constant of free space: $k = 8.987 \times 10^9$ N·m^2/C^2

Elementary unit of charge: $e = 1.602 \times 10^{-19}$ C

Magnetic constant of free space: $k' = 10^{-7}$ N/A^2 exactly

Mass of the earth = 5.976×10^{24} kg

Mass–energy conversion: 1 dalton = 931 Mev

Planck's quantum constant: $h = 6.624 \times 10^{-34}$ J·s
$$= 4.135 \times 10^{-15} \text{ ev·s}$$

Radius of the earth (average) = 6,370 km

Rest mass of electron = 9.110×10^{-31} kg
$$= 5.486 \times 10^{-4} \text{ Dl}$$
$$= 0.511 \text{ Mev}$$

Rest mass of proton = 1.673×10^{-27} kg

Speed of electromagnetic wave in vacuum: $c = 2.998 \times 10^8$ m/s

Universal gravitation constant: $G = 6.672 \times 10^{-11}$ N·m^2/kg^2

Answers to Try This

1-1. 3 min 30 s is 210 s:

$$v_{av} = \frac{\Delta s}{\Delta t} = \frac{1,500 \text{ m}}{210 \text{ s}} = 7.1 \text{ m/s}$$

1-2. Since east and south are at right angles to each other, their sum is the hypotenuse of a right triangle. Using the theorem of Pythagoras, we have $3.5^2 + 5.8^2 = s^2$, so $s = 6.8$ m. The direction will be south of east at an angle whose tangent is 5.8/3.5, or 59°.

1-3. To compensate for that easterly wind, he will have to head the plane in a direction east of north. His air velocity and the eastward wind velocity produce a sum directly north. He will achieve this if the angle has a sine equal to 110/620, so his heading is north 10° east.

1-4. a. Since the angle is 22° with the horizontal, the horizontal component is (40 mi/hr)(cos 22°) = 37 mi/hr.
  b. The other rectangular component is (40 mi/hr)(sin 22°) = 15 mi/hr.

1-5. Acceleration is change of speed per unit time; the change of velocity is (2,840 − 2,800) = 40 m/s. Then

$$a = \frac{\Delta v}{\Delta t} = \frac{40 \text{ m/s}}{25 \text{ s}} = 1.6 \text{ m/s}^2$$

1-6. Between 3 and 6 s, $\Delta t = (6 \text{ s} - 3 \text{ s}) = 3 \text{ s}$, and the velocity increases from 9.5 m/s to 14 m/s; $\Delta v = 4.5$ m/s. Therefore

$$a = \frac{\Delta v}{\Delta t} = \frac{4.5 \text{ m/s}}{3.0 \text{ s}} = 1.5 \text{ m/s}^2$$

1-7. **1.** Since it starts from rest and accelerates uniformly, $s = \frac{1}{2}at^2$ from which

$$a = \frac{2s}{t^2} = \frac{2(360 \text{ m})}{(5.0 \text{ s})^2} = 29 \text{ m/s}^2$$

    2. Again, acceleration is uniform from rest, so $v = \sqrt{2as} = \sqrt{2(2.8 \text{ m/s}^2)(220 \text{ m})} = 35$ m/s.

1-8. **a.** Starting from rest with uniform acceleration: $s = \frac{1}{2}at^2 = \frac{1}{2}(9.8 \text{ m/s}^2)(2.3 \text{ s})^2 = 26$ m.

    b. $v = at = (9.8 \text{ m/s}^2)(2.3 \text{ s}) = 23$ m/s.

1-9. **1.** Down; when slowing down, a is opposite v.
    2. West; when speeding up, a and v are in same direction.
    3. West.
    4. East, in the direction of the turn.

1-10. **a.** Since it travels horizontally at a uniform speed, $v = \Delta s/\Delta t$, so $\Delta s = v\,\Delta t = (20 \text{ m/s})(2.8 \text{ s}) = 56$ m.
    b. Accelerating downward at 9.8 m/s^2, it took 2.8 s to reach the ground: $s = \frac{1}{2}at^2 = \frac{1}{2}(9.8 \text{ m/s}^2)(2.8 \text{ s})^2 = 38$ m.

1-11. **a.** He travels the circumference of the circle in 30 s, so

$$v = \frac{2\pi r}{T} = \frac{2\pi(150 \text{ m})}{30 \text{ s}} = 31 \text{ m/s}$$

At the southernmost point, he is going west.

    b. $$a = \frac{v^2}{r} = \frac{(31.4 \text{ m/s})^2}{150 \text{ m}} = 6.6 \text{ m/s}^2$$

The direction of the acceleration is toward the center of the circle, northward.

1-12. The acceleration is due to gravity, since it is in free fall; $a = v^2/r$, so $v = \sqrt{ar} = \sqrt{(1.67 \text{ m/s}^2)(3.5 \times 10^6 \text{ m})} = 2{,}400$ m/s.

Chapter 2

2-1. **a.** Air pushes the ball forward to keep it in motion.
    b. Air retards the natural forward motion of the ball.

2-2. The motor's force must just balance the viscous drag of the water, which increases with speed.

2-3. $w = mg$, so

$$m = \frac{w}{g} = \frac{340 \text{ N}}{9.8 \text{ m/s}^2} = 35 \text{ kg}$$

2-4. 350 pounds.

2-5. 1. Its weight is $mg = (6 \text{ kg})(9.8 \text{ m/s}^2) = 59$ N. Since this is less than the buoyancy, the object will rise.

2. (a) 6; (b) 6; (c) 1; (d) 5; (e) 2; (f) 8.

2-6. The force of your tires pulling the earth along with the car.

2-7. a. The friction is the force that maintains horizontal equilibrium, so it must equal the scale reading, 4 N.

b. The elastic recoil of the tabletop maintains vertical equilibrium, so it must be equal to the weight of the brick; $w = mg = (1.6 \text{ kg})(9.8 \text{ m/s}^2) = 16$ N.

2-8. a. Since the 30° is the angle with the horizontal, the horizontal component is $(160 \text{ N})(\cos 30°) = 139$ N.

b. The vertical component is $(160 \text{ N})(\sin 30°) = 80$ N.

2-9. First, find the horizontal and vertical components of the tension in the string. Since the 20° angle is with the vertical, $T_{\text{vert}} = (16 \text{ N})(\cos 20°) = 15.0$ N. The horizontal component, acting toward the left, is $(16 \text{ N})(\sin 20°) = 5.5$ N.

Now set the upward forces equal to the downward forces: $B = 15.0 \text{ N} + 25 \text{ N} = 40$ N. And the forces toward the right equal those toward the left: $W = 5.5$ N.

2-10. a. The angle θ in Figure 2-13 is equal to the slope of the surface, which is 20°. Therefore, the component of weight parallel to the plane is $(120 \text{ kg})(9.8 \text{ m/s}^2)(\sin 20°) = 400$ N, and this is just balanced by the tension in the rope.

b. The component of weight perpendicular to the plane is $(120 \text{ kg})(9.8 \text{ m/s}^2)(\cos 20°) = 1100$ N.

2-11. The torque is her weight times the radial distance to the pivot: $\tau = (45 \text{ kg})(9.8 \text{ m/s}^2)(2.5 \text{ m}) = 1{,}100$ N·m.

2-12. Clockwise and counterclockwise torques around any selected pivot must be equal. All distances must be measured from that pivot. In this case, the rock is the most convenient pivot. The woman, her weight pushing (say) clockwise, is 1.0 m from the rock; then the scale must be pushing counterclockwise at a distance of 3.5 m from the rock. Therefore $w(1.0 \text{ m}) = (120 \text{ N})(3.5 \text{ m})$, and $w = 420$ N.

Chapter 3

3-1. a. Solid, liquid, gas.

b. Liquid, solid.

c. Solid, liquid, gas.

d. Solid.

e. Solid, liquid.

f. Solid.

g. Solid, liquid, gas.

h. Solid, liquid, gas.

i. Solid.

3-2. $D = m/V$, so

$$V = \frac{m}{D} = \frac{650 \text{ g}}{0.62 \text{ g/cm}^3} = 1{,}050 \text{ cm}^3$$

3-3. $p = \dfrac{F}{A} = \dfrac{7.5 \text{ lb}}{3(0.05 \text{ in}^2)} = 50 \text{ lb/in}^2$.

3-4. $p = F/A$, so

$$F = pA = \left(14.7 \ \frac{\text{lb}}{\text{in}^2}\right)(3 \text{ ft})(6 \text{ ft})\left(\frac{12 \text{ in}}{\text{ft}}\right)^2 = 38{,}000 \text{ lb}$$

3-5. $p = hDg = (6 \text{ m})(0.9 \ \times 10^3 \text{ kg/m}^3)(9.8 \text{ m/s}^2) = 5.3 \times 10^4 \text{ N/m}^2$.

3-6. $p = (0.89)^5(1 \text{ atm}) = 0.56 \text{ atm}$.

3-7. The mass of fluid displaced is the difference between the scale reading in air and in fluid, which is 10 g. The volume of fluid displaced is 8 cm³. Therefore

$$D = \frac{m}{V} = \frac{10 \text{ g}}{8 \text{ cm}^3} = 1.3 \text{ g/cm}^3$$

3-8. The mass of the ice is equal to the mass of the displaced water. The volume of ice submerged is (2 cm)(2 cm)(1.7 cm) = 6.8 cm³, so the mass of displaced water is 6.8 g—and this is also the mass of the ice cube. The volume of the ice cube is 8 cm³. Therefore

$$D = \frac{m}{V} = \frac{6.8 \text{ g}}{8 \text{ cm}^3} = 0.85 \text{ g/cm}^3$$

3-9. Try putting soap in the water.

3-10. a. The parachute presents a broad surface to the air, and causes the air to flow around it turbulently, thus increasing the drag.
b. A fish is streamlined; the water flows in laminar fashion around its body.

3.11. In a swollen artery, blood flow slows down, so that the pressure rises. The artery could burst.

3.12. The extra drag slows the plane down, and the extra lift allows it to remain airborne at the slower speed.

Chapter 4

4-1. A force of 520 N for 10 s provides an impulse of 5,200 N·s; 130 N for 4 s gives 520 N·s, only $\dfrac{1}{10}$ as much. Since the first blast slowed the craft down by 50 m/s, the second slows it by an additional 5 m/s $\left(\dfrac{1}{10}\right.$ as much$\left.\right)$ to bring it to 295 m/s.

4-2. $F = ma$, so

$$m = \frac{F}{a} = \frac{130 \text{ N}}{3.0 \text{ m/s}^2} = 43 \text{ kg}$$

4-3. The net force is 8,500 N − 6,200 N = 2,300 N. Then

$$a = \frac{F}{m} = \frac{2{,}300 \text{ N}}{1{,}200 \text{ kg}} = 1.9 \text{ m/s}^2$$

4-4. a. Momentum = mv = (0.30 kg)(35 m/s) = 10.5 kg·m/s.
b. Impulse is the change in momentum; since the ball started at rest, its change in momentum = 10.5 kg·m/s, and the impulse is 10.5 N·s.

c. Impulse $= F \, \Delta t$;

$$F = \frac{10.5 \text{ N} \cdot \text{s}}{0.5 \text{ s}} = 21 \text{ N}$$

4-5. Total momentum before the collision must equal total momentum after. Taking east as the positive direction:

$$(2{,}000 \text{ kg})(22 \text{ m/s}) + (1{,}200 \text{ kg})(-30 \text{ m/s}) = (2{,}000 \text{ kg} + 1{,}200 \text{ kg})v$$

from which $v = 2.5$ m/s.

4-6. $F = ma = \dfrac{mv^2}{r} = \dfrac{(55 \text{ kg})(6.0 \text{ m/s})^2}{30 \text{ m}} = 66 \text{ N}.$

4-7. a. $g \propto 1/r^2$, so gr^2 is a constant. Therefore, for any values of r and corresponding values of g, $g_1 r_1{}^2 = g_2 r_2{}^2$. This gives

$$g_2 = g_1 \left(\frac{r_1}{r_2}\right)^2 = (3.3 \text{ m/s}^2)\left(\frac{3.4 \times 10^6 \text{ m}}{8.8 \times 10^6 \text{ m}}\right)^2$$

$$g_2 = 0.49 \text{ m/s}^2$$

Since the meteorite is in free fall, that is its acceleration.
b. $w = mg = (300 \text{ kg})(0.49 \text{ m/s}^2) = 147 \text{ N}.$

4-8. $F_{\text{grav}} = \dfrac{Gm_1 m_2}{r^2} = \dfrac{(6.67 \times 10^{-11} \text{ N} \cdot \text{m}^2/\text{kg}^2)(5{,}700 \text{ kg})(14{,}000 \text{ kg})}{(4.0 \times 10^5 \text{ m})^2}$

$$= 3.3 \times 10^{-14} \text{ N}$$

4-9. $F_{\text{grav}} = Gm_1 m_2 / r^2$, so $m_2 = Fr^2/Gm_1$. If you had a kilogram at the surface of Mars, its weight (force of gravity on it) would be $mg = (1 \text{ kg})(3.3 \text{ m/s}^2) = 3.3 \text{ N}$. Using these values, we have

$$m_2 = \frac{(3.3 \text{ N})(3.4 \times 10^6 \text{ m})^2}{(6.67 \times 10^{-11} \text{ N} \cdot \text{m}^2/\text{kg}^2)(1 \text{ kg})} = 5.7 \times 10^{23} \text{ kg}$$

4-10. As long as the spacecraft is in free fall, whether or not it is near a planet, everything in it falls in the same way and everything is weightless. If you see something fall within the spaceship, that object is not falling the same way the spaceship is. Therefore, the ship is not in free fall. Its engines are on.

4-11. First, we need the value of g:

$$\frac{g}{9.8 \text{ m/s}^2} = \left(\frac{R}{3R}\right)^2; \qquad g = 1.1 \text{ m/s}^2$$

Then, from Equation 1-11b, $v = \sqrt{ar} = \sqrt{(1.1 \text{ m/s}^2)(3)(6.4 \times 10^6 \text{ m})} = 4{,}600$ m/s.

4-12. The circumference of the circle he travels in is $2\pi r = 2\pi (3.5 \text{ m}) = 22.0$ m. Since he travels this distance in 6.0 s, his speed is 22.0 m/6.0 s = 3.67 m/s. Then the angular momentum is $mvr = (40 \text{ kg})(3.67 \text{ m/s})(3.5 \text{ m}) = 510 \text{ kg} \cdot \text{m}^2/\text{s}$.

Chapter 5

5-1. The weight is divided between two strands; therefore
  a. Effort = 140 lb/2 = 70 lb.
  b. Effort distance = 32 ft × 2 = 64 ft.

5-2. The weight of the bureau is divided five times, so
  a. Effort distance = 22 m × 5 = 110 m.
  b. Effort = $mg/5 = (80 \text{ kg})(9.8 \text{ m/s}^2)/5 = 157$ N.

5-3. The work done is the vertical force (weight) times the vertical distance, or $W = mg\,\Delta s\cos\theta = (20\text{ kg})(9.8\text{ m/s}^2)(5.0\text{ m})(\cos 22°) = 910\text{ J}$.

5-4. The work input is $(30\text{ m})(220\text{ N}) = 6{,}600\text{ J}$. $W_{out}/W_{in} = 0.80$; $W_{out} = 0.80 \times 6{,}600\text{ J} = 5{,}300\text{ J}$.

5-5. a. $\text{MA} = \dfrac{\text{effort arm}}{\text{load arm}} = \dfrac{18\text{ cm}}{5\text{ cm}} = 3.6$.

   b. $\text{MA} = \dfrac{\text{load}}{\text{effort}}$; $\text{effort} = \dfrac{\text{load}}{\text{MA}} = \dfrac{220\text{ N}}{3.6} = 61\text{ N}$.

   c. $\text{MA} = \dfrac{\text{effort distance}}{\text{load distance}}$; effort distance = MA × load distance

   Effort distance = $3.6 \times 0.5\text{ cm} = 1.8\text{ cm}$

   d. W_{in} = effort × effort distance = $61\text{ N} \times 1.8\text{ cm} = 110\text{ N·cm}$.

   e. W_{out} = load × load distance = $220\text{ N} \times 0.5\text{ cm} = 110\text{ N·cm}$.

5-6. a. $\text{Ideal MA} = \dfrac{\text{load area}}{\text{effort area}} = \dfrac{350\text{ cm}^2}{25\text{ cm}^2} = 14$.

   b. $\text{Ideal effort} = \dfrac{\text{load}}{\text{ideal MA}} = \dfrac{(400\text{ kg})(9.8\text{ m/s}^2)}{14} = 280\text{ N}$.

   c. $\text{Actual effort} = \dfrac{\text{ideal effort}}{\text{efficiency}} = \dfrac{280\text{ N}}{0.90} = 310\text{ N}$.

5-7. W_{out} = efficiency × W_{in}, therefore $w(0.5\text{ m}) = 0.85(420\text{ N})(2.5\text{ m})$, and $w = 1{,}790\text{ N}$.

5-8. $\text{Ideal MA} = \dfrac{\text{effort distance}}{\text{load distance}} = \dfrac{2\pi(12\text{ in})}{0.25\text{ in}} = 302$

$$\text{Ideal effort} = \frac{1{,}500\text{ lb}}{302} = 5.0\text{ lb}$$

$$\text{Efficiency} = \frac{\text{ideal effort}}{\text{actual effort}} = \frac{5.0\text{ lb}}{45\text{ lb}} = 0.11,\text{ or }11\%$$

5-9. a. With 12 teeth on the large gear and 4 on the small one, the mechanical advantage is 3.

   b. The large gear produces three times the torque, or 180 N·m.

5-10. $P = \dfrac{W}{\Delta t} = \dfrac{mg\,\Delta s}{\Delta t} = \dfrac{(40\text{ kg})(9.8\text{ m/s}^2)(15\text{ m})}{25\text{ s}} = 240\text{ W}$.

5-11. The work input is $mg\,\Delta h = (220\text{ kg})(9.8\text{ m/s}^2)(5.0\text{ m}) = 10{,}780\text{ J}$. This must equal the work output, which is $F\,\Delta s$; therefore $F(0.50\text{ m}) = 10{,}780\text{ J}$ and $F = 22{,}000\text{ N}$.

5-12. $E_{grav} = mgh = (50\text{ kg})(9.8\text{ m/s}^2)(11\text{ m}) = 5{,}400\text{ J}$.

Chapter 6

6-1. Since it takes work to move the charges apart, their potential energy increases as they separate and decreases as they approach each other.

6-2. $E_{kin} = \dfrac{1}{2}mv^2 = \dfrac{1}{2}(1{,}200\text{ kg})(22\text{ m/s})^2 = 2.9 \times 10^5\text{ J}$.

6-3. a. $E_{kin} = \dfrac{1}{2}mv^2 = \dfrac{1}{2}(30\text{ kg})(6.0\text{ m/s})^2 = 540\text{ J}$.

   b. All this kinetic energy turns to gravitational energy ($= mgh$), so $540\text{ J} = (30\text{ kg})(9.8\text{ m/s}^2)h$, and $h = 1.8\text{ m}$.

6-4. Its energy drops from 0 to $-Gm_1m_2/r$, so the loss is

$$\frac{(6.67 \times 10^{-11} \text{ N·m}^2/\text{kg}^2)(6.0 \times 10^{24} \text{ kg})(60 \text{ kg})}{6.4 \times 10^6 \text{ m}} = 3.8 \times 10^9 \text{ J}$$

6-5. The escape velocity equation still holds:

$$v_{escape} = \sqrt{\frac{2Gm_2}{r}} = \sqrt{\frac{2(6.67 \times 10^{-11} \text{ N·m}^2/\text{kg}^2)(6.0 \times 10^{24} \text{ kg})}{2(6.4 \times 10^6 \text{ m})}} = 7,900 \text{ m/s}$$

6-6. The lost gravitational energy must equal the increase in both kinetic and thermal energy, so $(mgh)_{start} = \left(\frac{1}{2}mv^2\right)_{end} + \Delta H$, and

$$(30 \text{ kg})(9.8 \text{ m/s}^2)(6.0 \text{ m}) = \frac{1}{2}(30 \text{ kg})(7.2 \text{ m/s})^2 + \Delta H, \quad \text{from which}$$

$\Delta H = 990 \text{ J}$.

6-7. Since the hollow steel ball has its mass concentrated on the outside, further from the center of rotation, it has more rotational kinetic energy than the solid ball.

6-8. In falling, the water turns gravitational into translational kinetic energy, which becomes rotational kinetic energy in the turbine. The generator converts this to electric energy.

6-9. Its kinetic energy is $\frac{1}{2}mv^2 = \frac{1}{2}(1,400 \text{ kg})(30 \text{ m/s})^2 = 6.3 \times 10^5 \text{ J}$. This corresponds to an increase in mass:

$$\frac{E}{c^2} = \frac{6.3 \times 10^5 \text{ J}}{(3.0 \times 10^8 \text{ m/s})^2} = 7 \times 10^{-12} \text{ kg}$$

6-10. The mass converted is 0.1 g, or 1×10^{-4} kg. Then $E = mc^2 = (1 \times 10^{-4} \text{ kg})(3.0 \times 10^8 \text{ m/s})^2 = 9 \times 10^{12}$ J.

6-11. The engine continually turns chemical potential energy into thermal and kinetic energy. As the car goes uphill, the kinetic energy turns into thermal and gravitational energy. As the car goes downhill, the gravitational energy turns into thermal energy. As the car comes to rest, whatever kinetic energy remains turns to thermal energy.

Chapter 7

7-1. When the level of mercury in the thermometer stops changing, the thermometer is at the same temperature as your mouth.

7-2. It is silvered to prevent transfer of heat in or out by radiation. It is evacuated so that no convective transfer can take place.

7-3. The amount of column length above the zero mark is proportional to the Celsius temperature, so

$$\frac{11.5 \text{ cm}}{15.0 \text{ cm}} = \frac{T}{100°\text{C}}; \quad T = 76.7°\text{C}$$

7-4. $\Delta H = cm \, \Delta T = (0.90 \text{ J/g·C}°)(250 \text{ g})(65\text{C}°) = 14,600 \text{ J}$.

7-5. The heat lost by the water must equal the heat gained by the iron of the pot: $(mc \, \Delta T)_{water} = (mc \, \Delta T)_{iron}$, and $m(1 \text{ cal/g·C}°)(55\text{C}°) = (350 \text{ g})(0.115 \text{ cal/g·C}°)(120\text{C}°)$, so $m = 88$ g.

7-6. The iron ball will cool down to 0°C, and the heat it loses must be equal to the heat gained by the melting of the ice at that temperature: $(mc\,\Delta T)_{\text{iron}} = (mL)_{\text{ice}}$, and $(600\text{ g})(0.115\text{ cal/g·C°})(240\text{C°}) = m(80\text{ cal/g})$, so $m = 210$ g.

7-7. Pressure is proportional to *kelvin* temperature. $T_k = 40°\text{C} + 273\text{ K} = 313\text{ K}$ at the start. Dropping the pressure to half also reduces the temperature to half, which is 157 K. Then $157\text{ K} = T_C + 273\text{ K}$ and $T_C = -116°\text{C}$.

7-8. The ratio pV/T does not change, since no gas has been added or removed. The initial temperature is $20°\text{C} + 273\text{ K} = 293\text{ K}$, and the final temperature is $100°\text{C} + 273\text{ K} = 373\text{ K}$. Therefore

$$\frac{(1\text{ atm})(300\text{ cm}^3)}{293\text{ K}} = \frac{p(100\text{ cm}^3)}{373\text{ K}}; \qquad p = 3.8\text{ atm}$$

This is the absolute pressure. The gauge reads 2.8 atm.

7-9. $pV = nkT$, so $p = nkT/V$. To find the answer in SI units, all quantities in the equation must be in SI units. The temperature is $T_k = 30°\text{C} + 273\text{ K} = 303\text{ K}$. Since there are 10^3 liters in a cubic meter, the volume is $5.0 \times 10^{-4}\text{ m}^3$. Then

$$p = \frac{(5.0 \times 10^{22}\text{ mcl})(1.38 \times 10^{-23}\text{ J/mcl·K})(303\text{ K})}{5.0 \times 10^{-4}\text{ m}^3} = 4.2 \times 10^5\text{ Pa}$$

7-10. $E_{\text{kin}} = \dfrac{3}{2}kT$; $\qquad T = \dfrac{2}{3}\dfrac{E}{k} = \dfrac{2}{3}\dfrac{3.5 \times 10^{-21}\text{ J}}{1.38 \times 10^{-23}\text{ J/mcl·K}}$

$T = 170\text{ K}$

7-11. The temperature, and therefore mv^2, must be the same for both kinds of molecules. Therefore

$$(mv^2)_{\text{oxygen}} = (mv^2)_{\text{hydrogen}}$$

$$\frac{m_{\text{ox}}}{m_{\text{hy}}} = \left(\frac{v_{\text{hy}}}{v_{\text{ox}}}\right)^2 \quad \text{or} \quad \frac{32\text{ Dl}}{2\text{ Dl}} = \left(\frac{v_{\text{hy}}}{2{,}400\text{ m/s}}\right)^2$$

$$v_{\text{hy}} = (2{,}400\text{ m/s})\sqrt{16} = 9{,}600\text{ m/s}$$

Chapter 8

8-1. If the frequency is 7.5 Hz, the period is $1/f = 0.133$ s. The phase difference is therefore $(0.05/0.133)360° = 135°$.

8-2. Two points 90° out of phase are a quarter-wavelength apart so the wavelength is 100 cm. Then $v = f\lambda = (6.0\text{ Hz})(1.00\text{ m}) = 6.0\text{ m/s}$.

8-3. $v = 331\text{ m/s} + (0.6)24 = 345\text{ m/s}$. Then

$$\lambda = \frac{v}{f} = \frac{345\text{ m/s}}{320\text{ Hz}} = 1.08\text{ m}$$

8-4. The second overtone has three times the frequency of the fundamental, or $3 \times 264\text{ Hz} = 792\text{ Hz}$. The velocity is $331 + (0.6)22 = 344\text{ m/s}$. Then

$$\lambda = \frac{v}{f} = \frac{344\text{ m/s}}{792\text{ Hz}} = 0.43\text{ m}$$

8-5. You will not hear the high frequencies as well, since they do not bend around the edges of the doorway.

8-6. The speed of the sound is $331 + (0.6)10 = 337$ m/s. With the car to the left of the cop, both are traveling to the left, so both velocities are negative. Therefore

$$f' = f\left(\frac{v - v_0}{v - v_s}\right) = (240 \text{ Hz})\left(\frac{337 + 35}{337 + 25}\right) = 247 \text{ Hz}$$

8-7. If they arrive at the listener in phase, the path difference must be one wavelength, so the second speaker is 14.5 m from the listener.

8-8. The wavelength is

$$\frac{v}{f} = \frac{15 \text{ m/s}}{5.0 \text{ Hz}} = 3.0 \text{ m}$$

There must be a node at the attached end.
  a. The first node is a half-wavelength from the end, or 1.5 m.
  b. The first antinode is a quarter-wavelength from the end, or 0.75 m.

8-9. At the fundamental, the wavelength is twice the length of the string:

$$f = \frac{v}{\lambda} = \frac{280 \text{ m/s}}{0.60 \text{ m}} = 467 \text{ Hz}$$

The first overtone is twice this, or 933 Hz, and the second overtone is three times the fundamental, or 1,400 Hz.

8-10. The velocity is $331 + (0.6)22 = 344$ m/s. The wavelength of the fundamental is twice the length of the pipe, so

$$f = \frac{v}{\lambda} = \frac{344 \text{ m/s}}{0.44 \text{ m}} = 782 \text{ Hz}$$

The first overtone is twice this, or 1,565 Hz, and the second overtone is three times the fundamental, or 2,347 Hz.

8-11. At the fundamental, the fork is a $\frac{1}{2}$ wavelength long, so the wavelength is 72 cm. Therefore

$$f_1 = \frac{v}{\lambda_1} = \frac{420 \text{ m/s}}{0.72 \text{ m}} = 583 \text{ Hz}$$

At the first overtone, the fork is $1\frac{1}{2}$ wavelengths long; 36 cm = $1.5\ \lambda_2$, so $\lambda_2 = 24$ cm and

$$f_2 = \frac{420 \text{ m/s}}{0.24 \text{ m}} = 1,750 \text{ Hz}$$

At the second overtone, the fork is $2.5\ \lambda_3$, so 36 cm = $2.5\ \lambda_3$ and $\lambda_3 = 14.4$ cm:

$$f_3 = \frac{420 \text{ m/s}}{0.144 \text{ m}} = 2,917 \text{ Hz}$$

Chapter 9

9-1. It must be the same as the charge on the rubber, since it repels an object charged from the rubber rod.

9-2. Steel is a conductor, and the charges will repel each other and spread throughout the rod.

9-3. All the charge is outside, on the surface.

9-4. The charges will be equal. Use Coulomb's law: $F = kq_1q_2/r^2$. With $q_1 = q_2 = q$, this becomes

$$q = \sqrt{\frac{Fr^2}{k}} = \sqrt{\frac{(6.0 \times 10^{-5}\text{ N})(0.15\text{ m})^2}{9.0 \times 10^9\text{ N}\cdot\text{m}^2/\text{C}^2}} = 1.2 \times 10^{-8}\text{ C}$$

9-5. With three times the charge, the force will be three times as great, or 6×10^{-5} N. A negative charge is pushed the opposite way, to the east.

9-6. On a sphere, the curvature is the same everywhere, so the charges are uniformly spaced. The field lines are radial because they must be perpendicular to the surface. They point inward because they terminate on negative charges.

9-7. The field may induce a separation of the charges, but there will be just as much positive charge on one end as negative on the other. Thus, in a uniform field, the two forces on the object will be equal in magnitude and opposite in direction. Unless they are in line, there will be a torque.

9-8. Only motion in the direction of the field counts. So the work done in moving the charge is $F \Delta s = (6.5 \times 10^{-5}\text{ N})(0.05\text{ m}) = 3.3 \times 10^{-6}$ J.

9-9. From Equation 9-9,

$$q = \frac{E}{\Delta V} = \frac{850\text{ J}}{12\text{ V}} = 71\text{ C}$$

9-10. The number of electrons will be the total charge divided by the charge on each electron:

$$\frac{8.0 \times 10^{-19}\text{ C}}{1.60 \times 10^{-19}\text{ C}} = 5$$

9-11. **a.** Random velocity increases, according to the usual gas laws.
b. Drift velocity decreases, and the current drops.

Chapter 10

10-1. Each cell provides 1.5 V, so the number of cells is 22.5 V/1.5 V = 15.

10-2. From Equation 10-2, $\Delta q = I \Delta t = (50\text{ A})(300\text{ s}) = 15,000$ C.

10-3. From Equation 10-3,

$$I = \frac{P}{\Delta V} = \frac{600\text{ W}}{120\text{ V}} = 5\text{ A}$$

10-4. From Equation 10-4, $\Delta V = IR = (0.15\text{ A})(20\ \Omega) = 3$ V. The current is therefore flowing to a point 3 V lower than 45 V, or 42 V.

10-5. $R = \dfrac{\rho l}{A} = \dfrac{(2.6 \times 10^{-6}\ \Omega\cdot\text{cm})(1.2 \times 10^5\text{ cm})}{0.15\text{ cm}^2} = 2.1\ \Omega.$

10-6. The total resistance is $0.20\ \Omega + 1.0\ \Omega = 1.2\ \Omega$. Therefore the current is

$$\frac{\Delta V}{R} = \frac{3.0\text{ V}}{1.2\ \Omega} = 2.5\text{ A}$$

10-7. a. The potential drop across the internal resistance is 4.50 V − 4.35 V = 0.15 V. Therefore the internal resistance is

$$\frac{\Delta V}{I} = \frac{0.15 \text{ V}}{0.12 \text{ A}} = 1.3 \ \Omega$$

b. $R = \frac{\Delta V}{I} = \frac{4.35 \text{ V}}{0.12 \text{ A}} = 36 \ \Omega.$

10-8. I_1 and I_2 (the currents in R_1 and R_2) must both be 12 A, since there is no branching between them and A_2. Of the 20 A flowing out of the battery, 12 A go into R_1, so the other 8 A must be I_4. Of the 12 A in R_2, 9 A go into R_5 and R_6, so the other 3 A must be in R_3. I_4 (= 8 A) and R_3 (= 3 A) both flow into R_7, so $I_7 = 11$ A.

10-9. The potential differences through any path out of the positive end of the battery and back into the negative end must total 50 V. Thus, $\Delta V_6 + 20$ V = 50 V and $\Delta V_6 = 30$ V. Also, 20 V + ΔV_3 + 4 V + 18 V = 50 V, so $\Delta V_3 = 8$ V. And ΔV_7 + 5 V + 4 V + 18 V = 50 V, so $\Delta V_7 = 23$ V.

10-10. a. The total resistance is 20 Ω + 40 Ω + 60 Ω = 120 Ω. Then

$$I = \frac{\Delta V}{R} = \frac{24 \text{ V}}{120} = 0.20 \text{ A}$$

b. $\Delta V = IR = (0.20 \text{ A})(20 \ \Omega) = 4$ V.
c. $P = I^2 R = (0.20 \text{ A})^2 (60 \ \Omega) = 2.4$ W.

10-11. a. $\frac{1}{R} = \frac{1}{20 \ \Omega} + \frac{1}{40 \ \Omega} + \frac{1}{60 \ \Omega} = \frac{6 + 3 + 2}{120 \ \Omega}$
so $R = 120 \ \Omega/11 = 11 \ \Omega.$

b. $I = \frac{\Delta V}{R} = \frac{24 \text{ V}}{11 \ \Omega} = 2.2$ A.

c. The current in the 40-Ω resistance is

$$\frac{\Delta V}{R} = \frac{24 \text{ V}}{40 \ \Omega} = 0.60 \text{ A}$$

Then $P = I \ \Delta V = (0.60 \text{ A})(24 \text{ V}) = 14$ W.

10-12. First, find the equivalent resistance of each of the parallel combinations:

$$\frac{1}{R} = \frac{1}{40 \ \Omega} + \frac{1}{60 \ \Omega} = \frac{3 + 2}{120 \ \Omega}; \quad R = 24 \ \Omega$$

$$\frac{1}{R} = \frac{1}{10 \ \Omega} + \frac{1}{20 \ \Omega} = \frac{2 + 1}{20 \ \Omega}; \quad R = 6.7 \ \Omega$$

Since both parallel combinations are in series with the fifth resistor, their resistances can just be added: 24 Ω + 12 Ω + 6.7 Ω = 43 Ω.

Chapter 11

11-1. Away. The currents are in opposite directions.

11-2. Since the current is perpendicular to the field, the force is IlB = (10 A)(0.15 m)(2.0 × 10^{3} Ts) = 3 × 10^{3} N. Based on the right-hand rule, its direction is east.

11-3. **a.** $F = IlB = (5\text{ A})(0.10\text{ m})(30 \times 10^{-3}\text{ Ts}) = 0.015\text{ N}$.

   b. The torque on each side is $Fr = (0.015\text{ N})(0.10\text{ m})$, and the total torque is twice this, or 0.003 N·m.

11-4. From Equation 11-4,

$$B = \frac{F}{qv} = \frac{2.0 \times 10^{-12}\text{ N}}{(1.60 \times 10^{-19}\text{ C})(2.5 \times 10^{8}\text{ m/s})} = 0.050\text{ Ts}$$

11-5. **a.** The kinetic energy it gains equals the potential energy it lost; from Equation 9-9, $\Delta E = q\,\Delta V = (1.60 \times 10^{-19}\text{ C})(75\text{ V}) = 1.20 \times 10^{-17}\text{ J}$.

   b. From Equation 11-5, $mv = Bqr = (3.7 \times 10^{-3}\text{ Ts})(1.60 \times 10^{-19}\text{ C}) \times (0.30\text{ m}) = 1.78 \times 10^{-22}\text{ kg·m/s}$.

   c. $E_{\text{kin}} = \frac{1}{2}mv^2 = \frac{1}{2}(mv)^2/m$; so

$$m = \frac{(mv)^2}{2E_{\text{kin}}} = \frac{(1.78 \times 10^{-22}\text{ kg·m/s})^2}{2(1.20 \times 10^{-17}\text{ J})} = 1.3 \times 10^{-27}\text{ kg}$$

11-6. $B = \dfrac{2k'I}{r} = \dfrac{2(10^{-7}\text{ N·A}^2)(20\text{ A})}{0.050\text{ m}} = 8.0 \times 10^{-5}\text{ Ts}$

Based on the right-hand rule, it is directed northward.

11-7. From Equation 11-7,

$$I = \frac{B}{4\pi k'(n/l)}$$

If the diameter of the wire is 0.2 mm, there are 5 wires per millimeter, or 5,000 wires per meter, in each layer, for a total value $n/l = 20,000/\text{m}$. Then

$$I = \frac{1.5 \times 10^{-3}\text{ Ts}}{4\pi(10^{-7}\text{ N/A}^2)(20,000/\text{m})} = 0.060\text{ A}$$

11-8. See Figure A5-1.

FIG. A5-1

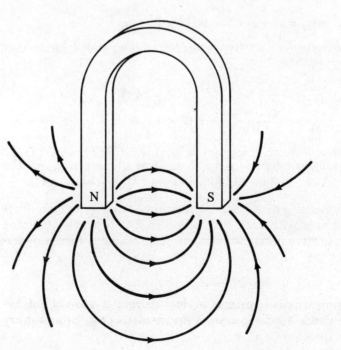

11-9. The dip of 76° means that the field vector points 76° from the horizontal. So, from Equation 1-4, $B_{hor} = (5.3 \times 10^{-5} \text{ Ts})(\cos 76°) = 1.3 \times 10^{-5}$ Ts.

11-10. $\mathscr{E} = Blv = (50 \times 10^{-3} \text{ Ts})(0.15 \text{ m})(2.5 \text{ m/s}) = 0.0188$ V. The current is

$$\frac{\mathscr{E}}{R} = \frac{0.0188 \text{ V}}{10} = 1.9 \times 10^{-3} \text{ A}$$

The right-hand rule shows that the force on positive charge in the rod is downward, and this is the way the current travels.

11-11. The maximum flux passing through the loop is $\phi = BA = (50 \times 10^{-3} \text{ Ts})(150 \times 10^{-4} \text{ m}^2) = 7.5 \times 10^{-4}$ Wb. Since this changes to 0 in 0.05 s, the average induced emf is

$$\frac{7.5 \times 10^{-4} \text{ Wb}}{0.05 \text{ s}} = 0.015 \text{ V}$$

for each turn, for a total of 0.75 V.

11-12. The electric energy output is $I \, \Delta V \, \Delta t = (6.0 \text{ A})(50 \text{ V})(30 \text{ s}) = 9,000$ J. This must be equal to the work done on the handle, which is $F \, \Delta s$. Therefore 9,000 J = F$(2\pi)(0.15 \text{ m})(20)$; so $F = 480$ N.

11-13. With the electromagnet moving toward the left, the left end of the solenoid must be an N-pole. The right-hand rule shows that, to produce this polarity, the solenoid current must run up across the front face, so it is going to the right through the galvanometer.

Chapter 12

12-1. As the loop rotates, the direction in which the magnetic flux passes through it changes every half-cycle.

12-2. **a.** Maximum current is $I_{eff}\sqrt{2} = (3.0 \text{ A})(\sqrt{2}) = 4.2$ A.
b. $P = I_{eff}^2 R = (3.0 \text{ A})^2(20 \text{ }\Omega) = 180$ W.

12-3. **a.** $\dfrac{V_P}{V_s} = \dfrac{N_P}{N_s} = \dfrac{120 \text{ V}}{15 \text{ V}} = 8$.
b. $I = P/\Delta V$. In the primary, $I = 3 \text{ W}/120 \text{ V} = 25$ mA. In the secondary, $I = 3 \text{ W}/15 \text{ V} = 0.20$ A.

12-4. The current entering the school is

$$\frac{P}{\Delta V} = \frac{140 \times 10^3 \text{ W}}{120 \text{ V}} = 1.17 \times 10^3 \text{ A}$$

This must pass through lines with a total resistance of 0.004 Ω, so the power lost in the lines is $I^2R = (1.17 \times 10^3 \text{ A})^2(0.004 \text{ }\Omega) = 5.5$ kW.

12-5. The three wires of a three-phase transmission line have potentials 120° out of phase with each other. Two of the three wires coming into the house have potentials 180° out of phase; the third is a neutral ground.

12-6. #8.

12-7. The grounding wire is to protect against current accidentally delivered to the case. A plastic case is an insulator and cannot carry current.

12-8. There are many—20 or more, probably.

12-9. When there is no current in the secondary, the current and potential difference in the primary are 90° out of phase with each other, so no power is delivered.

Chapter 13

13-1. $f = \dfrac{c}{\lambda} = \dfrac{3.00 \times 10^8 \text{ m/s}}{5.5 \times 10^{-7} \text{ m}} = 5.4 \times 10^{14}$ Hz.

13-2. The wavelength is

$$\lambda = \frac{c}{f} = \frac{3.00 \times 10^8 \text{ m/s}}{4,500 \times 10^6 \text{ Hz}} = 6.67 \times 10^{-2} \text{ m}$$

The antenna should be a half-wavelength long, or 3.3 cm.

13-3. Longer waves diffract around the object; since the object will cast no shadow, it cannot be seen.

13-4. From Equation 13-4b, with $n = 1$, $\lambda = d \sin \theta$. With 640 lines per millimeter the distance between slits is 1/640 mm, or m/(6.40×10^5). Then

$$\lambda = \left(\frac{\text{m}}{6.4 \times 10^5}\right)(\sin 37°) = 9.4 \times 10^{-7} \text{ m}$$

13-5. $f = \dfrac{c}{\lambda} = \dfrac{3.00 \times 10^8 \text{ m/s}}{6.9 \times 10^{-7} \text{ m}} = 4.3 \times 10^{14}$ Hz.

13-6. Observe it through a polarizing filter. Rotate the filter until the light is blocked out; its axis is then perpendicular to the direction of polarization of the light.

13-7. From Figure 13-13, the magenta filter seems to cut out everything from about 470 nm to 600 nm.

13-8. $E = hf = \dfrac{hc}{\lambda} = \dfrac{(6.62 \times 10^{-34} \text{ J·s})(3.00 \times 10^8 \text{ m/s})}{240 \times 10^{-9} \text{ m}} = 8.3 \times 10^{-19}$ J.

13-9. From Equation 13-9b, $W = hf - q \, \Delta V =$ $(6.63 \times 10^{-34} \text{ J·s})(3.51 \times 10^{15} \text{ Hz}) - (1.60 \times 10^{-19} \text{ C})(9.2 \text{ V}) =$ 8.6×10^{-19} J.

13-10. $E = hc/\lambda$, so

$$\lambda = \frac{hc}{E} = \frac{(4.14 \times 10^{-15} \text{ ev·s})(3.00 \times 10^8 \text{ m/s})}{22,000 \text{ ev}} = 5.6 \times 10^{-11} \text{ m}$$

Chapter 14

14-1. The rays from a bulb are arrayed radially outward; those from a laser are parallel.

14-2. The visible power produced by a 40-W bulb operating at 16% efficiency is $(0.16)(40 \text{ W}) = 6.4 \text{ W}$. This produces $(6.4 \text{ W})(450 \text{ lum/W}) = 2,880$ lum. The illumination at 3.5 m is

$$\frac{F}{4\pi R^2} = \frac{2,880 \text{ lum}}{4\pi(3.5 \text{ m})^2} = 19 \text{ lux}$$

14-3. The ray coming from your feet to the mirror and to your eye strikes the mirror at half the distance from your eye to the floor, and you do not use any of the mirror below that point. The smallest mirror is thus half your height.

14-4. The size of the image is always the same as the size of the object.

14-5. $\dfrac{\sin \theta_A}{\sin \theta_B} = \dfrac{n_B}{n_A};$ $\qquad \dfrac{\sin 55°}{\sin r} = \dfrac{2.42}{1.33};$ $\qquad r = 27°$

14-6. $\dfrac{\sin \theta_A}{\sin \theta_B} = \dfrac{n_B}{n_A};$ $\qquad \dfrac{\sin i_c}{\sin 90°} = \dfrac{1}{2.42};$ $\qquad i_c = 24°$

It sparkles because the light cannot escape unless it is coming to the surface at a small angle of incidence.

14-7. For red:

$$\frac{\sin 65.0°}{\sin r} = \frac{1.61}{1}; \qquad r = 34.3°$$

For violet:

$$\frac{\sin 65.0°}{\sin r} = \frac{1.66}{1}; \qquad r = 33.1°$$

14-8. In a $+20$ cm lens, rays entering parallel to the principal axis converge at 20 cm from the lens. In a -30 cm lens, rays entering parallel to the principal axis diverge as though coming from a point 30 cm behind the lens.

14-9. See Figure A5-2.

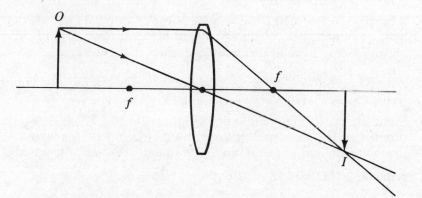

FIG. A5-2

14-10. See Figure A5-3.

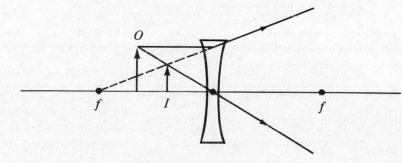

FIG. A5-3

14-11. **a.** First find the object distance:

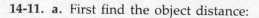

$$\frac{D_o}{D_i} = \frac{S_o}{S_i} \qquad \frac{D_o}{3.0 \text{ m}} = \frac{3.5 \text{ cm}}{50 \text{ cm}}; \qquad D_o = 0.21 \text{ m}$$

b. Now use this value to find the focal length:

$$\frac{1}{f} = \frac{1}{D_o} + \frac{1}{D_i} = \frac{1}{0.21 \text{ m}} + \frac{1}{3.0 \text{ m}}$$

$$\frac{1}{f} = \frac{(3.0 \text{ m}) + (0.21 \text{ m})}{(0.21 \text{ m})(3.0 \text{ m})}$$

$$f = 0.20 \text{ m}$$

14-12. This doubling of the diameter of the lens opening increases the area of the opening by a factor of 4, thus letting in four times as much light.

Chapter 15

15-1. This is 50 elementary charges, or $50(1.60 \times 10^{-19} \text{ C}) = 8.0 \times 10^{-18}$ C.

15-2. In the $n = 2$ state, its energy is $E_2 = -13.6 \text{ ev}/2^2 = -3.40$ ev. In the $n = 1$ state (ground state) it is -13.6 ev. The difference, 10.2 ev, is the energy of the photon emitted. Then

$$\lambda = \frac{hc}{E} = \frac{(4.14 \times 10^{-15} \text{ ev·s})(3.00 \times 10^8 \text{ m/s})}{10.2 \text{ ev}} = 1.22 \times 10^{-7} \text{ m}$$

15-3. **a.** The wavelength is the circumference of its orbit = $2\pi(5 \times 10^{-11} \text{ m}) = 3.1 \times 10^{-10}$ m.
   b. From Equation 15-4a, its angular momentum $mvr = h/2\pi = 1.05 \times 10^{-34}$ kg·m²/s.

15-4. 26 protons and 28 neutrons.

15-5. It bends into a circular path because of the centripetal force exerted by the magnetic field. It spirals inward because it is constantly giving up energy in collisions with hydrogen molecules.

15-6. $^{214}_{84}\text{Po} \rightarrow {}^{4}_{2}\text{He} + {}^{210}_{82}\text{Pb}$

$^{210}_{82}\text{Pb} \rightarrow {}^{210}_{83}\text{Bi} + {}^{0}_{-1}\text{e} + {}^{0}_{0}\nu$

15-7. From the chart, the half-life of polonium-214 is 1.6×10^{-4} s. The number of half-lives in 0.01 s is thus $0.01/(1.6 \times 10^{-4}) =$ about 6. Since the amount of polonium drops to half six times, the amount remaining is $(1 \text{ g})\left(\frac{1}{2}\right)^6 = 0.016$ g.

15-8. Mass of protons: 2×1.007825 2.01565
Mass of neutrons: 2×1.008665 2.01733
Mass of separate particles   4.03298
Mass of the helium nucleus   4.00260 Dl
Mass deficit of helium-4   0.03038 Dl
The binding energy is found by multiplying the mass deficit by 931 Mev/Dl, which gives 28 Mev.

15-9. $(500 \times 10^9 \text{ ev})\left(\frac{\text{Dl}}{931 \times 10^6 \text{ ev}}\right) = 540$ Dl

compared with about 1 Dl when it is at rest.

15-10. ${}^{4}_{2}\text{He} + {}^{39}_{19}\text{K} \rightarrow ({}^{43}_{21}\text{Sc}) \rightarrow {}^{1}_{1}\text{H} + {}^{42}_{20}\text{Ca}$

15-11. The energy released is $(25 \times 10^3 \text{ tons})(4 \times 10^9 \text{ J/ton}) = 1.0 \times 10^{14}$ J. The corresponding mass is

$$\frac{E}{c^2} = \frac{1.0 \times 10^{14} \text{ J}}{(3.00 \times 10^8 \text{ m/s})^2} = 1.1 \times 10^{-3} \text{ kg}$$

which is about 0.05% of the original mass.

15-12. $^{232}_{90}\text{Th} + ^{1}_{0}\text{n} \rightarrow ^{233}_{90}\text{Th}$

$^{233}_{90}\text{Th} \rightarrow ^{233}_{91}\text{Pa} + ^{0}_{-1}\text{e} + ^{0}_{0}\nu$

$^{233}_{91}\text{Pa} \rightarrow ^{233}_{92}\text{U} + ^{0}_{-1}\text{e} + ^{0}_{0}\nu$

15-13.
Mass of lithium-6		6.01512 Dl
Mass of deuterium		2.0140
Original total mass		8.02912
Mass of two helium nuclei:	2×4.00260	8.00520
Additional mass deficit		0.02392 Dl

Energy released per fusion reaction is (0.02392 Dl)(931 Mev/Dl) = 22 Mev.

Appendix 1

A1-1. **a.** The height of any ceiling is a constant.
b. All the heights constitute a variable set.
c. The number of inches in a foot is a universal constant.
d. Cars are at various speeds; variable.
e. All cars in a train have the same speed; constant.

A1-2. **a.** The constant is

$$\frac{\text{Distance traveled}}{\text{Fuel consumed}} = \frac{70 \text{ mi}}{2.5 \text{ gal}} = 28 \text{ mi/gal}$$

b. From the equation above,

$$\text{Fuel consumed} = \frac{\text{distance traveled}}{28 \text{ mi/gal}} = \frac{350 \text{ mi}}{28 \text{ mi/gal}} = 12.5 \text{ gal}$$

A1-3. **1.** The weight of the rod is proportional to the square of the thickness: w/T^2 = constant.

$$\frac{2.5 \text{ lb}}{(0.2 \text{ in})^2} = \frac{w}{(0.6 \text{ in})^2}$$

This can easily be converted to

$$w = (2.5 \text{ lb})\left(\frac{0.6 \text{ in}}{0.2 \text{ in}}\right)^2 = 22.5 \text{ lb, or } 23 \text{ lb}$$

2. The number of tiles is inversely proportional to the area of a single tile: $N \propto (1/s)$, or Ns = a constant. Therefore (1,500 tiles)(12 in^2) = N(30 in^2), so that N = (1,500 tiles)(12/30) = 600 tiles.

A1-4. **1.** opp/hyp = $\sin \theta$ = 35 cm/50 cm = 0.70, so θ = 44°.
2. adj/hyp = $\cos \theta$ − cos 65° = 0.4226

$$\text{hyp} = \frac{\text{adj}}{0.4226} = \frac{30 \text{ ft}}{0.4226} = 71 \text{ ft}$$

3. opp/adj = $\tan \theta$ = 20 in/35 in = 0.5714, so θ = 30°.

A1-5. **1.** $\frac{d}{4} - 6 = 22a$ Add 6 to both sides to get

$$\frac{d}{4} = 22a + 6$$ Now divide both sides by 4

$$d = \frac{22a + 6}{4}$$

2. $3r + t = 5t + 7$ Subtract t from both sides to get

$3r = 4t + 7$ Now divide both sides by 3

$$r = \frac{4t + 7}{3}$$

3. $aY + 3t^2 = 44$ Subtract $3t^2$ from both sides to get

$aY = 44 - 3t^2$ Now divide both sides by a

$$Y = \frac{44 - 3t^2}{a}$$

4. $p^2 + 6 = r - 3$ Subtract 6 from both sides

$p^2 = r - 9$ Now take the square root of both sides

$p = \sqrt{r - 9}$

A1-6. 1. First, convert 3 lb to ounces by multiplying by a unit fraction:

$$(3 \text{ lb})\left(\frac{16 \text{ oz}}{1 \text{ lb}}\right) = 48 \text{ oz}$$

Adding this to the ounces gives 75 oz.

2. First, convert the 65 cm to inches:

$$(65 \text{ cm})\left(\frac{1 \text{ in}}{2.54 \text{ cm}}\right) = 25.6 \text{ in}$$

Subtracting this from the original length of the board gives 31.4 in, which should be rounded off to 31 in.

A1-7. 1. 125 lb/30 in^2 = 4.2 lb/in^2, read "pounds per square inch."

2. (20 ft)(16 lb) = 320 ft·lb, read "foot-pounds."

3. $(100 \text{ cm})^3 = 100^3 \text{ cm}^3$, or 1,000,000 cm^3.

A1-8. Mass is in kg; time is in s, so mass per unit time is in kg/s, read "kilograms per second."

A1-9. 1. 14 million = $14 \times 10^6 = 1.4 \times 10^7$.

2. 234×10^{-3}; move the decimal point two places to the left, which is dividing by 10^2. To compensate, increase the order of magnitude by 2, to get 2.34×10^{-1}.

3. $4{,}190 = 4.190 \times 10^3$.

4. $0.000000398 = 3.98 \times 10^{-7}$.

A1-10. 1. $(3.5 \times 10^2)(4.6 \times 10^4) = 16.1 \times 10^{(4+2)} = 1.6 \times 10^7$.

2. $(5.7 \times 10^{-5})(6.0 \times 10^8) = 34.2 \times 10^3 = 3.4 \times 10^4$

3. $(2.6 \times 10^8) \div (1.3 \times 10^{12}) = 2.0 \times 10^{-4}$

4. $(3.75 \times 10^4) + (0.191 \times 10^4) = 3.94 \times 10^4$

5. $(3.95 \times 10^4) - (0.517 \times 10^4) = 3.43 \times 10^4$

A1-11. 121 mi.

Chapter 1

1. average speed
2. meter per second
3. scalar
4. vector
5. 4 m and 8 m
6. cosine
7. acceleration
8. meter per second squared
9. square
10. acceleration
11. south
12. west
13. vertical
14. 9.8 m/s^2
15. centripetal
16. earth
17. gravity

Chapter 2

1. velocity
2. opposite
3. equilibrium
4. newton
5. pound
6. viscous drag
7. weight
8. mass
9. weight
10. newton
11. elastic recoil
12. tension
13. buoyancy
14. southward; 30
15. elastic recoil
16. 25 lb
17. cosine
18. 200 N
19. torque
20. torque
21. newton-meter
22. 30 N·m
23. torques

Chapter 3

1. gas
2. shape
3. density
4. density
5. area
6. smaller
7. density
8. 760 nm
9. 3 atm
10. 2,200 mb
11. $\frac{1}{4}$
12. buoyancy
13. weight
14. weight
15. density
16. surface tension
17. viscosity
18. velocity
19. weight
20. laminar
21. streamlined
22. increases
23. decreases
24. faster
25. speed
26. angle of attack

Chapter 4

1. time
2. impulse
3. mass
4. newton
5. net unbalanced
6. velocity
7. impulse
8. momentum
9. perpendicular
10. centripetal
11. gravity
12. masses
13. distance between them
14. mass and radius
15. free fall
16. velocity
17. altitude
18. gravity
19. 24 h
20. torque
21. away from
22. conserved

Chapter 5

1. half
2. twice
3. effort distance
4. work
5. work, work
6. load
7. less than
8. efficiency
9. mechanical advantage
10. areas
11. distance, distance
12. length to height
13. friction
14. number of teeth
15. joule
16. power
17. weight
18. gravitational potential energy
19. weight and height
20. energy

Chapter 6

1. work
2. increases
3. kinetic
4. 4
5. scalar
6. gravitational potential; kinetic
7. total
8. zero
9. negative
10. kinetic
11. escape
12. thermal

13. energy
14. less than
15. chemical
16. electrical
17. mass

18. the speed of light
19. strong nuclear
20. mass
21. thermal

Chapter 7

1. temperature
2. heat
3. thermal equilibrium
4. a vacuum
5. metals
6. warmer
7. expands
8. ice
9. heat
10. specific heat
11. water
12. 4.19
13. phase

14. 540
15. pressure
16. 273 K
17. kelvin
18. 1 atm
19. pressure and volume
20. low P, high T
21. number of molecules
22. temperature
23. elastic
24. molecules
25. potential
26. specific

Chapter 8

1. period
2. frequency
3. frequency
4. 180°
5. amplitude
6. damping
7. wavelength
8. velocity
9. parallel
10. velocity
11. rarefactions
12. temperature
13. frequency
14. 110

15. tone quality
16. overtones
17. wavelength
18. point source
19. frequency
20. source
21. out of phase
22. difference
23. node
24. half-wavelength
25. node
26. one-half
27. node
28. natural or resonant frequencies

Chapter 9

1. repel
2. electric
3. negative
4. electrons
5. conductors
6. surface
7. insulator
8. curvature
9. coulomb
10. charges
11. one-fourth
12. closest together
13. negative

14. charge
15. positive
16. perpendicular
17. uniform
18. electric field
19. equipotential
20. field lines
21. volts
22. coulomb
23. electron
24. electrons
25. atom

Chapter 10

1. separation
2. emf
3. volt
4. current
5. ampere
6. ammeter
7. voltmeter
8. watts
9. power
10. resistance
11. ohm
12. potential difference
13. length
14. resistivity
15. voltmeter
16. internal resistance
17. current or internal resistance
18. current
19. zero
20. current
21. potential differences
22. series
23. largest
24. potential difference
25. current and power
26. smallest

Chapter 11

1. attract
2. magnetic
3. north
4. S
5. field and current
6. perpendicular
7. torque
8. downward
9. electric
10. circle
11. mass
12. currents
13. 2
14. clockwise
15. increase
16. ferromagnetic
17. electrons
18. alloys
19. north magnetic
20. moving
21. length and velocity
22. magnetic flux
23. rotating
24. work done
25. left
26. opened

Chapter 12

1. commutator
2. AC
3. zero
4. effective
5. smaller
6. magnetic field
7. potentials
8. transformer
9. two
10. 230,000
11. fuse or circuit breaker
12. increases
13. short circuits
14. green
15. induction
16. ballast
17. synchronous
18. increases
19. zero

Chapter 13

1. accelerated
2. speed
3. half-wavelength
4. perpendicular
5. X-rays or gamma rays
6. shorter
7. shorter
8. diffraction
9. wavelengths
10. antinode

11. wavelength
12. red
13. laser
14. modulating
15. polarized
16. perpendicular
17. three
18. yellow
19. short
20. infrared

21. frequency and power
22. higher
23. quantized
24. frequency
25. light
26. frequency
27. work function
28. energy
29. short

Chapter 14

1. diffraction
2. photon
3. lumens
4. green
5. lux
6. normal
7. angle of incidence
8. long
9. specular
10. virtual
11. the same
12. toward
13. index of refraction
14. velocity
15. sines
16. critical angle of incidence
17. higher

18. frequency
19. violet
20. converging
21. principal focus
22. principal focus
23. real
24. the focal length
25. two
26. diverging, or concave
27. focal length
28. size to size
29. virtual
30. infinity
31. dispersion
32. focal ratio or f-number
33. long
34. circle of confusion

Chapter 15

1. alpha particles
2. positive elementary
3. atomic number
4. hydrogen
5. photon
6. ionized
7. angular momentum
8. $h/2\pi$
9. standing waves
10. proton
11. neutrons
12. isotopes
13. daltons
14. ionizing
15. magnetic field
16. bubbles
17. gamma
18. electrons
19. helium

20. beta
21. 4
22. half-life
23. smaller than
24. binding
25. increases
26. electric
27. drift tubes
28. magnetic
29. electron
30. neutrons or protons
31. neutrons
32. neutrons
33. neutrons
34. plutonium-239
35. moderator
36. control rods
37. fusion
38. medium-sized

Solutions to End-of-Chapter Problems

Chapter 1

1. Since $v_{av} = \Delta s/\Delta t$, $\Delta s = v_{av}\,\Delta t = (4.5 \text{ m/s})(600 \text{ s}) = 2,700 \text{ m}$.

2. The time in SI units:

    $$(1 \text{ hr})\left(\frac{60 \text{ min}}{\text{hr}}\right)\left(\frac{60 \text{ s}}{\text{min}}\right) = 3,600 \text{ s}$$

    Then

    $$v_{av} = \frac{4.6 \times 10^4 \text{ m}}{3,600 \text{ s}} = 12.8 \text{ m/s}$$

3. For two velocities at right angles to each other, the vector sum is the hypotenuse of a triangle. Therefore $v = \sqrt{3.6^2 + 12.0^2} = 12.5$ m/s. The direction will be south of east at an angle whose tangent is 3.6/12.0, or 17°.

4. She must head her plane at an angle to the north such that the north-ward component of her velocity is 55 mi/hr, just enough to compensate for the wind. That angle will be the angle whose sine is 55/230, which is 13.8°.

5. Your displacement has two components: 4.5 mi east and $(6.2 - 2.0) =$ 4.2 mi south. The magnitude of this displacement is the vector sum of these components, $\sqrt{4.2^2 + 4.5^2} = 6.2$ mi. The direction is south of east at arctan 4.2/4.5 = 43°.

6. This direction is 60° north of west, so the westward component is (45 mi)(cos 60°) = 23 mi.

7. The 35 mi is the northward component of his displacement; 35 mi = Δs cos 55°, so his displacement is (35 mi)/cos 55° = 61 mi. Then

$$v_{av} = \frac{\Delta s}{\Delta t} = \frac{61 \text{ mi}}{6.0 \text{ hr}} = 10.1 \text{ mi/hr}$$

8. $a = \Delta v/\Delta t$. The increase in speed is (30 m/s − 12 m/s) = 18 m/s. Then

$$a = \frac{18 \text{ m/s}}{15 \text{ s}} = 1.2 \text{ m/s}^2$$

9. $a = \Delta v/\Delta t$, so

$$\Delta t = \frac{\Delta v}{a} = \frac{22 \text{ m/s} - 15 \text{ m/s}}{3.2 \text{ m/s}^2} = 2.2 \text{ s}$$

10. When starting from rest and accelerating uniformly, the average velocity is half the final velocity. Since $v_{av} = \Delta s/\Delta t$, $s = v_{av}\,\Delta t =$ (11 m/s)(15 s) = 165 m.

11. When starting from rest and accelerating uniformly, the standard formulas apply, so $v = \sqrt{2as}$. Then

$$a = \frac{v^2}{2s} = \frac{(18 \text{ m/s})^2}{2 \times 240 \text{ m}} = 0.68 \text{ m/s}^2$$

12. In free fall, the acceleration is 9.8 m/s²; there is uniform acceleration starting from rest, so $s = \frac{1}{2}at^2$ and

$$t = \sqrt{\frac{2s}{a}} = \sqrt{\frac{2 \times 24 \text{ m}}{9.8 \text{ m/s}^2}} - 2.2 \text{ s}$$

13. It comes to rest uniformly, so $v = \sqrt{2as}$ and

$$s = \frac{v^2}{2a} = \frac{(15 \text{ m/s})^2}{2 \times 9.8 \text{ m/s}^2} = 11.5 \text{ m}$$

14. There is uniform acceleration from rest, so $v = \sqrt{2as} = \sqrt{2(9.8 \text{ m/s}^2)(34 \text{ m})} = 26$ m/s.

15. $v = \sqrt{2as}$, so

$$s = \frac{v^2}{2a} = \frac{(15 \text{ m})^2}{2 \times 9.8 \text{ m/s}^2} = 11.5 \text{ m}$$

16. $a = \Delta v/\Delta t$, so $\Delta v = a\,\Delta t = (1.5 \text{ m/s}^2)(2.0 \text{ s}) = 3.0$ m/s. Upward acceleration for a descending elevator means it is slowing down, so its final speed is 4.4 − 3.0 = 1.4 m/s.

17. **a.** $v = \frac{2\pi r}{T} = \frac{2\pi(2.0 \text{ m})}{4.5 \text{ s}} = 2.8$ m/s.

 b. $a_c = \frac{v^2}{r} = \frac{(2.79 \text{ m/s})^2}{2.0 \text{ m}} = 3.9 \text{ m/s}^2$.

18. $a_c = v^2/r$, so $v = \sqrt{ar} = \sqrt{(3.0 \text{ m/s}^2)(6.5 \text{ m})} = 4.4$ m/s.

19. $a_c = v^2/r$, so

$$r = \frac{v^2}{a} = \frac{(20 \text{ m/s})^2}{3.5 \text{ m/s}^2} = 114 \text{ m}$$

20. $a_c = v^2/r$, so $v = \sqrt{ar} = \sqrt{(75 \text{ m/s}^2)(0.30 \text{ m})} = 4.74$ m/s. Then find its angular velocity: $v = \omega r$, so

$$\omega = \frac{v}{r} = \frac{4.74 \text{ m/s}}{0.30 \text{ m}} = 15.8 \text{ rad/s}.$$

Then convert the units:

$$\left(15.8 \frac{\text{rad}}{\text{s}}\right)\left(\frac{\text{rev}}{2\pi \text{ rad}}\right)\left(\frac{60 \text{ s}}{\text{min}}\right) = 150 \text{ rev/min}$$

21. **a.** Vertical motion is free fall starting from rest; $s = \frac{1}{2}at^2$, so

$$t = \sqrt{\frac{2s}{a}} = \sqrt{\frac{2 \times 22 \text{ m}}{9.8 \text{ m/s}^2}} = 2.12 \text{ s}$$

   b. Horizontal motion is uniform speed, so $v = \Delta s/\Delta t$ and $\Delta s = v\,\Delta t =$ (22 m/s)(2.12 s) = 47 m.

22. While in orbit, acceleration due to gravity is centripetal, so $g = v^2/r$, and $v = \sqrt{gr} = \sqrt{(3.3 \text{ m/s}^2)(3.4 \times 10^6 \text{ m})} = 3{,}350$ m/s.

23. $v = \dfrac{2\pi r}{T} = \dfrac{2\pi(2.2 \times 10^6 \text{ m})}{(40 \text{ min})(60 \text{ s/min})} = 5{,}760$ m/s

   Then

$$g = \frac{v^2}{r} = \frac{(5{,}760 \text{ m/s})^2}{2.2 \times 10^6 \text{ m}} = 15.1 \text{ m/s}^2$$

Chapter 2

1. $w = mg$, so

$$m = \frac{w}{g} = \frac{75 \text{ N}}{9.8 \text{ m/s}^2} = 7.7 \text{ kg}$$

2. $w = mg = (95 \text{ kg})(1.67 \text{ m/s}^2) = 159$ N.

3. Downward forces (weight and viscous drag) must equal upward forces (buoyancy and tension). Weight is $mg = (35 \text{ kg})(9.8 \text{ m/s}^2) = 343$ N. Therefore 343 N + 25 N = 50 N + T, and $T = 320$ N.

4. Force forward (engine thrust) = force backward (friction + viscous drag). Therefore, 31,000 N = 22,000 N + D, and $D = 9{,}000$ N.

5. The horizontal component of the tension in the rope is $T \cos 25°$. Forward force = retarding force, so 85 N = $T \cos 25°$, and $T = 94$ N.

6. **a.** Since the whole system is symmetrical, each half of the rope must be supporting half the weight, or $\frac{1}{2}$(32 kg)(9.8 m/s^2) = 157 N.

   b. The rope makes an angle of 50° with the vertical, so the vertical component of the tension in the rope is $T \cos 50°$ (or $T \sin 40°$). The vertical component of one side of the rope supports half the weight of the knapsack, so 157 N = $T \sin 40°$, and $T = 240$ N.

7. The vertical component of the tension is $T \cos 25°$, and each rope supports half the weight. Therefore $T \cos 25° = \frac{1}{2}$(35 kg)(9.8 m/s^2), and $T = 190$ N.

8. The angle the rope makes with the vertical is 40°, so the vertical component of the tension is $T \cos 40°$. Since this is the only upward force, it must be equal to the weight; $w = (350 \text{ N})(\cos 40°) = 270$ N.

9. The angle the weight vector makes with the perpendicular to the surface is 20°, so the force the books exert against the plank is $mg \cos 20° = (20 \text{ kg})(9.8 \text{ m/s}^2)(\cos 20°) = 180$ N.

10. The angle the weight vector makes with the surface is $90° - 35°$, so the component of weight acting parallel to the slope is $w \sin 35°$. This is the force needed to push the (frictionless) wagon up the slope, so $40 \text{ lb} = w \sin 35°$, and $w = 70$ lb.

11. **a.** $\tau = Fr = (250 \text{ N})(1.5 \text{ m})$, measuring r from the pivot. Then $Fr = 375 \text{ N·m}$.
    b. The torque applied by the rock must be the same, so
    $$F = \frac{\tau}{r} = \frac{375 \text{ N·m}}{0.5 \text{ m}} = 750 \text{ N}$$

12. When the left end is taken as the pivot, the torque applied by the weight of the painter is $Fr = (600 \text{ N})(1.0 \text{ m}) = 600 \text{ N·m}$, clockwise. The other rope exerts a counterclockwise torque of $T_R(2.5 \text{ m})$. Since there are no other torques, these must be equal, so $600 \text{ N·m} = T_R(2.5 \text{ m})$, and $T_R = 240 \text{ N}$. The upward forces must equal the weight of the painter, so $T_R + T_L = 600 \text{ N}$, and $T_L = 360 \text{ N}$.

13. The clockwise torque exerted by the rope must equal the counterclockwise torque at the handle, so $(1,200 \text{ N})(1.5 \text{ cm}) = F(45 \text{ cm})$, and $F = 40 \text{ N}$.

14. **a.** The component of weight parallel to the ramp is $mg \sin \theta = (7.0 \text{ kg})(9.8 \text{ m/s}^2)(\sin 15°) = 18 \text{ N}$.
    b. The component of weight normal to the surface is $mg \cos \theta = (7.0 \text{ kg})(9.8 \text{ m/s}^2)(\cos 15°) = 66 \text{ N}$.

15. The component of weight acting down the slope is $(450 \text{ N})(\sin 25°) = 190 \text{ N}$, and this must be equal to the sum of the uphill components. Two forces pull uphill, friction and the tension in the rope. Therefore, $190 \text{ N} = 75 \text{ N} + T$, and $T = 115 \text{ N}$.

16. Clockwise torque exerted by the girl around the rock must be equal to the counterclockwise torque exerted by the rope. Thus $(55 \text{ kg}) \times (9.8 \text{ m/s}^2)r = (350 \text{ N})(4.0 \text{ m})$, so $r = 2.6 \text{ m}$.

Chapter 3

1. The mass of the liquid is $265 \text{ g} - 120 \text{ g} = 145 \text{ g}$.
    $$D = \frac{m}{V} = \frac{145 \text{ g}}{100 \text{ cm}^3} = 1.45 \text{ g/cm}^3$$

2. $D = m/V$, so $m = DV = (0.83 \text{ g/cm}^3)(1,000 \text{ cm}^3) = 830$ g.

3. Its density must be 1.20 g/cm^3, and
    $$V = \frac{m}{D} = \frac{500 \text{ g}}{1.20 \text{ g/cm}^3} = 417 \text{ cm}^3$$

4. $p = \dfrac{F}{A} = \dfrac{2,500 \text{ lb}}{4 \times 4 \text{ in}^2} = 156 \text{ lb/in}^2$.

5. $F = pA = (8\frac{1}{2} \text{ in.})(11 \text{ in.})\left(14.7 \dfrac{\text{lb}}{\text{in}^2}\right)$
    $F = 1,370$ lb

6. $(62 \text{ lb/in}^2)\left(\dfrac{1 \text{ atm}}{14.7 \text{ lb/in}^2}\right) = 4.2$ atm.

7. $p = (0.89)^n(1 \text{ atm})$, where $n = 9.6$; $p = 0.32$ atm.

8. $p = hDg$. To put everything in SI units, we have to convert:
    $$D = \left(1\frac{\text{g}}{\text{cm}^3}\right)\left(\frac{10^2 \text{ cm}}{\text{m}}\right)^3\left(\frac{\text{kg}}{10^3 \text{ g}}\right) = 10^3 \text{ kg/m}^3$$

So $p = (65 \text{ m})(10^3 \text{ kg/m}^3)(9.8 \text{ m/s}^2) = 6.37 \times 10^5 \text{ N/m}^2$. Adding the pressure at the surface, $1.0 \times 10^5 \text{ N/m}^2$, gives an answer of $7.4 \times 10^5 \text{ N/m}^2$, or 7.4 atm.

9. $P = hDg = (110 \text{ m})(10^3 \text{ kg/m}^3)(9.8 \text{ m/s}^2) = 1.08 \times 10^6$ Pa. It is not necessary to add the atmospheric pressure, since the atmosphere is pressing on the other side of the dam.

10. $P = hDg = (7.5 \text{ m})(880 \text{ kg/m}^3)(9.8 \text{ m/s}^2) = 6.5 \times 10^4 \text{ N/m}^2 = 0.65$ atm. This is the gauge pressure.

11. It displaces 650 cm^3 of water, and its buoyancy is the weight of this water. Then $w = mg$ and $m = DV$, so

$$w = DVg = \left(1 \frac{\text{g}}{\text{cm}^3}\right)(650 \text{ cm}^3)(9.8 \text{ m/s}^2)\left(\frac{\text{kg}}{10^3 \text{ g}}\right) = 6.4 \text{ N}$$

12. Buoyancy is $(350 \text{ N} - 245 \text{ N}) = 105$ N. The volume of the anchor is the volume of water that weighs that much; $D = m/V$ and $w = mg$, so

$$V = \frac{w}{Dg} = \frac{105 \text{ N}}{(10^3 \text{ kg/m}^3)(9.8 \text{ m/s}^2)} = 1.1 \times 10^{-2} \text{ m}^3$$

13. The buoyancy on the rock is the weight of $(45 \text{ g} - 32 \text{ g}) = 13$ g of water, so its volume is 13 cm^3. Then

$$D = \frac{m}{V} = \frac{45 \text{ g}}{13 \text{ cm}^3} = 3.5 \text{ g/cm}^3$$

14. **a.** The mass of the 18 cm^3 of water is 18 g; since the block displaces its own weight, its mass is also 18 g.

b. $D = \dfrac{m}{V} = \dfrac{18 \text{ g}}{(3.0 \text{ cm})^3} = 0.67 \text{ g/cm}^3$.

15. The plank floats with 5/6.5 of its volume under water, so the water it displaces must have just 5/6.5 as much volume as the plank. Therefore, the density of the plank must be 5/6.5 that of water, or 0.77 g/cm^3.

16. Its density is 86% that of water, or 0.86 g/cm^3.

Chapter 4

1. Impulse $= F \, \Delta t$,

$$\Delta t = \frac{10^6 \text{ N·s}}{45{,}000 \text{ N}} = 22 \text{ s}$$

2. The same impulse is needed in both cases, so $F_1 \, \Delta t_1 = F_2 \, \Delta t_2$ and

$$F_2 = F_1\left(\frac{\Delta t_1}{\Delta t_2}\right) = (3{,}500 \text{ N})\left(\frac{5.0 \text{ s}}{2.0 \text{ s}}\right) = 8{,}750 \text{ N}$$

3. Acceleration is

$$\frac{\Delta v}{\Delta t} = \frac{6.5 \text{ m/s} - 4.0 \text{ m/s}}{3.0 \text{ s}} = 0.83 \text{ m/s}^2$$

Then $F = ma = (60 \text{ kg})(0.83 \text{ m/s}^2) = 50$ N.

4. **a.** $v = \sqrt{2as}$, so

$$a = \frac{v^2}{2s} = \frac{(25 \text{ m/s})^2}{2(0.40 \text{ m})} = 780 \text{ m/s}^2$$

b. $F = ma = (1,400 \text{ kg})(780 \text{ m/s}^2) = 1.1 \times 10^6 \text{ N}.$

5. $a = \dfrac{v}{t} = \dfrac{30 \text{ m/s}}{8.0 \text{ s}} = 3.75 \text{ m/s}^2$

   $F = ma = (2,800 \text{ kg})(3.75 \text{ m/s}^2) = 11,000 \text{ N}$

6. The downward force on it is its weight $= mg = (680 \text{ kg})(1.67 \text{ m/s}^2) = 1,136 \text{ N}$. If F is the thrust of the engines, the net upward force will be $F - 1,136 \text{ N} = ma$; $F = (680 \text{ kg})(2.0 \text{ m/s}^2) + 1,136 \text{ N} = 2,500 \text{ N}.$

7. The net force is $220 \text{ N} - F$ (the friction), so $220 \text{ N} - F = (75 \text{ kg})(2.0 \text{ m/s}^2)$; $F = 70 \text{ N}.$

8. The total momentum starts at zero and does not change, so $m_1v_1 + m_2v_2 = 0$; $(35 \text{ kg})v + (6 \text{ kg})(3.5 \text{ m/s}) = 0$, and $v = -0.6 \text{ m/s}.$

9. The total momentum is zero and does not change: $(7.5 \text{ kg})v + (0.0080 \text{ kg})(640 \text{ m/s}) = 0$; $v = 0.68 \text{ m/s}.$

10. **a.** $\Delta(mv) = (45 \text{ kg})(85 \text{ m/s}) = 3,825 \text{ kg·m/s}.$

    b. $\Delta(mv)$ for the rocket is $-3,825 \text{ kg·m/s}$; since it starts at rest, this is also its final mv. Then

$$v = \frac{3,825 \text{ kg·m/s}}{750 \text{ kg}} = 5.1 \text{ m/s}$$

11. **a.** From rest in the engine, the gas increases its momentum to $mv = (60 \text{ kg})(95 \text{ m/s}) = 5,700 \text{ kg·m/s}.$

    b. Impulse is change of momentum, both for the gas and for the ship. Therefore $F \, \Delta t = 5,700 \text{ N·s} - F(20 \text{ s})$, and $F - 290 \text{ N}.$

12. The total momentum before equals the total momentum after. Since they are going in opposite directions, one of them has negative momentum. Then $(62 \text{ kg})(3.5 \text{ m/s}) - (53 \text{ kg})(5.0 \text{ m/s}) = (53 \text{ kg} + 62 \text{ kg})v$ and $v = 0.42 \text{ m/s}.$

13. $F = ma = mv^2/r$, so

$$r = \frac{mv^2}{F} = \frac{(240 \text{ kg})(35 \text{ m/s})^2}{6,200 \text{ N}} = 47 \text{ m}$$

14. Friction is the centripetal force, so

$$F = \frac{mv^2}{r} = \frac{(45 \text{ kg})(4.5 \text{ m/s})^2}{2.5 \text{ m}} = 365 \text{ N}$$

15. The acceleration is due to gravity alone, so $g \propto 1/r^2$ and gr^2 is a constant. We know g and r at the surface of the earth. For the meteor r is its altitude plus the radius of the earth: $55,000 \text{ km} + 6,400 \text{ km} = 61,400 \text{ km}$. Then $g_1r_1{}^2 = g_2r_2{}^2$, or

$$g_2 = g_1\left(\frac{r_1}{r_2}\right)^2 = (9.8 \text{ m/s}^2)\left(\frac{6,400 \text{ km}}{61,400 \text{ km}}\right)^2 = 0.11 \text{ m/s}^2$$

16. First find the acceleration due to gravity, as in Problem 15:

$$g = (9.8 \text{ m/s}^2)\left(\frac{6,400 \text{ km}}{6,400 \text{ km} + 4,200 \text{ km}}\right)^2 = 3.57 \text{ m/s}^2$$

Then $w = mg = (3,500 \text{ kg})(3.57 \text{ m/s}^2) = 12,500 \text{ N}.$

17. $F = \dfrac{Gm_1m_2}{r^2} = \dfrac{(6.67 \times 10^{-11} \text{ N·m}^2/\text{kg}^2)(4.5 \times 10^5 \text{ kg})(7 \times 10^5 \text{ kg})}{(22,000 \text{ m})^2} = 4.3 \times 10^{-8} \text{ N}.$

Chapter 5

1. **a.** 2, since two strands support the load.
   b. 2 times load distance, or 24 m.
   c. $W = F \, \Delta s = (80 \text{ kg})(9.8 \text{ m/s}^2)(12 \text{ m}) = 9,400 \text{ J}$.
   d. $W = F \, \Delta s = (450 \text{ N})(24 \text{ m}) = 10,800 \text{ J}$.
   e. efficiency $= \dfrac{W_{out}}{W_{in}} = \dfrac{9,400 \text{ J}}{10,800 \text{ J}} = 0.87$, or 87%.

2. 4, since four strands support the load.

3. efficiency $= \dfrac{W_{out}}{W_{in}} = \dfrac{F_L \, \Delta s_L}{F_E \, \Delta s_E}$;
   but $\Delta s_E = 4 \, \Delta s_L$, so efficiency $= F_L / 4 F_E$. Therefore
   $$F_E = \frac{F_L}{4 \times \text{eff}} = \frac{750 \text{ lb}}{4 \times 0.80} = 230 \text{ lb}$$

4. $W = F \, \Delta s \cos \theta = (25 \text{ kg})(9.8 \text{ m/s}^2)(1.2 \text{ m})(\cos 0°) +$
   $(25 \text{ kg})(9.8 \text{ m/s}^2)(3.5 \text{ m})(\cos 90°) = 290 \text{ J}$.

5. $W = F \, \Delta s \cos \theta = (25 \text{ lb})(30 \text{ ft})(\cos 30°) = 650 \text{ ft·lb}$.

6. At best, $W_{out} = W_{in}$, so $(1,400 \text{ N})(2.5 \text{ m}) = F(50 \text{ m})$, and $F = 70 \text{ N}$.

7. **a.** With 26 teeth pushing 10, the ideal mechanical advantage is 10/26.
   b. Each turn of the crank rotates the wheel 2.6 times.
   c. When the crank makes a complete revolution, it travels $2\pi \times 14 \text{ cm}$, while the rim of the wheel travels $2\pi \times 33 \text{ cm} \times 2.6$. The ideal mechanical advantage is
   $$\frac{s_L}{s_E} = \frac{(2\pi)(14 \text{ cm})}{(2\pi)(33 \text{ cm})(2.6)} = 0.16$$

8. $\dfrac{F_E}{F_L} = \dfrac{r_L}{r_E}$; $\qquad \dfrac{F_E}{(22 \text{ kg})(9.8 \text{ m/s}^2)} = \dfrac{3.2 \text{ m}}{0.60 \text{ m}}$; $\qquad F_E = 1150 \text{ N}$.

9. The mechanical advantage is the ratio of the radii: 25.0 cm/4.0 cm = 6.3.

10. The ratio of the radii, measured from the pivot: 1.78 m/0.12 m = 15.

11. $\dfrac{F_E}{A_E} = \dfrac{F_L}{A_L}$; $\qquad \dfrac{F_E}{10 \text{ cm}^2} = \dfrac{(330 \text{ kg})(9.8 \text{ m/s}^2)}{120 \text{ cm}^2}$; $\qquad F_E = 270 \text{ N}$.

12. **a.** $W_{out} = F \, \Delta s = (240 \text{ kg})(9.8 \text{ m/s}^2)(1.0 \text{ m}) = 2,350 \text{ N}$.
    b. efficiency $= \dfrac{W_{out}}{W_{in}}$; $\qquad 0.75 = \dfrac{2,350 \text{ N}}{W_{in}}$; $\qquad W_{in} = 3,130 \text{ N}$.
    c. $F_E \, \Delta s = W_{in}$; $\qquad F_E(3.2 \text{ m}) = 3,130 \text{ N}$; $\qquad F_E = 980 \text{ N}$.

13. $\dfrac{\tau_E}{\tau_L} = \dfrac{n_L}{n_E}$; $\qquad \dfrac{50 \text{ N·m}}{\tau_L} = \dfrac{30}{12}$; $\qquad \tau_L = 20 \text{ N·m}$.

14. **a.** $P = \dfrac{F \, \Delta s}{\Delta t} = \dfrac{(40 \text{ kg})(9.8 \text{ m/s}^2)(10 \text{ m})}{8.0 \text{ s}} = 490 \text{ W}$.
    b. efficiency $= \dfrac{P_{out}}{P_{in}}$; $\qquad 0.70 = \dfrac{490 \text{ W}}{P_{in}}$ $\qquad P_{in} = 700 \text{ W}$.

15. $P = \dfrac{F \, \Delta s}{\Delta t}$ $\qquad \Delta t = \dfrac{F \, \Delta s}{P} = \dfrac{(100 \text{ kg})(9.8 \text{ m/s}^2)(12 \text{ m})}{240 \text{ W}} = 49 \text{ s}$.

16. $\Delta E = W = F \, \Delta s = (300 \text{ N})(60 \text{ m}) = 18,000 \text{ J}$.

17. $E_{grav} = mgh = (12 \text{ kg})(9.8 \text{ m/s}^2)(2.0 \text{ m}) = 240 \text{ J}$.

18. $E_{grav} = mgh$; $120,000 \text{ J} = mg(10 \text{ m})$; $w = mg = 12,000 \text{ N}$.

19. $E_{grav} = mgh = (60 \text{ kg})(9.8 \text{ m/s}^2)(250 \text{ m}) = 147,000 \text{ J}$.

1. $\Delta E = W = F\,\Delta s = (150\ \text{N})(22\ \text{m}) = 3{,}300\ \text{J}$.

2. **a.** The only form of energy added is kinetic, so $\Delta E_{\text{kin}} = W = F\,\Delta s =$
   $(75\ \text{N})(15\ \text{m}) = 1{,}130\ \text{J}$.
   b. $E_{\text{kin}} = \frac{1}{2}mv^2 \qquad v = \sqrt{\frac{2E_{\text{kin}}}{m}} = \sqrt{\frac{2 \times 1{,}130\ \text{J}}{380\ \text{kg}}} = 2.4\ \text{m/s}$.

3. All the kinetic energy it loses does work, so $W = \frac{1}{2}mv^2 = \frac{1}{2}(20\ \text{kg}) \times$
   $(10.0\ \text{m/s})^2 = 1{,}000\ \text{J}$.

4. The gravitational potential energy it loses turns into kinetic energy, so
   $(E_{\text{grav}})_{\text{top}} = (E_{\text{kin}})_{\text{bottom}}\colon \quad mgh_{\text{top}} = \frac{1}{2}mv^2_{\text{bottom}}, \quad \text{and} \quad v = \sqrt{2gh} =$
   $\sqrt{2(9.8\ \text{m/s}^2)(0.15\ \text{m})} = 1.7\ \text{m/s}$.

5. **a.** $W = F\,\Delta s = (6.5\ \text{N})(25\ \text{m}) = 163\ \text{J}$.
   b. 163 J.
   c. $E_{\text{grav}} = mgh$; $163\ \text{J} = (0.30\ \text{kg})(9.8\ \text{m/s}^2)h$, so $h = 55\ \text{m}$.

6. **a.** $mgh_1 = mgh_2 + E_{\text{th}}$; $E_{\text{th}} = mg(h_1 - h_2) =$
   $(0.16\ \text{kg})(9.8\ \text{m/s}^2)(5.0\ \text{m} - 4.2\ \text{m}) = 1.3\ \text{J}$.
   b. As it leaves the ground, its energy is all kinetic and equal to mgh_2.
   Therefore $\frac{1}{2}mv^2 = mgh_2$, and $v = \sqrt{2gh_2} = \sqrt{2(9.8\ \text{m/s}^2)(4.2\ \text{m})} =$
   $9.1\ \text{m/s}$.

7. **a.** $W = F\,\Delta s = (250\ \text{N})(12\ \text{m}) = 3{,}000\ \text{J}$.
   b. $E_{\text{grav}} = mgh = (30\ \text{kg})(9.8\ \text{m/s}^2)(6.0\ \text{m}) = 1{,}760\ \text{J}$.
   c. Thermal energy is the energy produced by friction, so it is 10% of 3,000 J, or 300 J.
   d. The work produced three kinds of energy: $W = E_{\text{th}} + E_{\text{kin}} + E_{\text{grav}}$;
   $3{,}000\ \text{J} = 300\ \text{J} + E_{\text{kin}} + 1{,}760\ \text{J}$, so $E_{\text{kin}} = 940\ \text{J}$.
   e. $E_{\text{kin}} = \frac{1}{2}mv^2$, so $940\ \text{J} = \frac{1}{2}(30\ \text{kg})v^2$; $v = 7.9\ \text{m/s}$.

8. **a.** $E = \frac{Gm_1m_2}{r} = \frac{(6.67 \times 10^{-11}\ \text{N·m}^2/\text{kg}^2)(6.0 \times 10^{24}\ \text{kg})(250\ \text{kg})}{6.4 \times 10^6\ \text{m}}$
   $E = 1.56 \times 10^{10}\ \text{J}$
   b. $W = F\,\Delta s$; $1.56 \times 10^{10}\ \text{J} = (640{,}000\ \text{N})(\Delta s)$, so the height $\Delta s = 24{,}000\ \text{m}$.

9. The gravitational energy it loses is
   $\frac{Gm_1m_2}{r} = \frac{(6.67 \times 10^{-11}\ \text{N·m}^2/\text{kg}^2)(7.48 \times 10^{22}\ \text{kg})(60\ \text{kg})}{1.73 \times 10^6\ \text{m}} = 1.73 \times 10^8\ \text{J}$

10. $v_{\text{escape}} = \sqrt{\frac{2Gm_2}{r}} = \sqrt{\frac{2(6.67 \times 10^{-11}\ \text{N·m}^2/\text{kg}^2)(7.48 \times 10^{22}\ \text{kg})}{1.73 \times 10^6\ \text{m}}} = 2{,}400\ \text{m/s}$

11. Thermal energy created = loss of kinetic energy = $\left(\frac{1}{2}mv^2\right)_{\text{initial}} -$
   $\left(\frac{1}{2}mv^2\right)_{\text{final}} = \frac{1}{2}(1{,}200\ \text{kg})(22\ \text{m/s})^2 - \frac{1}{2}(1{,}200\ \text{kg})(15\ \text{m/s})^2 =$
   $1.55 \times 10^5\ \text{J}$.

12. $E_{\text{grav}} = mgh = (1{,}800\ \text{kg})(9.8\ \text{m/s}^2)(15\ \text{m}) = 2.6 \times 10^5\ \text{J}$.

13. $m = \frac{E_{\text{grav}}}{gh} = \frac{20{,}000\ \text{J}}{(9.8\ \text{m/s}^2)(12\ \text{m})} = 170\ \text{kg}$.

14. a. $(mv)_{before} = (mv)_{after}$; $(1.5 \text{ kg})(4.0 \text{ m/s}) = (4.0 \text{ kg})v$, so v, after the impact, is 1.5 m/s.

b. The thermal energy created is the difference between the kinetic energy of the rock before the impact and the kinetic energy of the combination after: $\frac{1}{2}(1.5 \text{ kg})(4.0 \text{ m/s})^2 - \frac{1}{2}(4.0 \text{ kg})(1.5 \text{ m/s})^2 = E_{therm}$, so $E_{therm} = 7.5 \text{ J}$.

15. $E = W = F \, \Delta s = (60 \text{ N})(0.35 \text{ m}) = 21 \text{ J}$.

16. $P = E/\Delta t$, so

$$E = P \, \Delta t = (10^9 \text{ W})(365 \text{ days})\left(\frac{24 \text{ hr}}{\text{day}}\right)\left(\frac{3{,}600 \text{ s}}{\text{hr}}\right) = 3.2 \times 10^{16} \text{ J}$$

$E = mc^2$, so

$$m = \frac{E}{c^2} = \frac{3.2 \times 10^{16} \text{ J}}{(3.00 \times 10^8 \text{ m/s})^2} = 0.35 \text{ kg}$$

17. $E = mc^2 = (10^{-7} \text{ kg})(3.0 \times 10^8 \text{ m/s})^2 = 9 \times 10^{16} \text{ J}$.

Chapter 7

1. $\dfrac{T}{100°C} = \dfrac{12.2 \text{ cm} - 3.5 \text{ cm}}{19.0 \text{ cm} - 3.5 \text{ cm}}$ $T = 56°C$.

2. The problem must be solved in SI units, so

$$\left(0.031 \frac{\text{cal}}{\text{g} \cdot \text{C}°}\right)\left(\frac{4.19 \text{ J}}{\text{cal}}\right)\left(\frac{10^3 \text{ g}}{\text{kg}}\right) = 130 \text{ J/kg} \cdot \text{C}°$$

The lost kinetic energy is turned into thermal energy, so $\frac{1}{2}mv^2 = cm \, \Delta T$, and

$$\Delta T = \frac{v^2}{2c} = 2 \times \frac{(460 \text{ m/s})^2}{130 \text{ J/kg} \cdot \text{C}°} = 810 \text{C}°$$

3. Lost gravitational energy becomes thermal energy, so $mgh = cm \, \Delta T$; $\Delta T = gh/c$ and $c_{water} = 4{,}190 \text{ J/kg} \cdot \text{C}°$. Then

$$\Delta T = \frac{(9.8 \text{ m/s}^2)(110 \text{ m})}{4{,}190 \text{ J/kg} \cdot \text{C}°} = 0.26 \text{C}°$$

4. $\Delta H = cm \, \Delta T = (0.22 \text{ cal/g} \cdot \text{K})(580 \text{ g})(180°C - 20°C) = 20{,}000 \text{ cal}$.

5. The heat lost equals the heat gained, so $(cm \, \Delta T)_{metal} = (cm \, \Delta T)_{water}$. Thus, $c(760 \text{ g})(95°C - 28\text{C}°) = (1 \text{ cal/g} \cdot \text{C}°)(300 \text{ g})(28°C - 5°C)$, and $c = 0.14 \text{ cal/g} \cdot \text{C}°$.

6. $(cm \, \Delta T)_{iron} = (cm \, \Delta T)_{water}$:

$$\left(0.115 \frac{\text{cal}}{\text{g} \cdot \text{K}}\right) m(140°C - 65°C) = \left(1 \frac{\text{cal}}{\text{g} \cdot \text{K}}\right)(220 \text{ g})(65°C - 20°C)$$

$$m = 1{,}150 \text{ g}$$

(Note that $1\text{C}° = 1 \text{ K}$.)

7. First, melt the ice:

$$\Delta H = mL = (200 \text{ g})\left(80 \frac{\text{cal}}{\text{g}}\right) = 16{,}000 \text{ cal}$$

Then, heat it to the boiling point:

$$\Delta H = cm \, \Delta T = \left(1 \frac{\text{cal}}{\text{g} \cdot \text{K}}\right)(200 \text{ g})(100 \text{ K}) = 20{,}000 \text{ cal}$$

Then boil it:

$$\Delta H = mL = (200 \text{ g})\left(540\frac{\text{cal}}{\text{g}}\right) = 108{,}000 \text{ cal}$$

Total: 144,000 cal.

8. Heat lost in cooling the copper is equal to the heat gained in melting the ice

$$(cm\ \Delta T)_{\text{copper}} = (mL)_{\text{ice}}$$

$$\left(0.091\frac{\text{cal}}{\text{g·K}}\right)(350 \text{ g})(\Delta T) = (75 \text{ g})\left(80\frac{\text{cal}}{\text{g}}\right)$$

So $\Delta T = 190$ K; the copper cooled down from 190°C to 0°C.

9. $(cm\ \Delta T)_{\text{water}} = (cm\ \Delta T)_{\text{copper}}$. The equation must be written with the final temperature as the unknown:

$$\left(1\frac{\text{cal}}{\text{g·K}}\right)(300 \text{ g})(T - 15°\text{C}) = \left(0.091\frac{\text{cal}}{\text{g·K}}\right)(580 \text{ g})(145°\text{C} - T)$$
$$300T - 4{,}500°\text{C} = 7{,}650°\text{C} - 53T$$
$$T = 34°\text{C}$$

10. **a.** $p \propto T$, so p/T is constant. For ice water, $T = 0°\text{C} + 273 \text{ K} = 273 \text{ K}$. Then

$$\frac{p_1}{T_1} = \frac{p_2}{T_2} \qquad \frac{950 \text{ mb}}{273 \text{ K}} = \frac{620 \text{ mb}}{T_2} \qquad T_2 = 178 \text{ K}$$

b. $T_\text{K} = T_\text{C} + 273 \text{ K}$; $T_\text{C} = T_\text{K} - 273 \text{ K} = 178 \text{ K} - 273 \text{ K} = -95°\text{C}$.

11. For ice water, $T = 273$ K; for boiling water, $T = 373$ K.

$$\frac{p_1}{T_1} = \frac{p_2}{T_2} \qquad \frac{1{,}400 \text{ mb}}{273 \text{ K}} = \frac{p_2}{373 \text{ K}} \qquad p_2 = 1{,}910 \text{ mb}$$

12. $V \propto 1/p$, so pV is constant. Then $p_1V_1 = p_2V_2$: (1 atm)(2.5 l) = p_2(1.0 l); and $p_2 = 2.5$ atm. This is the gauge absolute pressure; the gauge pressure is 1.5 atm.

13. The initial temperature is 295 K. $pV \propto T$, so

$$\frac{p_1V_1}{T_1} = \frac{p_2V_2}{T_2} \qquad \frac{(1 \text{ atm})(600 \text{ cm}^3)}{295 \text{ K}} = \frac{(40 \text{ atm})(50 \text{ cm}^3)}{T_2}$$

and $T_2 = 980$ K.

14. p_1 (absolute) = 3 atm; $T_1 = 295$ K;

$$\frac{p_1V_1}{T_1} = \frac{p_2V_2}{T_2} \qquad \frac{(3 \text{ atm})(l)}{295 \text{ K}} = \frac{(6 \text{ atm})(2l)}{T_2}$$

and $T_2 = 1{,}180$ K.

15. $pV = nkT$, so $n = pV/kT$; 1 liter = 10^{-3} m³ and 20°C = 293 K.

$$n = \frac{(0.3 \times 10^5 \text{ N/m}^2)(10^{-3} \text{ m}^3)}{(1.39 \times 10^{-23} \text{ J/mcl·K})(293 \text{ K})} = 7.4 \times 10^{21} \text{ molecules}$$

16. At constant temperature, pV/n is constant;

$$\frac{p_1V_1}{n_1} = \frac{p_2V_2}{n_2} \qquad \frac{(1 \text{ atm})(500 \text{ cm}^3)}{n} = \frac{p_2(200 \text{ cm}^3)}{2n}$$

and $p_2 = 5$ atm.

17. $E_{kin} = \frac{3}{2}kT = \frac{3}{2}\left(1.38 \times 10^{-23}\frac{J}{mcl \cdot K}\right)(373\ K) = 7.7 \times 10^{-21}\ J.$

18. Since both have the same temperature, both have the same kinetic energy, and $m_1 v_1^2 = m_2 v_2^2$. Then

$$\frac{v_2}{v_1} = \sqrt{\frac{m_1}{m_2}} = \sqrt{\frac{32}{2}} = \frac{4}{1}$$

Chapter 8

1. a. $v = f\lambda$, so

$$f = \frac{v}{\lambda} = \frac{12\ m/s}{1.2\ m} = 10\ Hz$$

b. $T = \frac{1}{f} = \frac{1}{10\ s^{-1}} = 0.10\ s$

2. $12/30 = \phi/360°$; $\phi = 144°$.

3. From a compression to the nearest rarefaction is $\frac{1}{2}$ wavelength, so $\lambda = 0.70\ m$. Then $v = f\lambda = (4.0\ Hz)(0.70\ m) = 2.8\ m/s$.

4. $v = 331\ m/s + (0.6\ m/s)(-6°C) = 327\ m/s$.

5. $v = 331\ m/s + (0.6)(26°C) = 347\ m/s$. Then

$$\lambda = \frac{v}{f} = \frac{347\ m/s}{640\ Hz} = 0.54\ m$$

6. Divide by 2 twice to get 110 Hz.

7. C sharp is $\frac{5}{4}(440\ Hz) = 550\ Hz$.

8. $v = 331\ m/s + (0.6\ m/s)(18°C) = 342\ m/s$;

$f' = f\left(\frac{v - v_o}{v - v_s}\right)$. The observer is at rest, so $v_o = 0$.

With the police car on the left, it is traveling to the right, so v_s is positive. Therefore

$$f' = (380\ Hz)\left(\frac{342\ m/s}{342\ m/s - 28\ m/s}\right) = 414\ Hz$$

9. $v = 331\ m/s + (0.6\ m/s \cdot °C)(-12°C) = 324\ m/s$. With the source on the left, both are moving to the right, so both velocities are positive. Then

$$f' = (260\ Hz)\left(\frac{324\ m/s - 30\ m/s}{324\ m/s - 37\ m/s}\right) = 266\ Hz$$

10. $v = 331\ m/s + (0.6\ m/s \cdot °C)(10°C) = 337\ m/s$. Then

$$\lambda = \frac{v}{f} = \frac{337\ m/s}{680\ Hz} = 0.50\ m$$

The other speaker is one wavelength farther than the near one, at 16.3 m.

11. Tightening the matching string raises its frequency, so it must have been lower than 220 Hz to start. With 4 beats per second, it was 4 Hz lower, at 216 Hz.

12. $\lambda = \dfrac{v}{f} = \dfrac{8.0 \text{ m/s}}{6.0 \text{ Hz}} = 1.33$ m

The first two antinodes are $\dfrac{1}{4}$ and $\dfrac{3}{4}$ wavelength from the end, at 0.33 m and 1.00 m.

13. At the fundamental, it is $\dfrac{1}{2}$ wavelength long, so

$$\lambda = 0.70 \text{ m}; \qquad f = \dfrac{v}{\lambda} = \dfrac{180 \text{ m/s}}{0.70 \text{ m}} = 257 \text{ Hz}$$

At the first overtone, the string is one wavelength long;

$$f = \dfrac{180 \text{ m/s}}{0.35 \text{ m}} = 514 \text{ Hz}$$

At the second overtone, $0.35 \text{ m} = \dfrac{3}{2}\lambda$, so

$$\lambda = 0.233 \text{ m}; \qquad f = \dfrac{180 \text{ m/s}}{0.233 \text{ m}} = 771 \text{ Hz}$$

14. $v = 331 \text{ m/s} + (0.6 \text{ m/s})(20°\text{C}) = 343 \text{ m/s}$. At the fundamental, the pipe is $\dfrac{1}{4}$ wavelength, so $\lambda = 1.40$ m; then $f = v/\lambda = 245$ Hz. At the first overtone, $0.35 \text{ m} = \dfrac{3}{4}\lambda$, so $\lambda = 0.47$ m and $f = 735$ Hz. At the second overtone, $0.35 \text{ m} = \dfrac{5}{4}\lambda$ and $f = 1{,}225$ Hz.

15. At the fundamental, the fork is $\dfrac{1}{2}$ wavelength long, so

$$f = \dfrac{60 \text{ m/s}}{0.36 \text{ m}} = 167 \text{ Hz}$$

At the first overtone, $0.18 \text{ m} = \dfrac{3}{2}\lambda$ and $f = 500$ Hz.

16. The frequency of the second overtone is $3 \times 440 \text{ Hz} = 1{,}320 \text{ Hz}$, so the beat frequency is $1{,}318.5 - 1{,}320 = 1.5 \text{ Hz}$.

Chapter 9

1. The total charge is $+20$ nC; each has $+10$ nC.

2. **a.** $F = \dfrac{kq_1 q_2}{r^2} = \dfrac{(9.0 \times 10^9 \text{ N·m}^2/\text{C}^2)(25 \times 10^{-9} \text{ C})(15 \times 10^{-9} \text{ C})}{(0.15 \text{ m})^2} = 1.5 \times 10^{-4} \text{ N}.$
   b. Same.

3. Each ball is deflected 9 cm, so the force on it is

$$\dfrac{9}{60}(3.0 \text{ g})\left(\dfrac{\text{kg}}{1{,}000 \text{ g}}\right)\left(9.8 \dfrac{\text{m}}{\text{s}^2}\right) = 4.4 \times 10^{-3} \text{ N}$$

With equal charges, $F = kq^2/r^2$ and

$$q = \sqrt{\dfrac{Fr^2}{k}} = \sqrt{\dfrac{(4.4 \times 10^{-3} \text{ N})(0.18 \text{ m})^2}{9.0 \times 10^9 \text{ N·m}^2/\text{C}^2}} = 1.3 \times 10^{-7} \text{ C}$$

4. The ratio of force to charge is the field, so

$$\dfrac{3.0 \times 10^{-3} \text{ N}}{12 \text{ nC}} = \dfrac{2.5 \times 10^{-5} \text{ N}}{q}$$

and $q = 0.10$ nC.

5. Charge is conserved, so the other has $+12$ nC.

6. $\Delta E = q\,\Delta V = (220 \times 10^{-9}\text{ C})(30\text{ V} - 5\text{ V}) = 5.5 \times 10^{-6}\text{ J}$.

7. $\Delta E = q\,\Delta V = (6.0\text{ C})(24\text{ V}) = 144\text{ J}$.

8. $\Delta E = q\,\Delta V = (240\text{ nC})(110,000\text{ V}) = 0.026\text{ J}$.

9. **a.** $\Delta E = q\,\Delta V = (2.0 \times 10^{-9}\text{ C})(240\text{ V}) = 4.8 \times 10^{-7}\text{ J}$.
   b. $\Delta E = F\,\Delta s;\ 4.8 \times 10^{-7}\text{ J} = F(0.12\text{ m})$, so $F = 4.0 \times 10^{-6}\text{ N}$.

10. $\dfrac{8.0 \times 10^{-9}\text{ C}}{1.60 \times 10^{-19}\text{ C}} = 5 \times 10^{10}$ excess electrons.

11. $\Delta E = q\,\Delta V = (1.60 \times 10^{-19}\text{ C})(120,000\text{ V}) = 1.9 \times 10^{-14}\text{ J}$.

12. The length of time it takes a charge to travel the length of the wire is

$$\frac{1,300\text{ C}}{0.50\text{ C/s}} = 2,600\text{ s}$$

The drift velocity is therefore

$$\frac{150\text{ m}}{2,600\text{ s}} = 0.058\text{ m/s}$$

Chapter 10

1. $12 \times 2.2\text{ V} = 26\text{ V}$.

2. **a.** 26 J, by definition of potential difference.
   b. $I = q/\Delta t$, so $q = I\,\Delta t = (4.0\text{ A})(300\text{ s}) = 1,200\text{ C}$.
   c. $\Delta E = q\,\Delta V = (1,200\text{ C})(26\text{ V}) = 31,000\text{ J}$.
   d. $P = \dfrac{\Delta E}{\Delta t} = \dfrac{31,200\text{ J}}{300\text{ s}} = 104\text{ W}$.

3. **a.** P
   b. V
   c. T
   d. Q

4. Electric energy delivered $= I\,\Delta V\,\Delta t$, which must equal the heat energy generated. Then $I\,\Delta V\,\Delta t = cm\,\Delta T$. In SI units, $c = 4,190$ J/kg·K. Then

$$\Delta t = \frac{cm\,\Delta T}{I\,\Delta V} = \frac{(4,190\text{ J/kg·K})(0.25\text{ kg})(80\text{ K})}{(11\text{ A})(115\text{ V})} = 66\text{ s}$$

5. $P = I\,\Delta V$, so

$$I = \frac{P}{\Delta V} = \frac{60\text{ W}}{115\text{ V}} = 0.52\text{ A}$$

Then

$$R = \frac{\Delta V}{I} = \frac{115\text{ V}}{0.52\text{ A}} = 220\ \Omega$$

6. $P = I\,\Delta V = (4.0\text{ A})(120\text{ V}) = 480\text{ W}$. In running for 8 h, the energy it uses is $(480\text{ W})(8\text{ hr}) = 3,840$ watt-hours, or 3.84 kWh. Multiply by 8¢/kWh to get 31¢.

7. $I = \dfrac{\Delta V}{R} = \dfrac{(180\text{ V} - 30\text{ V})}{1,200\ \Omega} = 0.13\text{ A}$.

8. $\Delta V = IR = (90 \times 10^{-3}\text{ A})(40\ \Omega) = 3.6\text{ V}$.

9. $P = I\,\Delta V$; 900 W $= I(115$ V$)$, so $I = 7.8$ A.

10. $R = \dfrac{\rho l}{A}$, so

$$l = \frac{RA}{\rho} = \frac{(0.50\ \Omega)(0.10\ \text{cm}^2)}{(2.8 \times 10^{-6}\ \Omega\cdot\text{cm})} = 1{,}800\ \text{cm}$$

11. $R = \dfrac{\rho l}{A} = \dfrac{(1.72 \times 10^{-6}\ \Omega\cdot\text{cm})(1.50 \times 10^4\ \text{cm})}{0.05\ \text{cm}^2} = 0.52\ \Omega$

$I = \dfrac{\Delta V}{R} = \dfrac{12\ \text{V}}{0.52\ \Omega} = 23$ A

$P = I\,\Delta V = (23\ \text{A})(12\ \text{V}) = 280$ W

12. $R = \dfrac{\Delta V}{I} = \dfrac{12\ \text{V}}{140\ \text{A}} = 0.086\ \Omega$.

13. $I = \dfrac{\Delta V}{R} = \dfrac{22.5\ \text{V}}{(4.5\ \Omega + 0.8\ \Omega)} = 4.2$ A.

14. $\Delta V = IR = (4.2\ \text{A})(4.5\ \Omega) = 18.9$ V.

15. **a.** In the motor,

$$I = \frac{P}{\Delta V} = \frac{300\ \text{W}}{23.6\ \text{V}} = 12.7\ \text{A}$$

Then

$$R = \frac{\Delta V}{I} = \frac{23.6\ \text{V}}{12.7\ \text{A}} = 1.86\ \Omega$$

b. The potential drop in the battery is 24 V $-$ 23.6 V $= 0.4$ V. Then

$$R = \frac{\Delta V}{I} = \frac{0.4\ \text{V}}{12.7\ \text{A}} = 0.03\ \Omega$$

16. A_6 reads the same as A_2, 3 A. Also, $A_1 = A_5 + A_4 = 10$ A. And $A_1 = A_2 + A_3 + A_4$, so $A_3 = 5$ A.

17. $V_4 + V_5 = V_3$, so $V_4 = 5$ V. Also, $V_4 + V_5 + V_6 = V_7$, so $V_6 = 12$ V. And $V_1 = V_2 + V_7 = 26$ V.

18. **a.** $12\ \Omega + 18\ \Omega = 30\ \Omega$.

b. $I = \dfrac{\Delta V}{R} = \dfrac{24\ \text{V}}{30\ \Omega} = 0.80$ A.

c. $P = I^2R = (0.80\ \text{A})^2(12\ \Omega) = 7.7$ W.
$P = (0.80\ \text{A})^2(18\ \Omega) = 11.5$ W.

19. **a.** $P = I^2R$, so

$$I = \sqrt{\frac{P}{R}} = \sqrt{\frac{850\ \text{W}}{20\ \Omega}} = 6.5\ \text{A}$$

b. The total resistance needed is

$$R = \frac{\Delta V}{I} = \frac{160\ \text{V}}{6.5\ \text{A}} = 25\ \Omega$$

so the variable resistor must supply 5 Ω.

20. **a.** $I = P/\Delta V =$ (toaster) 600 W/120 V $= 5.0$ A; (lamp) 150 W/120 V $=$ 1.3 A; (radio) 40 W/120 V $= 0.33$ A.

b. Add them up to get 6.6 A.

21. $\dfrac{1}{R_P} = \dfrac{1}{R_1} + \dfrac{1}{R_2} + \dfrac{1}{R_3} = \dfrac{1}{100\ \Omega} + \dfrac{1}{250\ \Omega} + \dfrac{1}{400\ \Omega}$

$$\dfrac{1}{R_P} = \dfrac{60 + 24 + 15}{6{,}000\ \Omega}$$

so

$$R_P = \dfrac{6{,}000\ \Omega}{99} = 61\ \Omega$$

22. First determine the combined resistance of the parallel combination:

$$\dfrac{1}{R_P} = \dfrac{1}{20\ \Omega} + \dfrac{1}{60\ \Omega} + \dfrac{1}{80\ \Omega} = \dfrac{12 + 4 + 3}{240\ \Omega}; \qquad R_P = 13\ \Omega$$

Then add this to the others in series to get 50 Ω.

Chapter 11

1. a. $B = \dfrac{2k'I}{r} = \dfrac{2(10^{-7}\ \text{N/A}^2)(20\ \text{A})}{0.15\ \text{m}} = 2.7 \times 10^{-5}\ \text{Ts}$.

    b. According to the right-hand rule, east.

2. $I = \dfrac{Br}{2k'} = \dfrac{(0.050 \times 10^{-3}\ \text{Ts})(0.02\ \text{m})}{2 \times 10^{-7}\ \text{N/A}^2} = 5.0\ \text{A}$.

3. $B = \dfrac{F}{Il} = \dfrac{2.5 \times 10^{-4}\ \text{N}}{(5.0\ \text{A})(1\ \text{m})} = 5 \times 10^{-5}\ \text{Ts}$.

4. The force per unit length exerted by the field is the weight per unit length of the wire. Thus, $F/l = mg/l$. Since $B = F/Il$,

$$B = \dfrac{mg}{Il} = \left(1.50 \times 10^{-3}\ \dfrac{\text{kg}}{\text{m}}\right)\left(\dfrac{9.8\ \text{m/s}^2}{2.0\ \text{A}}\right) = 7.4\ \text{mTs}$$

5. a. East.

    b. $F = Bqv = (1.5\ \text{Ts})(1.60 \times 10^{-19}\ \text{C})(1.8 \times 10^{7}\ \text{m/s}) = 4.3 \times 10^{-12}\ \text{N}$.

6. $Bqv = mv^2/r$, so

$$m = \dfrac{Bqr}{v} = \dfrac{(0.75\ \text{Ts})(1.60 \times 10^{-19}\ \text{C})(2.4\ \text{m})}{2.6 \times 10^{7}\ \text{m/s}} = 1.10 \times 10^{-26}\ \text{kg}$$

7. $mv = Bqr$, so

$$v = \dfrac{Bqr}{m} = \dfrac{(0.25\ \text{Ts})(1.60 \times 10^{-19}\ \text{C})(0.40\ \text{m})}{1.67 \times 10^{-27}\ \text{kg}} = 9.6 \times 10^{6}\ \text{m/s}$$

8. $B = 4\pi k'I(n/l)$. There are 10 turns per millimeter in two layers, so $n/l = 20{,}000/\text{m}$. Then, $B = 4\pi(10^{-7}\ \text{N/A}^2)(0.50\ \text{A})(20{,}000/\text{m}) = 12.6\ \text{mTs}$.

9. a. $I = \dfrac{\Delta V}{R} = \dfrac{12\ \text{V}}{30\ \Omega} = 0.40\ \text{A}$.

Then

$$B = 4\pi k'I\dfrac{n}{l} = \dfrac{4\pi(10^{-7}\ \text{N/A}^2)(0.40\ \text{A})(200)}{0.15\ \text{m}} = 6.7 \times 10^{-4}\ \text{Ts}$$

    b. Nothing changes except that the number of turns per unit length doubles, so the field also doubles.

10. $F = BIl = (0.055 \times 10^{-3}\ \text{Ts})(1.0\ \text{m})(8.0\ \text{A}) = 4.4 \times 10^{-4}\ \text{N}$. The direction is northward, up at an angle of 18° from the horizontal.

11. $\mathcal{E} = Blv = (1.2 \text{ Ts})(0.20 \text{ m})(3.5 \text{ m/s}) = 0.84 \text{ V}.$

12. $\phi = BA = (350 \times 10^{-3} \text{ Ts})\pi(0.10 \text{ m})^2 = 1.1 \times 10^{-2} \text{ Wb}.$

13. $\mathcal{E} = \dfrac{\Delta\phi}{\Delta t} = \dfrac{(150)(1.1 \times 10^{-2} \text{ Wb})}{0.10 \text{ s}} = 17 \text{ V}.$

14. $I = \dfrac{\Delta V}{R} = \dfrac{65 \text{ V}}{(4.0 \ \Omega + 2.4 \ \Omega)} = 10.2 \text{ A}.$

 $P = I \ \Delta V = (10.2 \text{ A})(65 \text{ V}) = 660 \text{ W}.$

15. The easiest way is to consider whether the current across the front face of the solenoid is in the same direction or in direction opposite that in the electromagnet.
 a. As the current increases in the electromagnet, current in the solenoid must oppose it, so it is upward and then right through the galvanometer.
 b. Nothing happens until the electromagnet begins to leave the solenoid; then it must be attracted, so the two currents are in the same direction; right through the galvanometer.
 c. When the switch is opened, solenoid current must support the electromagnet current; left through the galvanometer.

Chapter 12

1. a. $P_{max} = I_{max}{}^2 R = (0.50 \text{ A})^2 (60 \ \Omega) = 15 \text{ W}.$
 b. $\Delta V_{eff} = \dfrac{\Delta V_{max}}{\sqrt{2}} = \dfrac{I_{max}R}{\sqrt{2}} = \dfrac{(0.50 \text{ A})(60 \ \Omega)}{\sqrt{2}} = 21 \text{ V}.$

2. a. $I_{eff} = \dfrac{P_{av}}{\Delta V_{eff}} = \dfrac{300 \text{ W}}{120 \text{ V}} = 2.5 \text{ A}.$
 b. Twice $P_{av} = 600 \text{ W}.$

3. $\dfrac{\Delta V_P}{\Delta V_S} = \dfrac{N_P}{N_S} \qquad \dfrac{120 \text{ V}}{6 \text{ V}} = \dfrac{2,400}{N_S} \qquad N_S = 200.$

4. a. $I = \dfrac{P}{\Delta V} = \dfrac{60 \text{ W}}{12 \text{ V}} = 5 \text{ A}.$
 b. $\dfrac{\Delta V_P}{\Delta V_S} = \dfrac{I_S}{I_P}; \qquad \dfrac{120 \text{ V}}{12 \text{ V}} = \dfrac{5 \text{ A}}{I_P}; \qquad I_P = 0.5 \text{ A}.$
 c. $\Delta V_{max} = \Delta V_{eff}\sqrt{2} = (120 \text{ V})\sqrt{2} = 170 \text{ V}.$

5. At 120 V:

$$I = \dfrac{6.5 \times 10^3 \text{ W}}{120 \text{ V}} = 54 \text{ A}$$

Power loss in the line $= I^2 R = (54 \text{ A})^2 (0.20 \ \Omega) = 580 \text{ W}.$
At 1,500 V:

$$I = \dfrac{6.5 \times 10^3 \text{ W}}{1,500 \text{ V}} = 4.3 \text{ A}.$$

Power loss in the line $= I^2 R = (4.3 \text{ A})^2 (0.20 \ \Omega) = 3.7 \text{ W}.$ The saving is 576 W.

6. $I = \dfrac{P}{V} = \dfrac{45 \times 10^3 \text{ W}}{560 \text{ V}} = 80 \text{ A}$

The power loss in the line $= I^2 R = (80 \text{ A})^2 (1.5 \ \Omega) = 9,600 \text{ W}.$

7. To produce 60 cycles with four poles, the armature must rotate 30 times a second.

8. #14 wire is limited to 15 A, so the most power that can be delivered is (15 A)(120 V) = 1,800 W.

9. In the primary:

$$I = \frac{P}{\Delta V} = \frac{2,500,000 \text{ W}}{325,000 \text{ V}} = 7.7 \text{ A}$$

In the secondary:

$$\frac{2,500,000 \text{ W}}{7,500 \text{ V}} = 330 \text{ A}$$

10. $\Delta V_{\text{max}} = \Delta V_{\text{eff}}\sqrt{2} = (30,000 \text{ V})\sqrt{2} = 43,000 \text{ V}.$

Chapter 13

1. **a.** Wave
   b. Electric field
   c. Nothing
   d. Wave
   e. Electric and magnetic fields.

2. **a.** $\lambda = \frac{c}{f} = \frac{3.00 \times 10^8 \text{ m/s}}{106 \times 10^6 \text{ Hz}} = 2.8 \text{ m}.$
   b. Half a wavelength, 1.4 m.

3. **a.** Near the antenna, the two fields alternate every quarter-wavelength, so the magnetic maximum is 1.1 m farther than the electric, or 1.6 m from the antenna.
   b. Horizontal antenna, vertical magnetic field.

4. $\lambda = \frac{c}{f} = \frac{3.00 \times 10^8 \text{ m/s}}{7.1 \times 10^{14} \text{ Hz}} = 4.2 \times 10^{-7} \text{ m, or 420 nm.}$

   Figure 13-13 shows this to be blue.

5. **a.** 10λ, or 5,300 nm.
   b. $\sin \theta = n\lambda/d$, so

   $$d = \frac{n\lambda}{\sin \theta} = \frac{5,300 \times 10^{-9} \text{ m}}{\sin 12°} = 2.5 \times 10^{-5} \text{ m}$$

6. $\sin \theta = \frac{\lambda}{d} \qquad d = \frac{1 \text{ m}}{580 \times 10^3}.$

   For the shortest wavelength,

   $$\sin \theta = \left(\frac{580 \times 10^3}{\text{m}}\right)(380 \times 10^{-9}) \qquad \theta = 12.7°$$

   For the longest wavelength,

   $$\sin \theta = \left(\frac{580 \times 10^3}{\text{m}}\right)(760 \times 10^{-9} \text{ m}) \qquad \theta = 26.2°$$

7. The shortest wavelength for red light is about 700 nm;

   $$\sin \theta = \left(\frac{530 \times 10^3}{\text{m}}\right)(3)(700 \times 10^{-9} \text{ m}) = 1.1$$

   No angles have sines greater than 1.

8. Those two frequencies act on the red and green pigments, with very little in the blue. It appears yellow.

9. Everything below about 650 nm.

10. About 600 nm.

11. $E = hf = (6.62 \times 10^{-34}\ \text{J}\cdot\text{s})(10^{14}\ \text{Hz}) = 6.6 \times 10^{-20}\ \text{J},$ or
$$(4.14 \times 10^{-15}\ \text{ev}\cdot\text{s})(10^{14}\ \text{Hz}) = 0.41\ \text{ev}.$$

12. $E = hf = \dfrac{hc}{\lambda} = \dfrac{(4.14 \times 10^{-15}\ \text{ev}\cdot\text{s})(3.00 \times 10^8\ \text{m/s})}{140 \times 10^{-9}\ \text{m}} - 8.9\ \text{ev}.$

13. This is the photon that has just that energy. $E - hc/\lambda$, so
$$\lambda = \frac{hc}{E} = \frac{(6.62 \times 10^{-34}\ \text{J}\cdot\text{s})(3.00 \times 10^8\ \text{m/s})}{3.1 \times 10^{-19}\ \text{J}} = 6.4 \times 10^{-7}\ \text{m, or 640 nm}$$

14. The photon energy is $E = hf = (4.14 \times 10^{-15}\ \text{ev}\cdot\text{s})(6.8 \times 10^{15}\ \text{Hz}) = 28.2\ \text{ev}$. Since the electrons leave the surface with 22.0 ev of energy, the work function is $28.2 - 22.0 = 6.2$ ev.

15. The photon energy is
$$\frac{hc}{\lambda} = \frac{(4.14 \times 10^{-15}\ \text{ev}\cdot\text{s})(3.00 \times 10^8\ \text{m/s})}{580 \times 10^{-9}\ \text{m}} = 2.1\ \text{ev}$$

The electron energy is 2.1 ev − 1.6 ev = 0.5 ev.

16. The photon energy is
$$\frac{hc}{\lambda} = \frac{(4.14 \times 10^{-15}\ \text{ev}\cdot\text{s})(3.00 \times 10^8\ \text{m/s})}{470 \times 10^{-9}\ \text{m}} = 2.6\ \text{ev}$$

The maximum electron energy is 2.6 − 2.2 ev = 0.4 ev, and the current will be cut off by 0.4 V.

Chapter 14

1. $I = F/4\pi r^2$, so $F = 4\pi r^2 I = 4\pi (1.5\ \text{m})^2 (20\ \text{lux}) = 570\ \text{lum}.$

2. The visible power output is 4% of 200 W = 8 W. Since a visible watt corresponds to roughly 450 lum, the luminous flux is (450 lum/W) × (8 W) = 3,600 lum.

3. Referring to Figure A7-1, angle 1 = angle 2, and angle 3 = angle 4. Since angle 2 + angle 3 = 90°, the sum of the four angles is 180°.

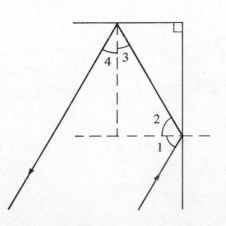

FIG. A7-1

4. The size does not change, but the distance drops to 6 m.

5. $n = \dfrac{c}{v} = \dfrac{3.00 \times 10^8 \text{ m/s}}{2.00 \times 10^8 \text{ m/s}} = 1.50$.

6. $n = \dfrac{\sin i}{\sin r} = \dfrac{\sin 45°}{\sin 32°} = 1.33$.

7. The first refraction occurs when the beam enters the benzene (B) from the air (A):

$$\frac{\sin A}{\sin B} = \frac{n_B}{n_A}; \qquad \frac{\sin 55°}{\sin B} = \frac{1.50}{1}$$

so $B = 33°$. This is the angle at which it enters the water, so

$$\frac{\sin 33°}{\sin W} = \frac{1.33}{1.50}; \qquad W = 38°$$

8. We need a material whose critical angle is 45°, so

$$\frac{\sin 45°}{\sin 90°} = \frac{1}{n}; \qquad n = 1.41$$

9. $n = \dfrac{\sin 62°}{\sin 47°} = \dfrac{\sin 90°}{\sin i_C}; \qquad i_C = 56°$.

10. Both make the same angle (A) in air, but different angles B_v and B_r in the glass. Then, since $n = \sin A/\sin B$, $\sin A = n \sin B$ is the same for both. Therefore $1.590 \sin 65.0° = 1.566 \sin B_r$, and $B_r = 67.0°$.

11. $\dfrac{1}{f} = \dfrac{1}{D_o} + \dfrac{1}{D_i}$, so

$$\frac{1}{D_i} = \frac{1}{f} - \frac{1}{D_o} = \frac{1}{5.0 \text{ cm}} - \frac{1}{2.0 \text{ cm}} = \frac{2-5}{10.0 \text{ cm}}$$

and $D_i = -3.33$ cm, a virtual image. To find the size:

$$\frac{D_o}{D_i} = \frac{S_o}{S_i} \qquad \frac{2.0 \text{ cm}}{3.33 \text{ cm}} = \frac{1.2 \text{ cm}}{S_i}$$

and $S_i = 2.0$ cm.

12. $\dfrac{1}{f} = \dfrac{1}{D_o} + \dfrac{1}{D_i} = \dfrac{1}{9.0 \text{ cm}} + \dfrac{1}{6.5 \text{ cm}} = \dfrac{6.5 + 9.0}{(9.0)(6.5) \text{ cm}}$

and $f = 3.8$ cm.

13. $\dfrac{1}{D_i} = \dfrac{1}{f} - \dfrac{1}{D_o} = \dfrac{1}{20 \text{ cm}} - \dfrac{1}{50 \text{ cm}} = \dfrac{5-2}{100 \text{ cm}}$

and $D_i = 33.3$ cm. To find its size:

$$\frac{D_o}{D_i} = \frac{S_o}{S_i} \qquad \frac{50 \text{ cm}}{33.3 \text{ cm}} = \frac{40 \text{ cm}}{S_i}$$

which gives $S_i = 27$ cm.

14. In this setup, $D_i = (3.0 \text{ m} - D_o)$, and we are trying to find values for D_o. Then

$$\frac{1}{f} = \frac{1}{D_o} + \frac{1}{D_i}$$

becomes

$$\frac{1}{0.50 \text{ m}} = \frac{1}{D_o} + \frac{1}{3.0 \text{ m} - D_o}$$

Multiplying through by $(0.50 \text{ m})(D_o)(3.0 \text{ m} - D_o)$ gives
$D_o(3.0 - D_o) = 0.50(3.0 - D_o) + 0.50 D_o$, which simplifies to
$D_o{}^2 - 3.0\, D_o + 1.5 = 0$. Now we apply the quadratic equation formula

$$x = \frac{-b \pm \sqrt{b^2 - 4ac}}{2a}$$

to get

$$D_o = \frac{3.0 \pm \sqrt{3.0^2 - 4(1.5)}}{2}$$

so $D_o = 2.36$ m or 0.63 m.

15. $\dfrac{D_o}{D_i} = \dfrac{S_o}{S_i} = \dfrac{1.00 \text{ m}}{0.065 \text{ m}}$ $D_i = 0.065\, D_o$

Therefore the $1/f$ equation becomes

$$\frac{1}{80 \text{ mm}} = \frac{1}{D_o} + \frac{1}{0.065\, D_o}$$

Then

$$\frac{1}{80 \text{ mm}} = \frac{0.065 + 1}{0.065\, D_o}$$

and $D_0 = 1{,}300$ mm, or 1.3 m.

16. $3.5 = \dfrac{\text{focal length}}{\text{diameter}} = \dfrac{80 \text{ mm}}{\text{diameter}}$.
So the diameter = 23 mm.

Chapter 15

1. a. 77.
   b. 113 (atomic mass number minus atomic number).
   c. 77 (atomic number).

2. a. $E = -13.6 \text{ ev}/n^2$. Since $n = 1$, $E = -13.6$ ev.
   b. With $n = 3$, $E = -1.51$ ev.
   c. n is indefinitely large and $E = 0$.

3. a. Using the answers to Problem 2, the energy drops from -1.51 ev to
    -13.6 ev, so $\Delta E = 12.1$ ev, and that is the energy of the photon.
   b. $E = hc/\lambda$, so

$$\lambda = \frac{hc}{E} = \frac{(4.14 \times 10^{-15} \text{ ev·s})(3.00 \times 10^8 \text{ m/s})}{12.1 \text{ ev}} = 1.03 \times 10^{-7} \text{ m}$$

4. Its total energy is then 18 ev $-$ 13.6 ev = 4.4 ev.

5. a. $mv = \dfrac{h}{\lambda} = \dfrac{6.62 \times 10^{-34} \text{ J·s}}{3.0 \times 10^{-10} \text{ m}} = 2.21 \times 10^{-24} \text{ kg·m/s}$.

   b. $v = mv/m$, and the mass is found in Appendix 3.

$$v = \frac{2.21 \times 10^{-24} \text{ kg·m/s}}{9.11 \times 10^{-31} \text{ kg}} = 2.43 \times 10^6 \text{ m/s}$$

   c. $E_{\text{kin}} = \dfrac{1}{2} mv^2 = \dfrac{1}{2}(9.11 \times 10^{-31} \text{ kg})(2.43 \times 10^6 \text{ m/s})^2 =$

    2.69×10^{-18} J. To put it into electron volts, divide by

    1.60×10^{-19} J/ev to get 16.8 ev.

6. **a.** Its wavelength is one-third the circumference of its orbit,

$$2\pi(1.5 \times 10^{-10}\,\text{m})/3 = 3.1 \times 10^{-10}\,\text{m}.$$

   b. $mv = \dfrac{h}{\lambda} = \dfrac{6.62 \times 10^{-34}\,\text{J·s}}{3.1 \times 10^{-10}\,\text{m}} = 2.1 \times 10^{-24}\,\text{kg·m/s}.$

7. From Appendix 3:

protons:	$15 \times 1.007825 = 15.11738$
neutrons:	$15 \times 1.008665 = 15.12998$
mass of separate particles	30.24735
mass of phosphorus-30	29.9783 Dl
mass deficit	0.2691 Dl

binding energy: 0.2691 Dl(931 Mev/Dl) = 250 Mev

8. **a.** $^{218}_{84}\text{Po} \rightarrow\ ^{214}_{82}\text{Pb} +\ ^{4}_{2}\text{He}.$

   b. The mass of polonium-218 is 218.0089. Subtract mass of helium-4 (4.00260) and lead-214 (213.9998) to find the additional mass deficit of 0.0065 Dl; multiply by 931 Mev/Dl to get 6.1 Mev.

9. It must drop to half four times; with a half-life of 3 min, this would take 12 min.

10. **a.** $^{210}_{82}\text{Pb} \rightarrow\ ^{210}_{83}\text{Bi} +\ ^{0}_{-1}\text{e} +\ ^{0}_{0}\nu.$

    b. The mass of lead-210 is 209.9848. Subtract the mass of bismuth-210 (209.9840) and an electron (0.00055) to get 0.0002 Dl; multiplying by 931 gives 0.2 Mev.

11. $E = q\,\Delta V =$ (two elementary charges)$(6.5 \times 10^5\,\text{V}) = 1.3$ Mev.

12. It must acquire an additional mass equal to its own mass, 1.67×10^{-27} kg. Then $E = mc^2 = (1.67 \times 10^{-27}\,\text{kg})(3.00 \times 10^8\,\text{m/s})^2 = 1.51 \times 10^{-10}$ J. To find the potential difference, $\Delta E = q\,\Delta V$ and

$$\Delta V = \frac{\Delta E}{q} = \frac{1.51 \times 10^{-10}\,\text{J}}{1.60 \times 10^{-19}\,\text{C}} = 9.4 \times 10^8\,\text{V}$$

which is not practical.

13. $^{4}_{2}\text{He} +\ ^{10}_{5}\text{B} \rightarrow (^{14}_{7}\text{N}) {-}^{13}_{7}\text{N} +\ ^{1}_{0}\text{n}.$

14. $^{22}_{11}\text{Na} \rightarrow\ ^{22}_{10}\text{Ne} +\ ^{0}_{+1}\text{e} +\ ^{0}_{0}\nu.$

15. $^{239}_{94}\text{Pu} +\ ^{1}_{0}\text{n} \rightarrow\ ^{142}_{54}\text{Xe} +\ ^{94}_{40}\text{Zr} + 3\ ^{1}_{0}\text{n}.$

16.

mass of hydrogen-1 (proton)	1.007825 Dl
mass of carbon-12	12.00000
	13.007825
mass of nitrogen-13	13.0057
mass deficit of reaction	0.0021 Dl

$\times$ 931 Mev/Dl = 1.96 Mev

Index

Tuning fork, 157
Turbulent flow, 59

Ultraviolet rays, 268, 288
Uniform acceleration, 12–14
Uniform field, 185
Universal gravitation, law of, 78, 354

Van de Graaff generator, 180, 197, 324
Vaporization, latent heat of, 140–141
Variables, 335, 336
Vector addition, 4, 5, 7
Vector diagram, 33–34
Vector sum, 3
Vectors, 3–10, 32
    components of, 8–9
    velocity, 5–7
Velocity, 3–7
    change in, 25, 26, 65–66
    drift, 193, 202
    escape, 120–121
    terminal, 58, 59
    of waves, 156–157
Velocity-time graph, 11
Velocity vectors, 5–7
Venturi meter, 60
Virtual image, 291–292, 303–304
Viscosity, 47, 58
Viscous drag, 27, 31, 32, 58–59
Vise, 102–103
Volt, 189, 196
    electron, 283
Voltmeters, 189, 190, 198, 204, 207, 209, 224

Water, specific heat of, 136
Watt, 105, 199
Wave front, 158
Wavelengths, 155, 156
    of light, 270
    measuring, 270–273
Waves, 153–174
    amplitude of, 155, 158
    diffraction and, 159–162
    electromagnetic, 266–269
    frequency of, 154, 158
    interference and, 165–167
    longitudinal, 156, 157
    monochromatic, 273
    period of, 154
    phase difference of, 154
    radio, 268, 269, 273
    sound, 157–158
    standing, 167–168, 169
    transverse, 156
    velocity of, 156–157
Weight, 27, 28, 31, 46
Weightlessness, 79–80
Weston galvanometer, 224
White hot, 278
Winch, 40, 103–105
Wind instruments, 171–172
Work, 92–93, 106, 108, 113
    in storage, 107–108
Work function of metal, 281, 283
Work input, 93–94, 95, 107
Work output, 93–94, 95, 107

X-rays, 268

Yoke, 232